MULTIGRID METHODS

FRONTIERS IN APPLIED MATHEMATICS

Richard E. Ewing, *Managing Editor*

MULTIGRID METHODS

Stephen F. McCormick, Editor

SOCIETY FOR INDUSTRIAL AND APPLIED MATHEMATICS
PHILADELPHIA, PENNSYLVANIA 1987

Library of Congress Catalog Card Number 87-60444
ISBN 0-89871-214-9

Typeset by The Universities Press, Ltd., Belfast, Ireland, and printed by Thompson-Shore, Inc., Dexter, Michigan.

CONTENTS

CONTRIBUTORS

R. Bank, Department of Mathematics, University of California at San Diego, La Jolla, California

K. Brand, Gesellschaft für Mathematik und Datenverarbeitung, St. Augustin, West Germany

W. Briggs, Computational Mathematics Group, The University of Colorado at Denver, Denver, Colorado

P. W. Hemker, Centrum voor Wiskunde en Informatica, Amsterdam, The Netherlands

G. M. Johnson, Institute for Computational Studies at Colorado State University, Fort Collins, Colorado

M. Lemke, Geselleschaft für Mathematik und Datenverarbeitung, St. Augustin, West Germany

J. Linden, Gesellschaft für Mathematik und Datenverarbeitung, St. Augustin, West Germany

J. Mandel, Computational Mathematics Group, The University of Colorado at Denver, Denver, Colorado

S. McCormick, Computational Mathematics Group, The University of Colorado at Denver, Denver, Colorado

J. W. Ruge, Computational Mathematics Group, The University of Colorado at Denver, Denver, Colorado

K. Stüben, Gesellschaft für Mathematik und Datenverarbeitung, St. Augustin, West Germany

P. Wesseling, Department of Mathematics, Delft University of Technology, Delft, The Netherlands

PREFACE

As all manuscripts of this type must be, this volume of the SIAM series *Frontiers in Applied Mathematics* was designed as much by compromise as by anything else. Decisions were carefully made about such things as notation, structure, content, style, purpose, and targeted audience, but, almost always, tradeoffs were involved.

We first decided on our basic purpose, the underlying formal structure we wanted, and what we expected of our audience. Our intent was a fairly cohesive book that developed, organized, and surveyed selected fields in the multigrid discipline. New material would be included, but only to fill gaps. The book would be directed toward readers who were well versed in related fields, especially the numerical solution of partial differential equations, but who were otherwise unfamiliar with multigrid methods.

Our biggest problem was designing a common system of notation and conventions. The system had to be simple, clean, versatile, unambiguous, and general enough to meet our diverse needs. At the end of this Preface, we summarize the design that emerged from our compromises.

It is crucial that the reader pay close attention to our notation. Most significantly, in lieu of iteration subscripts, we use *dynamic* variables, much as variables are used in computer languages. This puts the responsibility on the reader to always have in mind the sense of these variables, especially those that represent coarse grid quantities. It is a serious warning that *significant subtleties in our notation must not be overlooked.* We point out some of these subtleties in Chapter 1.

Since this volume is but one of several in the SIAM series *Frontiers in Applied Mathematics,* we have conformed to certain common objectives of the series including development of a basic theme and topic, incorporation of a survey of tutorial character, treatment of new developments in the field, and solicitation of individual contributions from experts in each area. However, because of the flexibility that this series allows, our volume departs a little from the others in character, primarily by tending toward a more unified text. From a pedagogical perspective, we have written the first chapter so that it may be used in the classroom as a basis for a module on multigrid methods; other chapters may provide resources for further topics of study.

After the section Notation and Convention and the introductory Chapter 1, Chapter 2 provides a practical development of multigrid for linear equations including a survey of the various forms of multigrid processes in use as well as the results of some numerical experiments. Chapter 3 surveys existing approaches for multigrid solution of Euler equations. Chapter 4 presents a state-of-the-art description of algebraic multigrid (AMG) and its applications. Chapter 5 develops a theoretical foundation for multigrid methods applied to essentially selfadjoint linear problems. A program that demonstrates the basic structure and simplicity of the multigrid algorithm is included as Appendix 1.

Finally, Appendix 2 contains the latest version of the Multigrid Bibliography, which is an attempt to provide references for almost all known multigrid publications. We also include a KWIC reference guide for the Multigrid Bibliography.

It is an understatement to say that many important topics are not treated in this volume. Because space forced us to limit our discussion, we tried to choose topics that have experienced recent and substantial progress, some almost to the point of unification (e.g., the variational theory of Chapter 5). This is evidenced by the fact that the problems we treated are predominantly of (nearly) elliptic type. We also excluded topics already represented in a unified way in the literature (cf. [101], [244], and [249] in the Multigrid Bibliography).

The authors wish to acknowledge their debt to Professor Hitoshe Ombe, whose many suggestions and editorial comments improved this book.

Notation and Convention

Following is a brief summary of the special notation used in this book. We will not attempt to explain our choices here, except to mention that we have allowed for certain ambiguities in exchange for simplicity. Thus, this notation must be considered always with the context in mind. Special attention should be given the variables that are dynamic (e.g., $\mathscr{u}$, u, $\mathscr{e}$, e, $\mathscr{r}$, and r) or generic (e.g., h and c).

A few, seemingly unavoidable notational exceptions appear in Chapters 3 and 4. Specifically, for AMG it seemed more natural to number the levels beginning with the finest grid. Also, it was convenient to use H to denote a level coarser than level h. This usage should not cause confusion with spaces or operators, since its context primarily involves operators such as I_h^H and I_H^h. Other notational exceptions were used in Chapter 4 to provide continuity with the AMG literature. In a similar way, we found it inappropriate to adopt all of our notation for Chapter 3 because of the prevailing conventions for fluid flow applications. Thus, Chapter 3 does not use script to denote continuous domain quantities or underlining to denote system variables; it uses Greek symbols for functions and vectors as well as constants, and capitals for certain functions and variables.

Each chapter will reference equations with numbers that indicate chapter sections, but not the chapters themselves. Thus, (3.1) will refer to equation 1 in subsection 3 of whatever chapter uses it. This should not cause any confusion, because no one chapter refers explicitly to equations in any other chapter.

References to the Multigrid Bibliography in the book are indicated by numbers in square brackets [ ]. Other references are compiled at the end of their respective chapters and are indicated in square brackets using the first two letters of the last name of the first author, with trailing numbers for resolving ambiguities. Thus, for example, in Chapter 3, [86] refers to Brandt's first paper cited in the Multigrid Bibliography, while [La2] refers to Lax's second paper cited at the end of the chapter.

GUIDE TO NOTATION

Notation	Explanation	Comment
$\mathbf{R}^d$, $\mathbf{C}^d$	Real, complex Euclidean space in d dimensions.	d is omitted when understood.
$\underset{\sim}{x} = (x_{[1]}, x_{[2]}, \cdots, x_{[d]})$	Independent variables in $\mathbf{R}^d$ or $\mathbf{C}^d$.	Underbar signifies that each entry of the vector is a function or variable. We may replace $x_{[1]}$ by x, $x_{[2]}$ by y, etc., so $\underset{\sim}{x} = (x, y, z)$ may be used, for example.
Ω	Open region in $\mathbf{R}^d$ or $\mathbf{C}^d$.	Its boundary is $\partial\Omega$, its closure is $\bar{\Omega}$. Ω^* may be used in place of Ω.
$\mathcal{H}(\Omega)$	Space of functions with domain Ω.	(Ω) is omitted when understood.
$\mathcal{H}'$	Dual space of $\mathcal{H}$.	
$\mathcal{H}^s(\Omega)$	Sobolev space of order s on domain Ω.	$\mathcal{H}^s$ may be a generic Hilbert scale.
$[\mathcal{H}^0, \mathcal{H}^1]_s$	Intermediate or interpolated spaces.	These scales are well defined when $\mathcal{H}^1 \subset \mathcal{H}^0$ is a dense, continuous embedding.

GUIDE TO NOTATION (*continued*)

Notation	Explanation	Comment
f	Function in $\mathscr{H}$ that is a right-hand side of a specific equation.	
$\boldsymbol{u}$	Function in $\mathscr{H}$ that is a solution of a specific equation.	
u	Function in $\mathscr{H}$ that is a dynamic approximation to $\boldsymbol{u}$.	
$\mathscr{A}$, $\mathscr{F}(\cdot)$	Linear, general operator on $\mathscr{H}$.	
$[\mathscr{H}]$	Space of continuous linear mappings from $\mathscr{H}$ into itself.	
$\langle u, v\rangle$, $\|u\|$	$\mathscr{L}_2$ inner product, norm when specified; generic inner product, norm otherwise.	
$(\mathscr{H})^l$	This denotes the cross product space $\underbrace{\mathscr{H}\times\mathscr{H}\times\cdots\times\mathscr{H}}_{l \text{ times}}$.	
$\langle u, v\rangle_{\mathscr{A}}$, $\|u\|_{\mathscr{A}}$	Energy inner product, norm when $\mathscr{A}$ is selfadjoint and positive definite (sa.p.d.) or symmetric positive definite (s.p.d.).	
h	Generic mesh size (assumes the mesh size is constant when h is a scalar, or constant in each coordinate direction when h is a vector).	h may represent a vector if $d>1$ so that $h=(h_{x_{[1]}}, h_{x_{[2]}})=(h_x, h_y)$, for example. In this case, we may use $\underline{h}$ for emphasis. (See the definition of $\underline{x}$ above.)
k	Level index (makes no assumption about mesh sizes).	$k=1$ is the coarsest, $k=q$ is the finest.
h_k	Level k mesh size (assumes the mesh size is constant on level k	$h_1 \geq h_2 \geq \cdots \geq h_q$. k replaces h_k when understood or when h_k is not

GUIDE TO NOTATION (*continued*)

Notation	Explanation	Comment
	when h_k is a scalar, or constant in each coordinate direction when h_k is a vector).	defined (e.g., the grid is irregular). h_k will be dropped altogether when understood.
Ω^h	Discrete region (grid) in $\mathbf{R}^d$ or $\mathbf{C}^d$ (Ω^h may or may not include points of $\partial\Omega$).	h is never dropped here except that h_k may be replaced by k. Note that Ω^2 refers to Ω^{h_2}, for example.
$H^h(\Omega^h)$	Grid function space (i.e., space of functions with domain Ω^h).	(Ω^h) may be dropped here when understood.
n^h	Number of points of Ω^h when n is understood to be a scalar; otherwise a vector $(n_{x_{[1]}}, n_{x_{[2]}}, \cdots, n_{x_{[d]}})$ where $n_{x_{[i]}}$ is the number of points of Ω^h in the $x_{[i]}$-coordinate direction (assumes Ω^h is logically rectangular). In either case, n may or may not count "boundary points" of Ω^h.	When n is a vector, we may use $n = (n_x, n_y, n_z)$ or $n = (l, m, n)$, for example. We may also use $\underset{\sim}{n}$ in this case for emphasis.
x_i^h, x_{ij}^h, or x_{ijk}^h	Grid point number i, ij, or ijk in Ω^h for $d = 1, 2,$ or 3, respectively.	x_i may also refer to grid point number i when $d > 1$ assuming a predefined ordering of all grid points in Ω^h. We may use i in place of x_i^h, ij in place of x_{ij}^h, etc.
f^h	Vector in H that is a right-hand side of a specific equation.	
$\mathbf{u}^h$	Vector in H that is a solution of a specific equation.	
u^h	Vector in H that is a dynamic approximation to $\mathbf{u}$.	
A^h, $F^h(\cdot)$	Linear (matrix), general operator on H.	h is never dropped here except that h_k may be replaced by k. Note that

GUIDE TO NOTATION (*continued*)

Notation	Explanation	Comment
		A^2 refers to A^{h_2}, for example. (The square of A^h is denoted $(A^h)^2$.)
a^h_{ij}	ijth entry of A.	$A = (a_{ij})$.
u^h_i, u^h_{ij}, or u^h_{ijk}	Value (entry) of u at the corresponding grid point.	
$\underline{u} = (u_{x[1]}, u_{x[2]}, \cdots, u_{x[d]})$	Vector function in the cross product space $(\mathcal{H})^d$ so each entry $u_{x[i]}$ is in $\mathcal{H}$.	Underbar signifies that each entry of the vector is a function or a variable. This is useful for systems of partial differential equations (PDEs). We may use $\underline{u} = (u_x, u_y, u_z)$ or $\underline{u} = (u, v, w)$, for example.
$\underline{u}^h = (u^h_{x[1]}, u^h_{x[2]}, \cdots, u^h_{x[d]})$	Same as $\underline{u}$ with H replacing $\mathcal{H}$.	Here, the entries of $\underline{u}^h$ are vectors. Allowing this use means that every vector could conceivably use the underbar. However, we reserve the underbar for vectors that really correspond to discrete systems. Thus, we imagine that the $u^h_{x[i]}$ are discrete versions of the distinct functions $u_{x[i]}$.
$\langle u, v\rangle_h$, $\Vert u\Vert_h$	Inner product, norm on H^h approximating its $\mathcal{L}_2$ counterpart on $\mathcal{H}$.	Here, h cannot be dropped. k may replace h_k, however.
$\langle u, v\rangle$, $\Vert u\Vert$	Euclidean inner product, norm on H when specified; generic inner product, norm otherwise.	
$\vert\vert\vert u\vert\vert\vert_s$	Denotes the norm $\langle (B^{-1}A)^s u, u\rangle_B^{1/2}$ where A and B are s.p.d.	
$\langle u, v\rangle_A$, $\Vert u\Vert_A$	Energy inner product, norm on H when A is sa.p.d. or s.p.d.	

GUIDE TO NOTATION (*continued*)

Notation	Explanation	Comment
$\mathscr{A}(u, v)$	Bilinear or sesquilinear form on $\mathscr{H}$.	Useful instead of energy forms when $\mathscr{A}$ is not sa.p.d. or s.p.d.
$A(u, v)$	Same as $\mathscr{A}(u, v)$ with H replacing $\mathscr{H}$.	
e	Analytic error in $\mathscr{H}$.	$e = \boldsymbol{u} - u$.
e^h	Algebraic error in H.	$e = \mathbf{u} - u$.
$\mathbf{e}^h$	Global error in H.	$\mathbf{e}_i^h = \mathbf{u}_i^h - \boldsymbol{u}(x_i^h)$.
$\varkappa$	Residual error in $\mathscr{H}$.	$\varkappa = \mathscr{A}e$.
r	Residual error in H.	$r = Ae$.
A^T	Matrix transpose.	
B^*	Matrix adjoint with respect to the energy inner product.	$B^* = A^{-1}B^TA$.
$S^\perp$	Orthogonal complement of S in H with inner product understood.	Similarly for $\mathscr{S}^\perp$ in $\mathscr{H}$.
$R(A)$	Range of A.	Similarly for $\mathscr{R}(\mathscr{A})$.
$RQ(u)$	Rayleigh quotient for A, B.	$RQ(u) = \langle Au, u\rangle / \langle Bu, u\rangle$. Similarly for $\mathscr{R}\mathscr{Q}(\mathscr{A})$.
$\rho(A)$	Spectral radius of A.	Similarly for $\rho(\mathscr{A})$.
$\lambda(A)$	(Discrete) spectrum of A or, generically, one of its eigenvalues.	Similarly for $\lambda(\mathscr{A})$. For specific values we may use $\lambda_i^h(A)$, $\lambda_i(\mathscr{A})$, λ_i^h, etc.
$N(A)$, ker A	Alternate representations for the null space or kernel of A.	
$\mathscr{J}^h$	Mapping from H to $\mathscr{H}$.	
J^h	Mapping from $\mathscr{H}$ to H.	
$\mathscr{I}$	Identity operator on $\mathscr{H}$.	
I	Identity operator on H.	
$G^h(u^h, f^h)$	General form of one sweep of a general algebraic smoother.	

GUIDE TO NOTATION (*continued*)

Notation	Explanation	Comment
$u - \omega B^{-1}(Au - f)$	General form of one sweep of a stationary linear smoother with preconditioner B^{-1} and relaxation parameter ω.	The linear part is $I - wB^{-1}A$.
$\leftarrow$	Denotes replacement; for use with dynamic variables.	$u \leftarrow G(u, f)$ denotes one smoothing sweep.
$u^h_{(k)}$	Specific assignment of the dynamic variable u.	May also use $u_{(\text{old})}$, $u_{(\text{new})}$, $u_{(\text{int})}$, etc.
$A^h = D^h - L^h - U^h$	Splitting of A into its strictly lower triangular, diagonal, and strictly upper triangular parts.	
$A^h = M^h - N^h$	General splitting.	
α	Specific constant independent of h, h_k, or k.	Any lowercase Greek letter other than those already reserved may be used here.
c	Generic positive constant independent of h, h_k, or k.	c and only c is used as an unspecified constant. The statement $\alpha = O(h^p)$ is equivalent to $\lvert\alpha\rvert \le ch^p$.
I^h_{2h}	Prolongation operator mapping H^{2h} to H^h.	$I^l_k : H^k \to H^l$ for $l > k$ is defined by $I^l_k = I^l_{l-1} I^{l-1}_{l-2} \cdots I^{k+1}_k$. Also, $I^h_h = I$.
I^{2h}_h	Restriction operator mapping H^h to H^{2h}.	I^l_k for $l < k$ is defined analogously.
$MG^h(u^h, f^h)$	Basic two-grid multigrid cycle when specified; generic otherwise.	$MG_\mu(u, f)$, $MV(u, f)$, and $MW(u, f)$ are μ-cycle, V-cycle ($\mu = 1$), and W-cycle ($\mu = 2$), respectively.
$FMG^h(u^h, f^h)$	Nested iteration based on $MG(u, f)$.	$FMG_\mu(u, f)$, $FMV(u, f)$, and $FMW(u, f)$ are defined analogously. $FMG(f) = FMG(0, f)$.
τ^h	Truncation error.	$\tau^h = A^h \mathbf{e}^h$.
τ^{2h}_h	Full approximation scheme (FAS) transfer.	$\tau^h_{2h} = A^{2h} I^{2h}_h u^h - I^{2h}_h A^h u^h$ (the linear case).

GUIDE TO NOTATION (*continued*)

Notation	Explanation	Comment
1-D, 2-D, 3-D	Abbreviations for the respective one-, two-, and three-dimensions (or dimensional).	
ODE, PDE	Abbreviations for the respective ordinary and partial differential equations.	
sa.p.d., s.p.d.	Abbreviations for the respective selfadjoint and symmetric positive definite.	

Stencils:

(a) $$\begin{bmatrix} a & a & a \\ a & a & a \\ a & a & a \end{bmatrix},$$

(b) $$\begin{bmatrix} a & a \\ a & a \end{bmatrix},$$

(c) $$\left] \begin{matrix} a & a & a \\ a & a & a \\ a & a & a \end{matrix} \right[,$$

(d) $$\left] \begin{matrix} a & a \\ a & a \end{matrix} \right[.$$

Used for 2-D operators (1-D is analogous). Stencils (a) and (b) are examples of *collection* stencils. Stencil (a) is typically used for A^h and, when coarsening is by vertices, I_h^{2h}. Stencil (b) is typically used for I_{2h}^h when coarsening is by cells. Both are to be centered at a grid point of the operator range and represent contributions to this point from neighboring points in the corresponding domain. Stencils (c) and (d) are examples of *distribution* stencils. They are typically used for I_{2h}^h when coarsening is by vertices and cells, respectively. Both are to be centered at a grid point of the domain and represent contributions of this point to neighboring points in the corresponding range.

CHAPTER 1

Introduction

W. BRIGGS AND S. McCORMICK

1.1. Purpose. This opening chapter is intended to serve several purposes. The primary one is to stand alone as a basic introduction to some of the essential principles of multigrid methods. Thus, the presentation has been kept rather basic in an attempt to express the underlying simplicity of the multigrid concept. The subtleties that inevitably arise in the implementation and analysis of these methods, especially for more complex applications, will not be treated in any significant way here but will be included in later chapters. Nevertheless, the principles and insights developed in this chapter should form an important basis for understanding multigrid methods, even in much more sophisticated settings.

An underlying objective of this chapter is to provide educational material that may be used, for example, as a module in a numerical analysis class. For this reason, we have included a discussion of some fundamental concepts of classical numerical analysis that are central to a full understanding of multigrid principles. In particular, some readers may find the extensive treatment of iterative methods to be unnecessary. However, if these methods and their convergence properties are well understood, then the motivation for multigrid methods seems very natural and becomes quite straightforward.

This chapter is intended not only to stand alone as a primer on multigrid methods, but also to lay the groundwork for the chapters that follow. For the remainder of the book, we will assume that the reader is familiar with the concepts developed in this opening chapter. Equally important, most of the notational conventions for the book will be explained here.

We begin by treating a few basic model problems, namely, simple linear second order elliptic boundary value problems in one and two dimensions. This choice was made primarily because our readers are perhaps most familiar with these equations and their discretizations, and because the discrete systems of linear equations that they produce lend themselves to simple analysis. This choice is also appropriate from a historical perspective because the early development of multigrid methods focused on such problems. However, the reader should not infer from this that multigrid techniques are limited in any way to these problems. In fact, as later chapters and cited references show, there is a very broad spectrum of problems to which these techniques have already been successfully applied, and this scope of applicability is expanding at a very rapid pace. Indeed, the very basic principles of multigrid techniques, such as the use of various discretization levels to resolve different components of error in the approximation, may well apply to most areas of numerical computation. We hope that this book will, in fact, help to inspire the reader to see new avenues where these basic concepts may play an important role.

As with other relatively new areas of research, there are many controversial issues in the multigrid field, many of which are fundamental. We cannot avoid choosing sides on certain of these issues, but we have tried to minimize this dilemma by staying with the "consensus" as much as possible. However, for this and other reasons, the reader should try to maintain a flexible, open-minded approach to this subject.

1.2. Model problems. A simple model that will serve to illustrate many important concepts is the familiar one-dimensional (1-D) boundary value problem

$$\begin{aligned} \mathcal{A}\boldsymbol{u} &= -\boldsymbol{u}''(x) = \mathcal{f} \quad \text{in } \Omega = \{x: 0 < x < 1\}, \\ \boldsymbol{u}(0) &= \boldsymbol{u}(1) = 0. \end{aligned} \tag{2.1}$$

(Functions defined on a continuum will be denoted by script letters; solutions of specific equations will be denoted by bold letters; the solution of (2.1) therefore uses both.) At times it may be useful to associate $\mathcal{f}$ or the exact solution $\boldsymbol{u}$ with a general real-valued function space $\mathcal{H} = \mathcal{H}(\Omega)$. This we equip with the usual $\mathcal{L}_2(\Omega)$ inner product and norm

$$\langle \mathcal{f}, \mathcal{g} \rangle = \int_\Omega \mathcal{f}\mathcal{g}\, d\Omega, \qquad \|\mathcal{f}\| = \langle \mathcal{f}, \mathcal{f} \rangle^{1/2},$$

or the *energy* inner product and norm

$$\langle \mathcal{f}, \mathcal{g} \rangle_{\mathcal{A}} = \langle \mathcal{A}\mathcal{f}, \mathcal{g} \rangle, \qquad \|\mathcal{f}\|_{\mathcal{A}} = \langle \mathcal{f}, \mathcal{f} \rangle_{\mathcal{A}}^{1/2}.$$

The discretization of (2.1) can be accomplished in many different ways. For simplicity, we choose the discrete domain Ω^h to be the set of uniformly spaced grid points $x_i^h = ih$, $1 \le i \le n^h$, where $h = 1/(n^h + 1)$. (When no ambiguity is possible, the superscript h will be dropped from the notation.) Replacing the derivatives $\boldsymbol{u}''(x_i)$ in (2.1) by a centered difference approximation, we have

$$\begin{aligned} \mathcal{A}\boldsymbol{u}(x_i) &= \frac{-\boldsymbol{u}(x_{i+1}) + 2\boldsymbol{u}(x_i) - \boldsymbol{u}(x_{i-1})}{h^2} + \tau_i \\ &= \mathcal{f}(x_i), \end{aligned} \tag{2.2}$$

where τ_i is the *discretization* (or *truncation*) error associated with this approximation. If we assume that the exact solution u admits a Taylor series expansion about x_i, then τ_i may be written as

$$\tau_i = \frac{h^2}{24}(u^{(iv)}(x_i^+) + u^{(iv)}(x_i^-)) = \frac{h^2}{12} u^{(iv)}(\bar{x}_i),$$

where $x_i^\pm$ are numbers between x_i and $x_{i\pm 1}$ and $\bar{x}_i$ is between x_{i-1} and x_{i+1}.

Clearly, the differentiability of u depends upon that of f. In this simple case, $u^{(iv)}(x) = -f''(x)$. Let $H^h = H^h(\Omega^h)$ denote the space of grid functions on Ω^h. Then the inner product $\langle \cdot, \cdot \rangle_h$ on H^h that approximates the $\mathscr{L}_2(\Omega)$ inner product may be defined by $\langle u, v\rangle_h = hu^Tv$ for u, v in H^h. The associated norm is then given by $\|u\|_h = \langle u, u\rangle_h^{1/2}$. Let τ_i and $f_i = f(x_i)$ be components of the n-vectors τ and f, respectively. Finally, suppose β is a bound on the relative size of the second derivatives of f, that is,

$$\left(h \sum_{i=1}^{n} (f''(\bar{x}_i))^2\right)^{1/2} \leq \beta \, \|f\|_h.$$

(We assume $\beta < \infty$ uniformly in h, which is true if f is sufficiently smooth. This may be recognized as a relationship between discrete *Sobolev* and $\mathscr{L}_2$ norms of f.) Then it is possible to bound the truncation error by

$$\|\tau\|_h \leq \frac{\beta \, \|f\|_h}{12} h^2.$$

This means that a *second order* approximation has been used in (2.2) or, alternatively, that the discretization error is $O(h^2)$. (We define a quantity $q(h)$ to be $O(h^p)$ if $\limsup_{h\to 0} (q(h)/h^p)$ is finite.)

If the truncation error is dropped from (2.2), the result is a set of difference equations that are satisfied by the discrete approximations $\mathbf{u}_i$, $1 \leq i \leq n$, i.e.,

$$\text{(2.3)} \qquad \frac{-\mathbf{u}_{i+1} + 2\mathbf{u}_i - \mathbf{u}_{i-1}}{h^2} = f_i, \qquad 1 \leq i \leq n.$$

If $\mathbf{u}_i$ and f_i are regarded as components of the n-vectors $\mathbf{u}$ and f, then we may express (2.3) in matrix form as $A\mathbf{u} = f$, where A is the $n \times n$, symmetric and positive definite (s.p.d.) tridiagonal matrix with diagonal entries $2h^{-2}$ and super- and sub-diagonal entries $-h^{-2}$. Notice that the boundary conditions $\mathbf{u}_0 = \mathbf{u}_{n+1} = 0$ have been incorporated into this system of equations.

When we compare equations (2.2) and (2.3), it is evident that the discretization error τ may be interpreted as the amount by which the exact solution u fails to satisfy the discrete equations. Thus, the discretization error is a measure of how well the discrete equations represent the continuous problem. However, this does not directly determine how well the discrete solution $\mathbf{u}$ approximates the exact solution u at the grid points. To do this, we first define the *global* error to be the n-vector $\mathbf{e}$ with components

$$\mathbf{e}_i = \mathbf{u}_i - u(x_i).$$

Now subtracting (2.2) from (2.3), we are led to a relationship between the global and the truncation errors given by

$$A\mathbf{e} = \tau.$$

This gives a bound on $\mathbf{e}$ of the form

$$\|\mathbf{e}\|_h \leq \|A^{-1}\|_h \, \|\tau\|_h \leq \frac{\beta \, \|f\|_h}{12} h^2 \, \|A^{-1}\|_h,$$

where β was defined earlier.

To obtain a more quantitative estimate, note that the symmetry of A implies that

$$\|A^{-1}\|_h = \sup_{\|u\|_h=1} \|A^{-1}u\|_h = \rho(A^{-1}),$$

where $\rho(\cdot)$ denotes spectral radius. It is easy to verify that the eigenfunctions of $\mathscr{A}$ are $w_{(k)}(x) = \sin(k\pi x)$, $k = 1, 2, 3, \cdots$, and the associated eigenvalues are $\lambda_k = k^2\pi^2$. Analogously, the matrix A has eigenvectors whose components are given by

$$w^h_{i(k)} = w_{(k)}(x_i), \qquad 1 \leq i, k \leq n,$$

while its eigenvalues range from $\lambda^h_1 \cong \lambda_1 = \pi^2$ to $\lambda^h_n \cong 4/h^2 - \pi^2$. (By relations of the form $\cong$ and $\lesssim$, we mean that their respective relations $=$ and $<$ hold up to a relative error of $O(h^2)$.) Thus, the eigenvalues of A^{-1} range from $\lambda^h_1 \cong h^2/4$ to $\lambda^h_n \cong 1/\pi^2$ so that $\|A^{-1}\|_h \lesssim 1/\pi^2$. A bound on the global error is therefore given by

$$\|\mathbf{e}\|_h \leq \frac{\bar{\beta} \, \|f\|_h \, h^2}{12\pi^2}, \tag{2.4}$$

where $\bar{\beta} \cong \beta$. Thus, the global error $\mathbf{e}^h$ is second order in h, which means that if h is halved, the error *bound* decreases roughly by a factor of four.

These and many of our later estimates are expressible in other norms. In fact, we will later refer to the *Euclidean* inner product and norm on H^h given by

$$\langle u, v \rangle = u^T v, \qquad \|u\| = \langle u, u \rangle^{1/2}.$$

We will also use the *energy* inner product and norm on H given by

$$\langle u, v \rangle_A = \langle u, Av \rangle, \qquad \|u\|_A = \langle u, u \rangle_A^{1/2}.$$

Note that the energy quantities generalize to any s.p.d. matrix A.

The general objective in applying an iterative method to a system of linear equations is to compute a vector u^h that approximates the exact solution $\mathscr{u}$ at the grid points x_i^h to within a prescribed accuracy $\varepsilon > 0$. More precisely, choosing the norm$\|\cdot\|_h$, for instance, and letting w^h be the vector with components $w_i^h = \mathscr{u}(x_i^h)$, then we want

$$\|u^h - w^h\|_h \leq \varepsilon. \tag{2.5}$$

Since the immediate numerical problem is to approximate $\mathbf{u}^h$, we will satisfy (2.5) by way of the following:

$$\|u^h - \mathbf{u}^h\|_h + \|\mathbf{u}^h - w^h\|_h \leq \varepsilon. \tag{2.6}$$

The first term in (2.6) is controlled by the number of iterations, becoming smaller the more iterations we perform. The second term in (2.6) is controlled by the size of h and decreases with h according to (2.4). In general it seems best to roughly balance these errors: why go to extreme measures to reduce the *algebraic* error, $e^h = \mathbf{u}^h - u^h$, when the global error is comparatively large; or conversely, why have a poor approximation to a very accurate discrete solution? Thus, we will attempt to satisfy (2.5) by way of the conditions

$$\|u^h - \mathbf{u}^h\|_h \leq \frac{\varepsilon}{2}, \tag{2.7a}$$

$$\|\mathbf{u}^h - w^h\|_h \leq \frac{\varepsilon}{2}. \tag{2.7b}$$

Although the error norm in (2.7a) is not generally computable, this condition can lead to practical stopping criteria for the iterative solution of (2.3). We will discuss this issue later. The immediate concern is (2.7b), which dictates the choice of h. Specifically, using (2.4), we can replace (2.7b) by the condition

$$h \leq \frac{6\pi^2 \varepsilon}{\bar{\beta}\, \|f\|_h} \equiv h_\varepsilon. \tag{2.7b'}$$

Thus, if we are given $\bar{\beta}$ (or, more realistically, a bound for $\bar{\beta}$), then the step size $h = h_\varepsilon$ is small enough to guarantee that the discretization is sufficiently accurate to meet our objective.

Another model that will be considered in this chapter is the two-dimensional (2-D) boundary value problem

$$\begin{aligned} -\Delta \mathscr{u} &= f(x, y) \quad \text{in } \Omega = \{(x, y)\colon 0 < x, y < 1\}, \\ \mathscr{u} &= 0 \quad \text{on } \partial\Omega. \end{aligned} \tag{2.8}$$

Here, $\Delta \mathscr{u} = \mathscr{u}_{xx} + \mathscr{u}_{yy}$ is the usual 2-D *Laplacian* and $\partial\Omega$ denotes the boundary of Ω.

As in the 1-D case, (2.8) may be discretized on a uniform grid given by

$$\Omega^h = \{(x_i, y_j): x_i = ih_x,\ y_j = jh_y,\ 1 \le i \le n_x,\ 1 \le j \le n_y\}$$

where $h_x = 1/(n_x + 1)$, $h_y = 1/(n_y + 1)$, and n_x and n_y are integers. By h we would generally mean the vector $h = (h_x, h_y)$, but for simplicity we assume $n_x = n_y = n$ and instead use the liberal notation $h = h_x = h_y$. Now, replacing the derivatives in (2.8) by centered difference approximations at each grid point (x_i, y_j), we have

$$\begin{aligned}(2.9)\qquad &\frac{1}{h^2}[-u(x_i + h, y_j) - u(x_i - h, y_j)\\ &\quad -u(x_i, y_j + h) - u(x_i, y_j - h) + 4u(x_i, y_j)] + \tau_{ij} = f(x_i, y_j),\end{aligned}$$

where again we can show that $\tau_{ij} = O(h^2)$. (These bounds now depend in a complicated way on the existence and magnitude of various partial derivatives of f.) When we drop the truncation error τ_{ij} from (2.9), the result is a set of difference equations for the discrete approximation $\mathbf{u}_{ij}$ to $u(x_i, y_j)$, $1 \le i, j \le n$. Assuming that the unknowns $\mathbf{u}_{ij}$ are ordered *lexicographically* by rows (or columns), we then find that this system of discrete equations may be represented by $A\mathbf{u} = f$, where the $n^2 \times n^2$ matrix A has the following block tridiagonal form:

$$(2.10)\qquad A = \frac{1}{h^2}\begin{pmatrix} T & -I & & 0 \\ -I & T & -I \ddots & \\ & \ddots & \ddots & -I \\ 0 & & -I & T \end{pmatrix}.$$

Here, T is the $n \times n$ tridiagonal matrix with diagonal entries 4 and super- and sub-diagonal entries -1, and I is the appropriate identity matrix. The equation of this system centered at the grid point (x_i, y_j) may be represented using the 5-point *stencil*

$$\frac{1}{h^2}\begin{bmatrix} & -1 & \\ -1 & 4 & -1 \\ & -1 & \end{bmatrix} u_{ij} = f_{ij},$$

which displays the interconnections between the unknowns associated with neighboring grid points. When we proceed as in the 1-D case, it is possible to show that the global error and truncation error are related by $A\mathbf{e} = \tau$ and that the error vector satisfies a bound of the form $\|\mathbf{e}\|_h \le \kappa h^2$, where κ is a constant which depends only upon f and certain of its partial derivatives.

An important observation is that the matrix of (2.10) is sparse and, in many practical applications, quite large. The sparseness, though highly structured, can cause difficulties for elimination methods because of the substantial fill-in that occurs. Therefore, conventional elimination methods can be quite inefficient and iterative methods should be considered. As will be shown, traditional iterative (or *relaxation*) methods tend to become ineffective as the problem size grows. Nevertheless, these basic iterative methods provide the foundation of multigrid methods, so it is important to discuss them and understand where their limitations lie.

1.3. Basic iterative schemes (relaxation). In the following treatment of iterative methods, the notation will be simplified by the use of *dynamic* variables to represent approximations that are successively updated, thereby taking on various *assignments* during the solution procedure. The replacement process of a particular iterative scheme will be indicated by a left arrow ($\leftarrow$). Thus, if u is an approximation to $\mathbf{u}$, the solution of (2.3), then an update of u will be represented by

$$u \leftarrow G(u),$$

where G is some expression involving the old (current) assignment of u. When it becomes absolutely necessary to distinguish between different specific assignments, notation such as $u_{(\text{old})}$, $u_{(\text{new})}$, or $u_{(\lambda)}$ will be used. When $u_{(\text{new})}$ is used, then u will be understood to mean $u_{(\text{old})}$.

An important word of caution: the reader should be very careful to understand precisely what each dynamic variable means, that is, what its assignment is at each point in the discussion. We will try to emphasize this at appropriate times.

Among the simplest iterative methods for solving matrix equations, such as (2.3), are the relaxation schemes that change u one entry at a time to meet certain criteria for each corresponding equation. Specifically, the ith step of the *Jacobi* method involves changing u_i so that the ith equation in (2.3) is satisfied. Thus, one full *sweep* of Jacobi applied to (2.3) can be expressed as follows:

$$u_{i_{(\text{new})}} \leftarrow \frac{u_{i-1} + u_{i+1} + h^2 f_i}{2}, \qquad i = 1, 2, \cdots, n. \tag{3.1}$$

When we let D, $-L$, and $-U$ denote the diagonal, strictly lower triangular, and strictly upper triangular parts of A, respectively, then $A = D - L - U$, and we may rewrite (3.1) in the following convenient matrix form:

$$u \leftarrow D^{-1}(L + U)u + D^{-1}f.$$

As we will show, the convergence of this scheme depends upon properties of the *iteration matrix* $R = D^{-1}(L + U)$.

The closely related *Gauss–Seidel* method results from the Jacobi method by using the updated approximations as soon as they are computed. In component form, this may be expressed as follows:

$$u_{i_{(\text{new})}} \leftarrow \frac{u_{i+1} + u_{i-1_{(\text{new})}} + h^2 f_i}{2}, \qquad i = 1, 2, \cdots, n. \tag{3.2}$$

If we again use the splitting $A = D - L - U$, then a Gauss–Seidel relaxation sweep appears in matrix form as follows:

$$u \leftarrow (D - L)^{-1} U u + (D - L)^{-1} f.$$

Now the iteration matrix $R = (D - L)^{-1} U$ determines the convergence properties of the scheme.

The Jacobi and Gauss–Seidel methods are just two of many *stationary linear* iteration schemes that can be written in the general form

$$u \leftarrow Ru + g, \tag{3.3}$$

where R is the iteration matrix. We will analyze the Jacobi method in some detail but will first make several key observations on the general scheme (3.3).

Convergence can occur in (3.3) only if the solution of the original problem $A\mathbf{u} = f$ is unchanged by the iteration. We therefore assume that

$$\mathbf{u} = R\mathbf{u} + g. \tag{3.4}$$

When we recall that $e = \mathbf{u} - u$ and subtract (3.3) from (3.4), then (3.3) becomes

$$e \leftarrow Re. \tag{3.5}$$

This means that the evolution of the error is determined only by the iteration matrix and the initial error.

A simple but important theoretical and practical observation can now be made concerning the homogeneous system $A\mathbf{u} = 0$. Specifically, note that the right-hand side f enters into (3.5) only by way of the initial error e. Since it is usual to assume that this error is arbitrary, the analysis of stationary linear iterative methods on general linear equations can usually be restricted to the homogeneous equations $A\mathbf{u} = 0$ if we assume an arbitrary initial guess u. Doing so provides the added advantage that the error is known: $\mathbf{u} = 0$ implies that $e = -u$, that is, the current approximation is also the actual error. We will use this approach in the following convergence analysis. (It can also be useful for numerical tests of performance, especially since it usually avoids difficulties with machine accuracy.)

In practice, the error e is generally not computable. However, it is often possible to determine useful information about the error by computing the *residual*, $r = f - Au$, which is analogous to the discretization error in the sense that r is the amount by which an iterate u fails to satisfy the original discrete problem. Furthermore, the definitions of the residual r and the error e combine to give the following important relationship:

$$Ae = r. \tag{3.6}$$

Another simple but very useful observation that arises from this relationship is that solving $A\mathbf{u} = f$ with an initial guess u is equivalent to solving $Ae = r$ with the initial guess $e = 0$. The two problems are connected by the relationship $e = \mathbf{u} - u$. This observation is fundamental to multigrid analysis and underlies the development that is given in the remainder of this chapter.

At this point, it is instructive to analyze the convergence properties of Jacobi relaxation. Recall that the error is governed by the relation

$$e_{(\text{new})} = D^{-1}(L + U)e = \left(I - \frac{h^2}{2}A\right)e,$$

where the splitting $A = D - L - U$ has been used. Using the symmetry of $R = I - (h^2/2)A$, we can bound the Euclidean norm of the new error in terms of that for the old according to the following:

$$\|e_{(\text{new})}\| \le \|R\| \, \|e\| = \rho(R) \, \|e\|.$$

(In fact, for the case considered here, this inequality is also valid in the norms $\|\cdot\|_h$ and $\|\cdot\|_A$.) Thus, $\rho(R)$ is a bound for the *convergence factor* $\|e_{(\text{new})}\|/\|e\|$ and the error is therefore decreased by each relaxation sweep provided $\rho(R) < 1$. Recall that $\lambda_1^h(A) \cong \pi^2$ and $\lambda_n^h(A) \cong 4/h^2 - \pi^2$. Since R is a polynomial in A, its *spectrum*, $\lambda(R)$, consists of the eigenvalues μ_i given approximately by

$$\mu_1 \cong 1 - \frac{\pi^2 h^2}{2}, \quad \mu_2 \cong 1 - 2\pi^2 h^2, \cdots, \quad \mu_n \cong -1 + \frac{\pi^2 h^2}{2}.$$

We see that the inequality $\rho(R) < 1$ is satisfied. Notice that the eigenvalues associated with low frequency (or long wave) modes ($i \ll n$) are close to 1, while the eigenvalues associated with the high frequency (or short wave) modes ($i \cong n$) are close to -1. A spectral radius very near 1 is attained in practice and does, in fact, result in poor convergence. More accurately, it implies slow attenuation for those eigencomponents of the error associated with eigenvalues near 1 in absolute value. Because these components are invariant under R (since R is a polynomial in A), they quickly begin to dominate the evolving error. This almost always guarantees slow asymptotic convergence. (It is possible but highly unlikely that the initial, and hence all subsequent, errors are orthogonal to the slowly damped components.) This is the reason that practitioners observe good reduction of error in the early iterations with Jacobi (and other relaxation) schemes, but much poorer performance as relaxation proceeds.

It is useful to ask whether classical modifications to Jacobi relaxation can improve convergence in any significant way. Consider, for example, the procedure in which a Jacobi update $u_{(\text{int})}$ is computed as an intermediate step from the current approximation u and then used to form the new iterate

$$u_{(\text{new})} \leftarrow u + \omega(u_{(\text{int})} - u),$$

where ω is a prescribed scalar parameter. Notice that $\omega = 1$ corresponds to a Jacobi update. The iteration matrix for this *damped* (or *underrelaxed*) Jacobi process is just

$$R_\omega = I - \omega \frac{h^2}{2} A.$$

The matrix R_ω can be analyzed in much the same way as the Jacobi iteration matrix, R_1. Its spectrum, $\lambda(R_\omega)$, consists of the values

$$\mu_1 \cong 1 - \frac{\omega h^2 \pi^2}{2}, \quad \mu_2 \cong 1 - 2\omega h^2 \pi^2, \cdots, \quad \mu_n \simeq 1 - 2\omega.$$

If we now attempt to choose ω to damp the high frequencies (for example, $\omega = \frac{1}{2}$ so that $\mu_n \cong 0$), then we still have $\mu_1 = 1 - O(h^2)$ and slow convergence remains for the low frequency error components. On the other hand, choosing ω to damp the low frequencies (for example, $\omega = 2/h^2\pi^2$ so that $\mu_1 \cong 0$) actually leads to magnification of the high frequency components, since then $\mu_n = 1 - O(h^{-2})$.

A choice of ω which approximately balances the extreme eigenvalues ($\mu_1 \cong \mu_n$) is

$$\omega_{\text{opt}} = \frac{2}{2 + h^2\pi^2/2},$$

for which

$$\rho(R_{\omega_{\text{opt}}}) \cong \frac{4 - h^2\pi^2}{4 + h^2\pi^2} = 1 - O(h^2).$$

(Note that $\omega_{\text{opt}} \cong 1$. Thus, Jacobi's method without damping already approximately balances its effect on the extreme ends of the spectrum of A.) Unfortunately, now *both* high and low frequency error components have slow decay rates.

Having determined that Jacobi relaxation, even with damping, cannot effectively reduce *all* components of the error, can we still use it to advantage in reducing some subset of the components? Suppose the spectrum of A is contained in the interval $[a, c]$, which we partition as $[a, c] = [a, b) \cup [b, c]$. Consider the interval $[b, c]$ corresponding to the high frequency components and let the "optimal" ω for this interval be denoted ω_{bc}; that is, ω_{bc} balances the eigenvalues of $R_{\omega_{bc}}$ over $[b, c]$. This implies that

$$1 - \omega_{bc}\frac{h^2}{2}b = \omega_{bc}\frac{h^2}{2}c - 1$$

so that

$$\omega_{bc} = \frac{4}{h^2(b+c)}.$$

With $c = 4/h^2$, suppose we choose $b = 2/h^2$ so that $[b, c]$ captures the top half of $\lambda(A)$. Then $\omega_{bc} = \frac{2}{3}$ and

$$\rho(R_{\omega_{bc}}/S_{[b,c]}) \approx \left|1 - \omega_{bc}\frac{h^2}{2}b\right| = \frac{1}{3},$$

where $R_{\omega_{bc}}/S_{[b,c]}$ is the iteration matrix $R_{\omega_{bc}}$ restricted to $S_{[b,c]} = \{\text{span } w \text{ in } H\colon Aw = \lambda w,\ b \le \lambda \le c\}$. Thus, the Jacobi method can be damped to give a satisfactory *convergence factor* for the high frequency components of the error. In fact, this factor depends directly on the *condition number* of $A/S_{[b,c]}$ given by $\text{cond}\,(A/S_{[b,c]}) = \max\,\{\lambda/\mu\colon \lambda,\ \mu \text{ in } \lambda(A) \cap [b, c]\} \cong 2$. Because this condition number is small and independent of h, so is the convergence factor for damped Jacobi.

Figures 1.1(a)–(c) illustrate how the damped Jacobi method behaves on certain error components. Specifically, the method was used with $\omega = \frac{2}{3}$ to solve (2.3) with $h = 1/64$, $f = 0$, and various initial guesses (errors). Figure 1.1(a) shows that the oscillatory (high frequency) error is quickly attenuated after 10 iterations, while the *smooth* (low frequency) error of Fig. 1.1(b) is not. The error of Fig. 1.1(c) with mixed frequencies shows a good initial reduction because the oscillatory components are eliminated. However, further sweeps do little to eliminate the remaining smooth components.

The above argument suggests that a *spectral* (or *Fourier mode*) analysis might be useful in understanding relaxation methods. Indeed, much of the original analysis of the basic multigrid schemes (cf. [88], [538][1]) used precisely this point of view. Although spectral analysis will not be pursued in this chapter, this might be a good place to present some of the essential ideas. Specifically, if we are working in 1-D on a grid Ω^h, then an arbitrary grid function $u \in H^h$ may be represented in terms of a discrete Fourier series of the form

$$u_j = \sum_k c_k e^{i\theta_k j}.$$

[1] Numbers in square brackets refer to the Multigrid Bibliography included as Appendix 2. Other references are compiled at the end of their respective chapters and are referred to by letters.

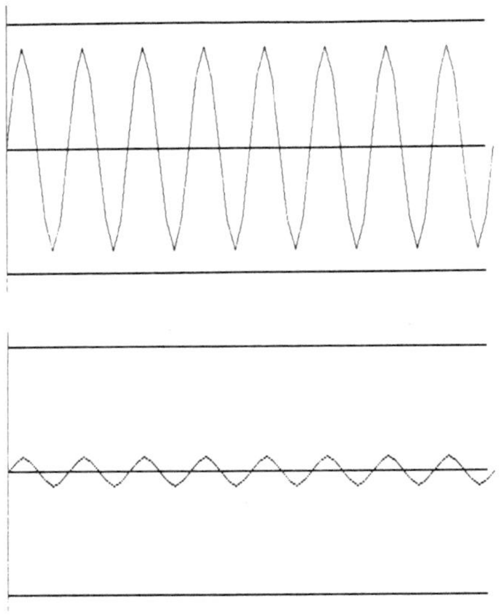

FIG. 1.1(a). *The damped Jacobi method* ($\omega = 2/3$) *applied to equation* (2.3) *of the text with* $h = 1/64$ *and* $f = 0$. *The initial guess (which is also the initial error) has a short wavelength of* $\lambda = 1/8$. *In this case* 10 *iterations of the damped Jacobi method produce a significant decrease in the solution.*

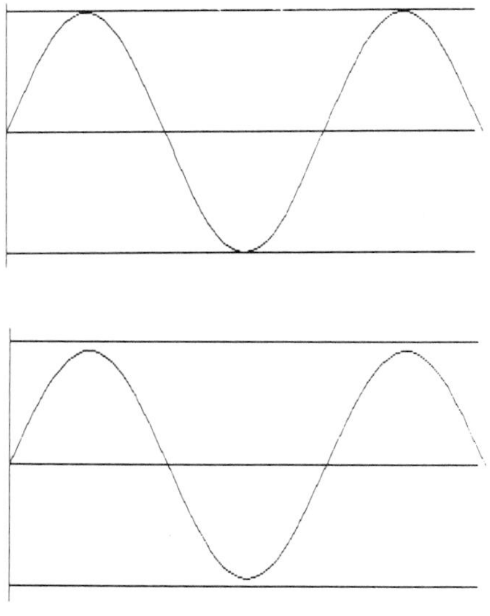

FIG. 1.1(b). *The same experiment as shown in Fig.* 1.1(a) *except that the initial solution has a relative long wavelength of* $\lambda = 2/3$. *The upper figure shows the solution after one iteration. The lower figure shows very little damping after* 10 *iterations.*

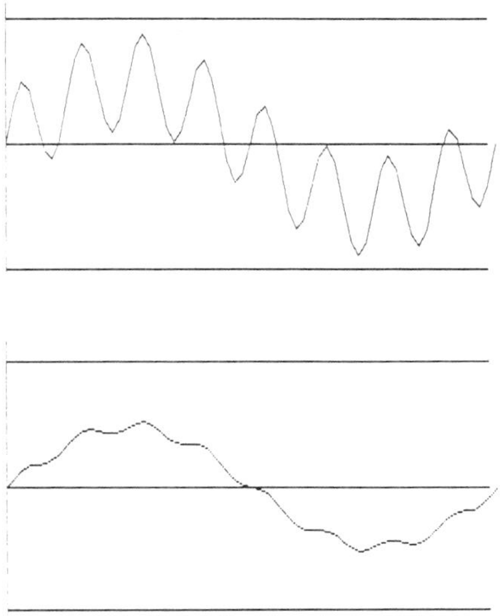

FIG. 1.1(c). *The damped Jacobi method applied to an inital guess consisting of a long wave* ($\lambda = 1$) *and a short wave* ($\lambda = 1/8$). *After* 10 *iterations, the wave is significantly reduced in amplitude, while the long wave persists.*

Here, $i = \sqrt{-1}$. (The precise form of the boundary conditions does not affect the following discussion.) Unlike the continuous case, however, only roughly n modes are needed for this representation. The *wave numbers* (or *spatial frequencies*) θ_k have the form $2\pi k/(n+1)$ for $-n/2 \lesssim k \lesssim n/2$ and correspond to the Fourier modes with wavelengths $\lambda_k = \infty, 1, \frac{1}{2}, \cdots, 4h, 2h$. This leads to the important observation that variations on a scale less than $2h$ cannot be resolved on Ω^h. (In fact, these variations will undergo *aliasing*, appearing spuriously on a scale that *can* be resolved by the grid, e.g., the mode $e^{i4\pi j/3}$, which has a wavelength of $3h/2 < 2h$, will appear as the mode $e^{i(2\pi - 4/3\pi)j} = e^{-i2\pi j/3}$ with wavelength $3h > 2h$.) In this setting, we see that $S_{[b,c]}$ (defined above) is the space of grid functions spanned by the modes with wavelengths $4h \leq \lambda \leq 2h$. This is roughly half of the Fourier modes present on Ω^h.

In anticipation of later developments, we might ask what the Fourier modes of Ω^h look like on the coarser grid Ω^{2h}. Figure 1.2 shows Ω^h marked by x's and $\cdot$'s. The dashed line represents the $\theta_{n/2}$ mode on Ω^h with a wavelength of $4h$. Seen on grid Ω^{2h} marked by x's, however, this mode (marked by open circles) has a wavelength of $2(2h)$ which means that it is the mode with the shortest possible wavelength on Ω^{2h}. Modes with a wavelength of less than $4h$ on Ω^h cannot be represented on Ω^{2h}. One of the

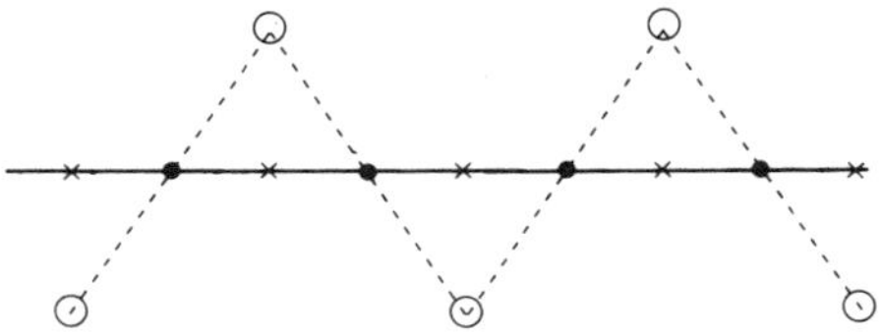

FIG. 1.2. *The coarse grid Ω^{2h} is marked by x's. The fine grid Ω^h is marked by x's and ·'s. A wave with wavelength 4h (dashed line) on Ω^h appears as the mode with the shortest possible wavelength on Ω^{2h} (open circles).*

useful interpretations of this observation is that a grid function with moderate smoothness on Ω^h appears as an oscillatory function on Ω^{2h}. It then follows that a relaxation scheme such as weighted Jacobi, in eliminating the oscillatory modes on Ω^h, generally leaves modes that are still oscillatory on Ω^{2h}. This suggests that a repeated application of relaxation on successively coarse grids, if we can decide how to do this, should progressively eliminate all of the modes in the error. These ideas will be taken up again in a more systematic way in the next section.

The damped Jacobi method's ability to selectively eliminate the high frequency components of the error will be important in all that follows; we will call it the *smoothing property*. This property is certainly not confined to damped Jacobi. Many common relaxation methods, when applied to s.p.d. systems, also possess it. The property appears to be a limitation of these methods, but this limitation provides our first clue to how the method can be salvaged. This is one of the entry points to multigrid methods. Before pursuing this clue, however, we present two additional arguments that offer alternative ways to approach and interpret multigrid methods.

One natural way to improve the performance of classical iterative methods is to try to obtain a good initial guess at as little cost as possible. A well-known strategy involves a coarser grid, say, Ω^{2h}. Computation of u^{2h}, *an approximation to* $\mathbf{u}^{2h}$, is less expensive, not only because there are fewer unknowns to update, but also because the convergence factor $1 - O(h^2)$ is improved. However, if the coarse grid does not satisfy the bound (2.7b′), then the global error bound (2.7b) may not be satisfied, so we cannot expect that u^{2h} by itself will be accurate enough. At the same time, though, u^{2h} may still be valuable as an initial approximation in a relaxation method for finding $\mathbf{u}^h$.

In order to use u^{2h} as an initial guess for $\mathbf{u}^h$, we must first devise an effective scheme for mapping *coarse* grid vectors in H^{2h} into *fine* grid vectors in H^h. The most natural way to do this is with an operation called *interpolation* (or *prolongation*). We restrict ourselves to *piecewise multilinear interpolation* (although there are many alternatives) and define the

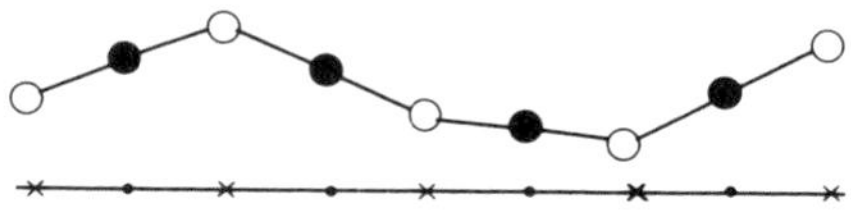

FIG. 1.3. *A coarse grid vector u^{2h} (open circles) interpolated to a fine grid vector u^h (open and closed circles) by the operation $u^h = I_{2h}^h u^{2h}$ defined in the text.*

operator $I_{2h}^h: H^{2h} \to H^h$, with $u^h = I_{2h}^h u^{2h}$, in 1-D by the following (see Fig. 1.3):

$$\begin{aligned} u_{2i}^h &= u_i^{2h}, \\ u_{2i+1}^h &= (u_i^{2h} + u_{i+1}^{2h})/2, \qquad 0 \le i \le n^{2h}. \end{aligned}$$

In 2-D this becomes

$$\begin{aligned} u_{2i,2j}^h &= u_{ij}^{2h}, \\ u_{2i+1,2j}^h &= (u_{ij}^{2h} + u_{i+1,j}^{2h})/2, \qquad 0 \le i, j \le n^{2h}, \\ u_{2i,2j+1}^h &= (u_{ij}^{2h} + u_{i,j+1}^{2h})/2, \\ u_{2i+1,2j+1}^h &= (u_{ij}^{2h} + u_{i+1,j}^{2h} + u_{i,j+1}^{2h} + u_{i+1,j+1}^{2h})/4. \end{aligned}$$

(Notice that when these formulas refer to values of u at boundary points, we assume that these values are zero on both $2h$ and h.)

These operators may be represented in *stencil* form:

one dimension $\quad]\tfrac{1}{2} \;\; 1 \;\; \tfrac{1}{2}[$

two dimensions $\quad \left]\begin{matrix} \tfrac{1}{4} & \tfrac{1}{2} & \tfrac{1}{4} \\ \tfrac{1}{2} & 1 & \tfrac{1}{2} \\ \tfrac{1}{4} & \tfrac{1}{2} & \tfrac{1}{4} \end{matrix}\right[.$

Stencils representing maps of H^{2h} into H^h use reverse square brackets. The stencil is defined on the fine grid and is meant to be centered at a point corresponding to the coarse grid. Each entry indicates the contribution of the coarse grid point to the three (or nine) corresponding fine grid points. The central entry is the coefficient of u_{2i}^h (or $u_{2i,2j}^h$) in the formula for u_i^{2h} (or u_{ij}^{2h}).

The idea now is that we can form an initial guess on grid h by determining a rough approximation, u^{2h}, to $\mathbf{u}^{2h}$ and computing $u^h = I_{2h}^h u^{2h}$. The issue, of course, is how well this strategy works. If we examine the smooth function of Fig. 1.4(a), it appears that, for such $\mathbf{u}^h$, this process could be very effective. In fact, numerical experience shows that such an initial guess can eliminate many of the early relaxation sweeps generally required by *naive* guesses (e.g., zero or random vectors). Thus, the (residual) error usually starts off much smaller than it would with the initial guess $u^h = 0$. However,

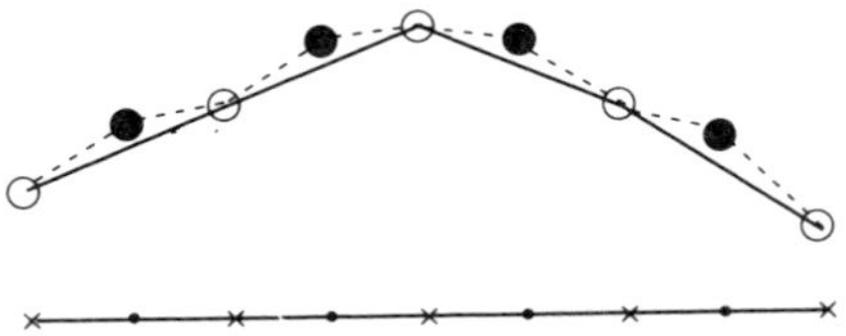

FIG. 1.4(a). *A smooth fine grid vector (dashed line through open and closed circles) may be well approximated by a coarse grid interpolant (solid line through open circles).*

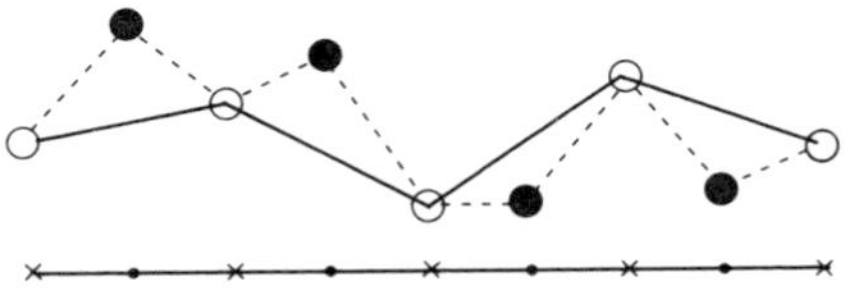

FIG. 1.4(b). *However, an oscillatory fine grid vector (dashed line through open and closed circles) may be poorly approximated by its coarse grid interpolant (solid line through open circles).*

asymptotic convergence rates are generally independent of the initial guess, so slow rates will quickly begin to appear. Thus this strategy does have some limitations, particularly when a high degree of accuracy is required, which is just where efficient computation is really needed.

Notice, however, that this initial guess strategy would not be very useful if $\mathbf{u}^h$ were highly oscillatory as depicted in Fig. 1.4(b). This figure shows that, while smooth errors *can* be well approximated by coarse grid interpolants, oscillatory errors *cannot*. This means that the initial guess for oscillatory errors might just as well be naive.

We will call this strategy of generating an informed initial guess *nested iteration,* since it uses a nested sequence of grids Ω^{2h}, Ω^h to compute better approximations to $\mathbf{u}$. At this point (and in later sections), it is important to think *recursively*. Specifically, while it may be effective to use grid $2h$ to get a good initial guess for grid h, we may find that iteration on grid $2h$ is still so slow that a good initial guess is desirable here as well. We are thus led to using grid $4h$ just as we used grid $2h$. This suggests the use of a nested sequence of $q \geq 1$ grids Ω^{h_k}, $h_k = 2^{q-k}h$, $1 \leq k \leq q$, where the approximation u^{h_1} is easy to compute and where

$$u^{h_{k+1}} = I_{h_k}^{h_{k+1}} u^{h_k}$$

is used to start the grid h_{k+1} iteration.

The idea of nested iteration has some intuitive appeal, since it is natural to think that a coarse grid solution should give an improved initial guess for a fine grid problem. Yet the idea does have its limitations, as we have observed. Once again, however, the drawbacks of nested iteration also hold a clue to salvaging it, as we shall see in the next section.

We have seen that nested iteration *can* be an effective method of obtaining a good initial guess for relaxation when none is available. Our third line of thinking arises when we ask what can be done if an initial guess is already given. Recall an earlier observation that iterating on the original equation (2.3), with an arbitrary initial guess, is equivalent to iterating on the residual equation (3.6) with an initial guess of zero. Thus, in situations where some initial guess is given, it is important to think about the residual equation and the possibility of using relaxation to solve for the error starting with a naive initial guess of zero. This approximation to the error can be used to correct the current approximation to $\mathbf{u}^h$. But now, all of our previous discussions of relaxation ideas pertain to the solution of the residual equation. In particular, we are led to consider using nested iteration as a means of obtaining an initial guess *for the error.* To do this, however, we must first have a way of writing the coarse grid (say, Ω^{2h}) counterpart of the residual equation (3.6). This means that we need to find a sensible representation of r^h on Ω^{2h}. (This question does not arise when nested iteration is applied directly to the original problem (2.3), since f^{2h} is presumably a sensible approximation to f^h.)

Suppose that u^h is given on Ω^h and let $r^h = f^h - A^h u^h$, which is easily computed. In order to represent r^h on Ω^{2h}, we need a way to map fine grid vectors in H^h to coarse grid vectors in H^{2h}. This may be accomplished using a *restriction operator,* $I_h^{2h}: H^h \to H^{2h}$. We will consider the two most common ones, the first of which is *injection.* In 1-D, injection is defined by $u^{2h} = I_h^{2h} u^h$, where $u_i^{2h} = u_{2i}^h$. In 2-D, it is given by $u_{ij}^{2h} = u_{2i,2j}^h$. The *full-weighting* restriction operator is defined in 1-D by

$$u_i^{2h} = \tfrac{1}{4}(u_{2i-1}^h + 2u_{2i}^h + u_{2i+1}^h)$$

and in 2-D by

$$\begin{aligned} u_{ij}^{2h} = \tfrac{1}{16}[&u_{2i+1,2j+1}^h + u_{2i+1,2j-1}^h + u_{2i-1,2j+1}^h + u_{2i-1,2j-1}^h \\ &+ 2(u_{2i,2j+1}^h + u_{2i,2j-1}^h + u_{2i+1,2j}^h + u_{2i-1,2j}^h) + 4u_{2i,2j}^h]. \end{aligned}$$

The stencils for these operators are as follows:

	One Dimension	*Two Dimensions*
Injection:	$[0 \quad 1 \quad 0]$	$\begin{bmatrix} 0 & 0 & 0 \\ 0 & 1 & 0 \\ 0 & 0 & 0 \end{bmatrix}$,
Full Weighting:	$[\tfrac{1}{4} \quad \tfrac{1}{2} \quad \tfrac{1}{4}]$	$\begin{bmatrix} \frac{1}{16} & \frac{1}{8} & \frac{1}{16} \\ \frac{1}{8} & \frac{1}{4} & \frac{1}{8} \\ \frac{1}{16} & \frac{1}{8} & \frac{1}{16} \end{bmatrix}$.

Stencils representing maps of H^h into H^{2h} use square brackets. The stencil covers the fine grid and is meant to be centered at a corresponding coarse grid point. Each entry indicates the contribution of the neighboring fine grid points to the central coarse grid point. Thus, for injection, the value at the coarse grid points is just the value at the corresponding fine grid point. Full weighting uses a weighted average of the values at the corresponding fine grid point and its nearest neighbors.

The idea is now fairly simple, though somewhat subtle. Suppose that an initial guess, u^h, is given. Then instead of applying nested iteration to (2.3) in ignorance of u^h, we apply it to (3.6) as follows:

Step 1: Form $r^h = f^h - A^h u^h$.
Step 2: Set $f^{2h} = I_h^{2h} r^h$.
Step 3: Compute an approximation, u^{2h}, to the solution of $A^{2h}\mathbf{u}^{2h} = f^{2h}$.
Step 4: Set $u^h \leftarrow u^h + I_{2h}^h u^{2h}$.

A few comments are certainly in order. First of all, f^{2h} is now being used as a dynamic variable. In the above scheme, it is used as a grid $2h$ approximation to r^h, *not* f^h. In a similar way, u^{2h} is a grid $2h$ approximation to e^h, *not* $\mathbf{u}^h$. (This is a very important subtlety of the notation.) Therefore, $I_{2h}^h u^{2h}$ is not an approximation to the solution $\mathbf{u}^h$, but rather to the error e^h. The correction to the solution thus takes the form $u^h \leftarrow u^h + I_{2h}^h u^{2h}$. Finally, we should mention that this strategy could also be used recursively. An initial guess for $\mathbf{u}^{2h}$ (an approximation to the error) could be obtained from grid Ω^{4h} by solving the grid $4h$ counterpart of the residual equation. This grid $4h$ problem could also be started by way of grid $8h$, and so on.

How good is this strategy? The assessment of nested iteration applied to the original equation (2.3) can be applied equally well to the residual equation (3.6). The resulting conclusion is that nested iteration applied to the residual equation is effective for, at most, improving the early stages of relaxation, and then only when the *error* is smooth. There is an encouraging complementarity here: nested iteration will do well if the error is smooth, but if the error is not smooth, then relaxation should be effective.

Before continuing, it may be helpful to summarize the major points of this section. Taking damped Jacobi relaxation as a fairly representative stationary linear iterative scheme, we find that an analysis of its convergence properties leads to three observations:

(i) *Relaxation*. The damped Jacobi method has the *smoothing property* that ensures rapid convergence during early stages of relaxation when high frequency components of the error are quickly eliminated. Unfortunately, this scheme is usually very slow to eliminate low frequency error components.

(ii) *Nested iteration*. When a good initial approximation for relaxation on a grid Ω^h is unavailable, one may often be found by solving the corresponding problem on a coarser grid Ω^{2h}. Recursive application of this idea leads to the strategy of *nested iteration,* in which a sequence of nested grids is used to obtain informed initial guesses. Unfortunately, this has no effect on the asymptotic convergence rate of relaxation, so when high accuracy is required, many relaxation sweeps must still be done. Moreover, this can be effective only to the extent that the *solution* $\mathbf{u}^h$ is smooth.

(iii) *Residual equation.* When an initial guess is available, it is advisable to think in terms of the *residual equation* $Ae = r$ using zero as an initial guess. Thus, nested iteration may be used even here to obtain an improved initial guess for e. This too has the limitation that it can improve only early stages of relaxation, and then only when the *error* e is smooth.

1.4. Multigrid methods. We will now follow the clues in the previous section to see how the limitations of relaxation and nested iteration can be overcome. In fact, we will tie these ideas together to form one effective process, namely, multigrid. From what we already know, this should seem like a natural possibility. Since relaxation is successful in eliminating oscillatory error components, but not smooth ones—and since, quite the reverse, nested iteration on the residual equation can be effective on smooth components but not oscillatory ones—there should be a way to combine them to mutual advantage.

Toward this end, imagine that we are faced with solving the original equation (2.3) from scratch with no good initial guess. In this instance, it would seem worthwhile to try nested iteration to obtain a better initial guess: the worst that could happen is that $\mathbf{u}^h$ would be so oscillatory that the initial guess $u^h = I_{2h}^h u^{2h}$ would be of little value. In the best case, $\mathbf{u}^h$ might be very smooth and nested iteration would provide an improved initial guess. In either case, the error $e^h = \mathbf{u}^h - u^h$ after nested iteration should be oscillatory, since the smooth part of $\mathbf{u}^h$ is well approximated by $u^h = I_{2h}^h \mathbf{u}^{2h}$. Thus, if relaxation is used just after nested iteration, it should work very well.

Unfortunately, there is still the persistent problem of slow asymptotic convergence of the relaxation. Suppose, for example, that the error tolerance ε is very small so that a small mesh size h is needed to satisfy (2.7b). Then relaxation on Ω^h will usually slow to its asymptotic rate well before the convergence criterion (2.7a) is met. But such slow convergence is a clear indication that the error e^h is smooth. According to the third clue, this is just the point at which nested iteration should be most effective. However, now we should use nested iteration, not on (2.3), but *on the residual equation* (3.6), to obtain an estimate for the *error.* This estimate of the error can then be transferred to Ω^h to correct the current approximation u^h. We can then apply further relaxation sweeps to the corrected u^h. When the error once again becomes smooth, nested iteration may be applied again. In summary, we can use nested iteration on the residual equation (3.6) to reduce smooth components of e^h and relaxation on the original equation (2.3) to reduce the oscillatory ones. The continued use of nested iteration interspersed with relaxation on Ω^h is the essence of multigrid.

The basic *two-grid* cycling scheme can now be described as follows. Let ν_1 and ν_2 be integers that determine the number of Ω^h relaxation sweeps to be done before and after nested iteration, respectively. Then one two-grid cycle of multigrid beginning with an initial guess u^h and right-hand side f^h is denoted by

$$u^h \leftarrow MG^h(u^h, f^h)$$

and defined by the following steps:

Step 1: (*pre-relaxation*): Perform ν_1 relaxation sweeps (damped Jacobi, for example) on $A^h\mathbf{u}^h = f^h$ starting with initial guess u^h.
Step 2: (*coarse grid solution*): Form $r^h = f^h - A^h u^h$ and $f^{2h} = I_h^{2h} r^h$. Compute an approximation, u^{2h}, to the solution of $A^{2h}\mathbf{u}^{2h} = f^{2h}$.
Step 3: (*coarse grid correction*): Let $u^h \leftarrow u^h + I_{2h}^h u^{2h}$.
Step 4: (*post-relaxation*): Perform ν_2 relaxation sweeps on $A^h\mathbf{u}^h = f^h$ with initial guess u^h.

The development of this *basic* multigrid scheme has been highly heuristic, and it leaves many issues open. In the remaining sections and in other chapters, both practical and theoretical aspects of multigrid methods will be treated in a more rigorous way. We first finish this section by introducing more practical multigrid algorithms that use several grids in the cycling process. Again, we are asked to think recursively: The second step of the above two-level scheme requires solving a linear system on the grid Ω^{2h}. If an iterative scheme is applied to this problem, it should not be surprising that the slow convergence that plagued the Ω^h problem will also occur on Ω^{2h}. In order to overcome this, we are led to repeat the two-level scheme on a yet coarser grid Ω^{4h}. This suggests a recursive scheme in which successively coarser grids are used, where the descent to coarser levels continues until the convergence rate of relaxation is acceptable. (We could also stop at some coarse level and solve the system directly or by some other iterative process.) From the coarsest grid, the algorithm then would ascend through finer grids, correcting the solution at each level. Perhaps the simplest such *multi-level* scheme is called the *V-cycle*: Given h, f^h, u^h, ν_1, and ν_2 as before, let h_1 be the coarsest grid which is to be used. The *V*-cycle scheme is then denoted by

$$u^h \leftarrow MV^h(u^h, f^h)$$

and is defined recursively by the following four steps:

Step 1: Relax ν_1 times on $A^h\mathbf{u}^h = f^h$ starting with initial guess u^h.
Step 2: If $h = h_1$, go to Step 4. Otherwise, set $r^h = f^h - A^h u^h$, $f^{2h} = I_h^{2h} r^h$, $u^{2h} \leftarrow 0$, and $u^{2h} \leftarrow MV^{2h}(u^{2h}, f^{2h})$.
Step 3: Set $u^h \leftarrow u^h + I_{2h}^h u^{2h}$.
Step 4: Relax ν_2 times on $A^h\mathbf{u}^h = f^h$ starting with initial guess u^h.

Figure 1.5(a) illustrates the grid schedule of the V-cycle where $h_1 = 8h$.

Another multi-level scheme is the *W-cycle*. This can be useful for more general applications because it expends more effort on coarser levels, hopefully producing better coarse grid corrections. Denoted by

$$u^h \leftarrow MW^h(u^h, f^h),$$

the W-cycle is a special case of the *μ-cycle* scheme, which is defined as follows. Let μ be a positive integer. With h, f^h, u^h, ν_1, ν_2, and h_1 as before, the μ-cycle is denoted by

$$u^h \leftarrow MG_\mu^h(u^h, f^h)$$

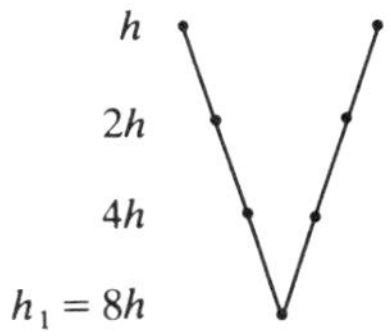

FIG. 1.5(a). *Grid schedule for a V-cycle with four levels.*

FIG. 1.5(b). *Grid schedule for a W-cycle with four levels.*

FIG. 1.5(c). *Grid schedule for an FMV cycle on four levels with $\nu_0 = 1$ applications of the basic V-cycle.*

and defined recursively by the following four steps:

Step 1: Relax ν_1 times on $A^h\mathbf{u}^h = f^h$ starting with initial guess u^h.
Step 2: If $h = h_1$, go to Step 4. Otherwise, set $r^h = f^h - A^h u^h$, $f^{2h} = I_h^{2h} r^h$, and $u^{2h} \leftarrow 0$, and perform $u^{2h} \leftarrow MG_\mu^{2h}(u^{2h}, f^{2h})$ μ times.
Step 3: Set $u^h \leftarrow u^h + I_{2h}^h u^{2h}$.
Step 4: Relax ν_2 times on $A^h\mathbf{u}^h = f^h$ starting with initial guess u^h.

Figure 1.5(b) illustrates the grid schedule of the W-cycle which is defined by $MW = MG_2$. (Note that the V-cycle is also a special case of the μ-cycle because $MV = MG_1$.)

Nested iteration was used to develop the coarse grid correction process of multigrid. In essence, we capitalized on the *dual* advantage that coarse grid relaxation is both cheaper (because there are fewer points) and more effective (because the asymptotic convergence factor $1 - O(h^2)$ is smaller). Thus, multigrid *basic* cycling schemes start with relaxations on the fine grid, then proceed to coarser levels to reduce smooth error components. Yet, if coarse grid computation has such advantages, it is perhaps better to start on coarser levels and proceed to finer grids only when sufficiently good approximations to the solution of the original equation have been achieved. In effect, we are thus led to using nested iteration again, but this time on the multigrid scheme itself. This yields the so-called *full multigrid* (*FMG*) cycling scheme. With the same quantities as those used with MV and an additional integer ν_0 that dictates the number of basic cycles, the full multigrid algorithm for the V-cycle is denoted by

$$u^h \leftarrow FMV^h(u^h, f^h)$$

and defined recursively by the following two steps:

Step 1: If $h = h_1$, then go to Step 2. Otherwise, set $r^h = f^h - A^h u^h$, $f^{2h} = I_h^{2h} r^h$, $u^{2h} \leftarrow 0$, $u^{2h} \leftarrow FMV^{2h}(u^{2h}, f^{2h})$, and $u^h \leftarrow I_{2h}^h u^{2h}$.
Step 2: Perform $u^h \leftarrow MV^h(u^h, f^h)$ ν_0 times.

Figure 1.5(c) illustrates the FMV cycle with $\nu_0 = 1$. Similar definitions can be made for FMW and FMG_μ.

The basic aim of FMG is to guarantee a good initial guess for level h before any processing is done there. Thus, FMG will often be used without a starting guess or, more precisely, with $u^h = 0$. In this case, we write $FMG_\mu^h(f^h)$ in place of $FMG_\mu^h(u^h, f^h)$, for example. If we initially have $f^{2h} = f^h$ for all h, then this algorithm simplifies so that $u^h \leftarrow FMV^h(f^h)$ now becomes the following:

Step 1: If $h = h_1$, go to Step 2. Otherwise, set $u^{2h} \leftarrow FMV^{2h}(f^{2h})$ and $u^h \leftarrow I_{2h}^h u^{2h}$.
Step 2: Perform $u^h \leftarrow MV^h(u^h, f^h)$ ν_0 times.

For each of the cycling schemes just introduced, there are many issues and alternatives including the number of different grids that should be used, the number of cycles that should be made through those grids, and the number of iterations that should be made on each level. These are some of the more immediate issues that arise, but there are many more, some of which will be considered in the remaining sections and chapters. The reader is also directed to other references (cf. [101], [538]) for further study.

1.5. Performance (implementation, complexity, results). Most of the discussion up to this point has dealt with the model problem in $d=1$ dimension. With little extra work, implementation and complexity can be discussed in the general case $d \geq 1$. Therefore, in this section we consider the solution of the model problems (2.2) ($d=1$) and (2.8) ($d=2$) and their analogues in higher dimensions.

A simple but representative calculation concerns the storage costs of the various multigrid methods presented in § 4. On each grid Ω^h, it is necessary to store u^h and f^h, each requiring about n^d locations. If Ω^h is coarsened in powers of two, as in our previous discussion, then the total storage requirement is about

$$2n^d(1+2^{-d}+2^{-2d}+\cdots)=\frac{2n^d}{1-2^{-d}}.$$

This means that multigrid requires only a marginal increase in storage over fine grid relaxation.

For programming multigrid methods in 1-D, it is easiest to store the unknowns for the various grids sequentially in one long vector. In two or more dimensions, there are several possible data structures. One common practice is to store the unknowns for the various grids sequentially in a single vector but to reference them as d-dimensional arrays in subroutines by passing initial indices for the appropriate level. The sample program given at the end of the book as Appendix 1 illustrates this approach.

We can make a similar calculation of the computational cost of the various cycling schemes. If we assume that relaxation requires c arithmetic operations to update a single unknown, then one sweep of relaxation on Ω^h costs about cn^d operations. (Throughout this book, c is taken to be a generic constant and will be used for different purposes.) This cost is called a *work unit* (WU). Experience indicates that intergrid transfers and other noniterative operations typically add only 10–20% to the computational load, so such costs will be neglected. The cost of performing one V-cycle, MV^h, with $\nu_1=\nu_2=1$ is therefore about

$$2cn^d(1+2^{-d}+2^{-2d}+\cdots)=\frac{2cn^d}{1-2^{-d}}$$

or $2/(1-2^{-d})$ WUs. Similarly, the cost of one complete *FMV* cycle with $\nu_0 = 1$ is about

$$\frac{2}{1-2^{-d}} cn^d(1 + 2^{-d} + 2^{-2d} + \cdots) = \frac{2cn^d}{(1-2^{-d})^2}$$

or $2/(1-2^{-d})^2$ WUs. This means that one *FMV* cycle costs roughly the same as three or four fine grid relaxation sweeps.

Of course, the question is how well the *MV* and *FMV* cycling schemes work. It is often said that multigrid processes converge at a rate independent of h. The reasoning is that the convergence factor for most relaxation schemes *is* independent of h for the oscillatory modes. Through the use of coarse grids, multigrid methods use relaxation essentially as a means of attenuating the oscillatory modes. Therefore, the overall convergence factor for a good multigrid scheme is usually close to the convergence factor of relaxation restricted to the oscillatory modes. This is usually small and bounded independent of h.

A more precise statement about multigrid convergence is difficult to make. First of all, the analysis of any multigrid algorithm and application can be quite complex, since it must account for many factors (e.g., relaxation and its ordering, prolongation, restriction, and operator coefficients). Second, a priori analysis can provide valuable qualitative insight, but is almost always deficient in giving quantitative results. Most of all, different philosophical perspectives lead to substantially different approaches to analysis. Three popular forms of analysis have emerged; we will discuss each briefly.

Fourier Mode Analysis. The basic approach here is to make several simplifying but presumably reasonable assumptions regarding each of the various processes in a particular algorithm as well as features of the underlying problem. This allows us to avoid such difficulties as nonlinearities, boundary effects, and mixing of the various modes, and to concentrate on a narrow range of frequencies. For example, analysis of MG^h with damped Jacobi applied to the Poisson equation (2.2) might assume that the coarse grid correction exactly eliminates smooth modes and does not influence oscillatory modes. It is then enough to analyze the worst-case convergence factor of damped Jacobi for modes that do not appear on grid $2h$. For 1-D, we may define the *smoothing factors* to be

$$\mu = \max_{\pi/2 \le |\theta| \le \pi} \mu(\theta),$$

where $\mu(\theta)$ is the convergence factor of the damped Jacobi method applied to the mode with wave number θ. (The Jacobi method as applied to uniformly discretized Poisson equations leaves Fourier modes invariant, so the Fourier modes are also eigenvectors of the iteration matrix R and the

$\mu(\theta)$ are really the corresponding eigenvalues of R.) Our earlier analysis showed that, for the damped Jacobi method, $\mu \cong \frac{1}{3}$. We may then suppose that the convergence rate of MG^h is essentially $\mu^{\nu_1+\nu_2}$ which, for small ν_1 and ν_2, is close to the rate of convergence observed in practice.

The major limitation to this approach is that, while it can be a very effective predictor of performance, it is usually not rigorous. Mode analysis may be made rigorous under certain circumstances and for small enough h, but the question is still unsettled for practical, general applications.

Rigorous Theory. As we shall see in the next and subsequent chapters, there is now an abundant and rigorous theory of multigrid methods applied to an increasingly broad class of problems. In fact, rigorous theoretical results establish in a very general setting that these methods have convergence factors bounded independent of h. In some cases, the provided estimates depend on various details of the method and application and can become quantitative only when these details are specified. Thus, although theoretical work has led to many practical algorithms and principles, it is not yet very useful for predicting convergence factors.

Numerical Tests. The proof is in the pudding, so careful numerical experiments can be used to assess performance for particular algorithms and applications. In fact, worst-case convergence factors of the basic cycling schemes can usually be determined independently of the right-hand side. This can be done by setting $f = 0$, starting with a random initial guess u^h and carrying out several cycles until the ratio of successive discrete errors $e^h = u^h$ has stabilized. (This approach generally does not apply to FMG algorithms.) There are general limitations on such tests; the obvious one is that they give only a posteriori results. Nevertheless, multigrid methods are now surviving the test of time in many numerical environments, so this form of analysis is essential.

Rather than pursuing these three lines of analysis further, we will assume that the basic multigrid cycle under consideration has a per-cycle convergence factor bounded by $\gamma < 1$, which is independent of h. In particular, we will assume that MV^h with fixed choices for ν_1 and ν_2 has this property. The real question, then, concerns the cost of achieving acceptable numerical results by multigrid. More precisely, what is the order of complexity of a multigrid scheme that achieves the computational objective expressed in (2.5)?

To answer this question, we first need some assumptions on discretization error. In analogy to (2.4), we suppose the discretization is $O(h^p)$ for some positive integer p, which means that there exists a constant $\kappa < \infty$ independent of h so that

$$\|\mathbf{e}\|_h \leq \kappa h^p. \tag{5.1}$$

Now suppose the tolerance $\varepsilon > 0$ is given and that h is chosen to meet the objective in (2.7b) (i.e., $h \leq (\varepsilon/2\kappa)^{1/p}$). Because, as we remember, $e = \mathbf{u} - u$, the criterion that our numerical approximation must meet is simply

$$\|e\| \leq \kappa h^p. \tag{5.2}$$

When we have satisfied (5.2), we will say that we have solved the grid h problem *to the level of truncation.*

For the standard multigrid algorithm MV^h, the error in the initial guess $u^h \leftarrow 0$ is typically $O(1)$ in norm. This means that in order to satisfy (5.2), we must reduce the error at least by a factor of $O(h^p)$. Since $h = 1/(n+1)$, we must perform ν cycles where $\gamma^\nu = O(n^{-p})$. Letting $n_d = n^d$ denote the number of grid points in Ω^h, we then find this requires $O(\log n_d)$ cycles. Since each MV^h cycle costs $O(n_d)$ arithmetic operations, then the total computational complexity is $O(n_d \log n_d)$.

This is essentially the complexity of specialized direct Poisson solvers. Yet multigrid can do much better. The key is to use a better initial guess, something we have already achieved with *FMG*. Specifically, suppose we have solved the grid $2h$ problem to the level of truncation, i.e.,

$$\|u^{2h} - \mathbf{u}^{2h}\|_{2h} \leq \kappa(2h)^p.$$

Then by (5.1) and our choice of h, we will have

$$\|u^{2h} - w^{2h}\|_{2h} \leq 2\kappa(2h)^p \leq 2^p\varepsilon,$$

where again we have used w to denote the vector with components $w_i = u(x_i)$. Thus, u^{2h} is an approximation to w^{2h} with error no larger than a factor of 2^p times the tolerance ε. Under reasonable conditions, we can then easily show that the approximation $u^h = I_{2h}^h u^{2h}$ satisfies

$$\|u^h - w^h\|_h \leq \varepsilon\theta 2^p,$$

where $\theta \cong 2$. (In fact, if we use a natural *variational* approach when A is s.p.d., then $\theta = 1$.) In other words, if we start by solving grid $2h$ to within the level of truncation, then we need to further reduce the error by a factor of only $\theta 2^p$. This requires ν_0 basic cycles where $\gamma^{\nu_0} = \theta 2^p$; that is, $\nu_0 = O(1)$. If FMV^h (or some other appropriate *FMG* scheme) is implemented with this ν_0, then we are assured that each grid, including the finest, is solved to within the level of truncation. Thus, the real computational complexity of FMV^h is just $O(n^d)$.

The convergence bound $\gamma = \gamma(\nu_1, \nu_2)$ for the basic cycling process tends to zero as $\nu_1 + \nu_2$ tends to infinity. Thus, we can achieve the relation $\gamma^{\nu_0} = \theta 2^p$ by increasing the number of relaxation sweeps (ν_1, ν_2), the number of cycles (ν_0), or both. However, for many practical problems and for proper implementations, $\nu_0 = 1$ and $\nu_1 = \nu_2 = 1$ or 2 suffice. We are led to the conclusion that (full) multigrid can solve many problems to the level of truncation at a cost equivalent to three to five relaxation sweeps.

Another practical issue concerns the scheduling procedure that determines the ordering of the grids during the multigrid cycles. The algorithms that we have developed are based on one general scheduling technique called the *fixed* scheme, where the schedule is predetermined, in this case by the choices of ν_0, ν_1, and ν_2. This now seems to be the most popular method, partly because it is possible to determine its performance by one of the analytical approaches mentioned above and to use this analysis to "optimize" choices of the parameters. A second approach, the *accommodative* scheme, is a selfadjusting strategy that monitors the convergence of relaxation in order to determine where to use coarser or finer levels. For example, if on grid h a criterion of the form $\|r^{2h}\|_{2h} \leq \sigma \|r^h\|_h$ is met, this signals acceptable convergence and the scheme can proceed to level h. If this criterion is not yet satisfied and one of the form $\|r^{2h}_{(\text{new})}\|_{2h} \geq (1-\delta) \|r^{2h}\|_{2h}$ is met, this signals slow convergence and the scheme can proceed to level $4h$. A danger with this scheme is that the residual may be inappropriate for measuring the behavior of the evolving error. (Compared to Euclidean and energy norms of the error, the residual norm magnifies oscillatory effects at the expense of smooth ones.)

Both scheduling schemes require termination criteria. If the objective is to converge to within truncation error, the advantage of the fixed scheme is that a priori analysis may be able to predetermine the proper *FMG* scheme that ensures such accuracy. However, both schemes have advantages over most iterative methods in that the various grid levels can be used to estimate the actual accuracy achieved in the finest grid. In any case, multigrid can also use conventional stopping criteria. This includes tests of the form $\|r^h\| < \varepsilon$ where, without information afforded by the various multigrid levels, we would take ε very small (say, on the order of machine accuracy).

We close this chapter with some numerical experiments illustrating the performance of the methods that we have just discussed. Consider the 2-D boundary value problem

$$(5.3)\qquad \begin{aligned} -\Delta u &= 2[(1-6x^2)y^2(1-y^2) + (1-6y^2)x^2(1-x^2)], \\ &\qquad\qquad \text{in } \Omega = (0,1) \times (0,1), \\ u &= 0 \quad \text{on } \partial\Omega, \end{aligned}$$

which has the exact solution $u = x^2y^2(1-x^2)(1-y^2)$. This problem will be treated by the V-cycle method and the full multigrid (FMV) method, both of which use red-black Gauss–Seidel relaxation. (This is a specially ordered Gauss–Seidel method that labels a point ij as red if $i+j$ is even and as black otherwise. One red-black sweep now involves relaxing first on the red points, then on the black. Note that the ordering within a red sweep—or

TABLE 1.1
Numerical results for V-cycle with red-black Gauss–Seidel applied to the model problem (5.3) *on* $n \times n$ *grids with* $n = 15$, 31 *and* 63.

n	Grid size	$\|r\|_h$	Factor	error_h	error_∞
17	17	.71 (−3)	–	.60 (−3)	.49 (−3)
17	17	.31 (−4)	.04	.41 (−3)	.19 (−3)
17	17	.21 (−5)	.07	.41 (−3)	.20 (−3)
17	17	.17 (−6)	.08	.41 (−3)	.20 (−3)
17	17	.20 (−7)	.12	.41 (−3)	.20 (−3)
17	17	.14 (−7)	.71	.41 (−3)	.20 (−3)
33	33	.23 (−3)	–	.74 (−3)	.34 (−3)
33	33	.13 (−4)	.06	.15 (−3)	.48 (−4)
33	33	.11 (−5)	.08	.15 (−3)	.50 (−4)
33	33	.12 (−6)	.08	.15 (−3)	.49 (−4)
33	33	.23 (−7)	.18	.15 (−3)	.49 (−4)
33	33	.18 (−7)	.80	.15 (−3)	.49 (−4)
65	65	.70 (−4)	–	.11 (−2)	.30 (−3)
65	65	.47 (−5)	.07	.59 (−4)	.15 (−4)
65	65	.49 (−6)	.10	.48 (−4)	.13 (−4)
65	65	.70 (−7)	.14	.47 (−4)	.11 (−4)
65	65	.28 (−7)	.40	.47 (−4)	.11 (−4)
65	65	.26 (−7)	.93	.47 (−4)	.11 (−4)

Note. One relaxation sweep is performed on each level. Results are given after each fine grid relaxation. Factor is the ratio of $\|r\|_h$ on two successive cycles. $\text{Error}_h = \|u^h - w^h\|_h$ and $\text{error}_\infty = \|u^h - w^h\|_\infty$, where $w_i^h = \mathscr{u}(x_i^h)$.

within a black one—is irrelevant for five-point stencils, since changing a red point variable here has no effect on any other red point equation.)

Table 1.1 summarizes the results of using the V-cycle with one relaxation sweep on each level. For each of the $n \times n$ grids ($n = 15$, 31, and 63), the norms of the residual and global errors are printed after each fine grid relaxation. The discrete $\mathscr{L}_2$ norm of the residual, $\|r\|_h$, and the factor by which it decreases (i.e., the ratio of successive values) are given. Notice that the norm decreases at an asymptotic rate of roughly 0.1 until the effects of roundoff error (in the computation of the residual) enter. At this point, the convergence degrades abruptly. On all three grids, this occurs after five V-cycles. The discrete $\mathscr{L}_2$ norm of the error $\mathscr{u} - u$ decreases with each V-cycle and quickly reaches the level of the truncation error. The maximum norm of this error also reaches the level of truncation error quickly, showing the expected decrease by a factor of four as the grid size is halved.

The V-cycle solutions clearly show the values of the various norms when

TABLE 1.2
Numerical results for FMV with red-black Gauss–Seidel applied to the model problem (5.3) *on a* 63×63 *grid.*

Grid size	$\|r\|_h^{1,1}$	$\text{error}_h^{1,1}$	$\|r\|_h^{2,1}$	$\text{error}_h^{2,1}$
3	0.0	.83 (−2)	0.0	.83 (−2)
5	.27 (−2)	.35 (−2)	.33 (−3)	.33 (−2)
9	.17 (−2)	.17 (−2)	.36 (−3)	.12 (−2)
17	.62 (−3)	.81 (−3)	.13 (−3)	.41 (−3)
33	.18 (−3)	.36 (−3)	.37 (−4)	.14 (−3)
65	.47 (−4)	.15 (−3)	.96 (−5)	.48 (−4)

Note. Results are printed after each fine grid relaxation. Scheme (1, 1) uses one relaxation on each level. Scheme (2, 1) uses two relaxation sweeps before injection and one sweep before interpolation. Error_h is as defined in Table 1.1. Note that $r = 0$ on the coarsest level because it contains only one interior point, making relaxation a direct solver.

the method converges to the discrete solution $\mathbf{u}$. These results can be used for comparison when we turn to *FMV* approximations. Table 1.2 shows the results of applying two different *FMV* cycling schemes to the model problem (5.3). Both schemes use red-black Gauss–Seidel relaxation. The scheme denoted by (1, 1) performs one relaxation sweep on each level. As Table 1.2 indicates, after a single (1, 1) *FMV* cycle, the norm of the error $\mathscr{u} - u$ has not reached the level of the truncation error as determined by the V-cycle results. The *FMV* errors are 2.0, 2.5, and 3.1 times the V-cycle errors on the respective grids. Clearly, additional relaxation sweeps are needed in the *FMV* cycle. Specifically, the scheme denoted by (2, 1) performs two relaxation sweeps before each residual injection and one sweep before each interpolation. Table 1.2 shows the effect of the extra sweeps. The norms of the error for a single (2, 1) *FMV* cycle are virtually identical to the norms of the converged approximation obtained by the V-cycle.

A final rough comparison of the work involved in these calculations might be useful. Let one work unit be the amount of work required to do one relaxation sweep on the finest grid. Ignoring the work involved in intergrid transfers, we find (see § 1.5) that one V-cycle costs roughly 8/3 work units. It can also be shown that a (1, 1) *FMV* cycle costs about 4/3 of a V-cycle or 32/9 work units. Since the (2, 1) *FMV* cycle requires half again as many sweeps as the (1, 1) *FMV* cycle, 16/3 work units are required to converge to truncation error for the model problem. Table 1.1 indicates that two to four V-cycles suffice to converge to within truncation error, which costs 16/3 to

32/3 work units. This suggests that, while the two schemes may have comparable costs for small problems, the V-cycle may become relatively more expensive as the problem size grows. (Remember that FMV complexity in terms of work units is essentially bounded independent of h.) We emphasize, however, that the objective of a V-cycle should be reduction of the algebraic error (which is usually intermediate to some grander objective), not convergence to within truncation error, so strict comparisons are not really appropriate here.

A listing of a FORTRAN program for the FMV scheme is given at the end of the book as Appendix 1.

Acknowledgment. This work was supported by the U.S. Air Force Office of Scientific Research under grant AFOSR-86-0126 and the Department of Energy under grant DE-AC03-84-ER80155.

CHAPTER 2

Linear Multigrid Methods

P. WESSELING

2.1. Introduction. This chapter is devoted to linear multigrid (MG) methods, by which we mean MG methods especially designed for linear partial differential equations. However, some examples of applications of linear MG to nonlinear problems will be given. The practical aspects of the present state of the art will also be discussed.

MG principles have a much wider scope than the class of problems described here; see Brandt [102] for a bird's-eye view.

The focus of this chapter is on finite difference and finite volume discretization. For finite elements, see Chapter 5.

We restrict ourselves to two-dimensional (2-D) problems because this simplifies our exposition. Most of the concepts and methods discussed here carry over to 3-D. Also, not much work has been done yet on 3-D MG (see, however, Behie and Forsyth [61], Caughey [123], Fuchs and Zhao [181], Gary et al. [189], Kettler and Wesseling [330], Scott [511], Thole and Trottenberg [552], van der Wees [Va2],[1] and van der Wees et al. [Va3]).

2.2. Equations and discretizations. Application of linear MG to two types of equations will be considered, namely, the single second order equation (note that Cartesian tensor notation is used with conventional summation over repeated indices; Greek subscripts stand for dimension indices [i], and the subscript $,\alpha$ denotes the derivative with respect to x_α)

$$-(d_{\alpha\beta} u_{,\alpha})_{,\beta} + (b_\alpha u)_{,\alpha} + au = s, \tag{2.1}$$

and the system of first order equations or conservation laws

$$\frac{\partial \underline{u}}{\partial t} + \frac{\partial \underline{f}(\underline{u})}{\partial x} + \frac{\partial \underline{g}(\underline{u})}{\partial y} = \underline{s}, \tag{2.2}$$

where

$$\underline{u}, \underline{s} : \mathbf{R}^+ \times \mathbf{R}^2 \to \mathbf{R}^p, \qquad \underline{f}, \underline{g} : \mathbf{R}^p \to \mathbf{R}^p. \tag{2.3}$$

Here $\mathbf{R}^+$ denotes the set of nonnegative reals.

[1] References using letters are compiled at the end of each chapter.

The application of MG to general systems of second order equations is not yet well developed, although some general considerations are given by Hackbusch [244, Chap. 11]. Progress has been made for the special cases of the Stokes and Navier–Stokes equations (Brandt [99], [101], Brandt and Dinar [105], Fuchs and Zhao [181], Hackbusch [244, § 9.4], Johnson [316], Maitre et al. [366], Mol [418], Pitkäranta and Saarinen [458], Verfürth [569], [570], Wesseling [594], and Wesseling and Sonneveld [595]). However, we will not review systems of second order equations here, but rather will discuss applications of linear MG to nonlinear versions of (2.1) and (2.2). For the application of nonlinear MG to (2.2), see Chapter 3.

It is convenient to use Cartesian tensor notation for (2.1) but not for (2.2). For equation (2.1) ellipticity is assumed:

$$d_{\alpha\beta}\xi_\alpha\xi_\beta > \gamma\xi_\alpha\xi_\alpha, \quad \gamma > 0 \quad \text{for all } \xi_\alpha \in \mathbf{R}. \tag{2.4}$$

If, furthermore, $a \geq 0$, then linear MG can be made to work efficiently, as will be seen.

Equation (2.2) is assumed to be hyperbolic. Important examples of (2.2) in fluid dynamics are the Euler equations of inviscid gas dynamics and the shallow water (or de Saint-Venant) equations; see Courant and Friedrichs [Co1]. We will focus on the Euler equations in this chapter, because implicit schemes are more important to this application than to shallow water equations, where MG has not been applied.

The region Ω, in which (2.1) and (2.2) are to be solved, is assumed to be the unit square covered by a uniform computational grid Ω^h with mesh size h in both directions. These restrictions on Ω and Ω^h are not severe because, for ease of programming in MG applications, the computational grid is usually chosen to be topologically equivalent to a uniform rectangular mesh; furthermore, a boundary-fitted coordinate transformation to a rectangular uniform grid is usually employed. In fact, one of the strengths of MG as compared with older methods is that the restrictions to be placed on the coefficients in (2.1) and (2.2) are very weak, so that one has great freedom in applying coordinate transformations. Another way to obtain a rectangular computational domain is to embed the physical domain in a rectangle by "padding" the equations. (This will be discussed later.) The assumption of equal mesh size in all directions is merely for ease of exposition.

Equation (2.1) may, for example, be discretized with finite differences as follows. Define forward and backward divided differences in the x_α-direction on Ω^h by the following:

$$\begin{aligned} \Delta_\alpha u(x) &= (u(x + e_\alpha h) - u(x))/h, \\ \nabla_\alpha u(x) &= (u(x) - u(x - e_\alpha h))/h, \end{aligned} \tag{2.5}$$

with $x = (x_{[1]}, x_{[2]})$ and e_α the unit vector in the x_α-direction. Then a *finite*

$$\begin{bmatrix} \times & \times & \times \\ \times & \times & \times \\ \times & \times & \times \end{bmatrix} \quad \begin{bmatrix} \times & \times & \\ \times & \times & \times \\ & \times & \times \end{bmatrix} \quad \begin{bmatrix} & \times & \\ \times & \times & \times \\ & \times & \end{bmatrix}$$

FIG. 2.1. *Finite difference stencils.*

difference approximation of equation (2.1) is given by

$$(2.6) \qquad -\tfrac{1}{2}(\nabla_\alpha d_{\alpha\beta}\Delta_\beta + \Delta_\alpha d_{\alpha\beta}\nabla_\beta)u + \tfrac{1}{2}(\nabla_\alpha + \Delta_\alpha)(b_\alpha u) + a u = s.$$

The difference stencil corresponding to this finite difference discretization is the 7-point stencil in Fig. 2.1. Throughout this chapter, the grid points will be numbered in the usual way, which is as follows. We have $h = 1/(n-1)$, $n - 1 = 2^q$ for some positive integer q, in the finite difference case. The grid points have coordinates (ih, jh), $i, j = 0, 1, 2, \cdots, n-1$, and receive a corresponding number $1 + i + jn$. The number of points in the $x_{[1]}$-direction is n.

To clarify whether one might expect to obtain effective MG methods for solving (2.6), we define the following property.

DEFINITION 2.1. A matrix A is called a *K-matrix* if $a_{ii} > 0$, $a_{ij} \le 0$, $j \ne i$, $a_{ii} \ge -\sum_{j \ne i} a_{ij}$, with strict inequality for at least one i.

If A is an *irreducible* K-matrix, K is an M-*matrix* (Varga [Va6, Chap. 3]). In the present case, A is irreducible because the computational grid is sufficiently connected. If some $a_{ij} > 0$, $j \ne i$, then A is not an M-matrix. Generally, if A is an M-matrix, good smoothing methods can be found, and the existence of incomplete factorizations is guaranteed (Meijerink and van der Vorst [Me1] and Varga et al. [Va7]). If A is not an M-matrix, good smoothing methods are generally harder to find.

Let us take, for example, $d_{[1],[1]} = d_{[2],[2]} = \varepsilon = \text{constant}$, $d_{[1],[2]} = d_{[2],[1]} = 0$, $b_\alpha = b_\alpha = \text{constant}$, $a = 0$ in (2.6). Then, the matrix A corresponding to the system of equations (2.6), with the grid point numbering introduced above, has the following elements (after multiplication by h^2):

$$(2.7) \qquad \begin{aligned} &a_{i,i-n} = -\varepsilon - b_{[2]}h/2, \quad a_{i,i-1} = -\varepsilon - b_{[1]}h/2, \quad a_{i,i} = 4\varepsilon, \\ &a_{i,i+1} = -\varepsilon + b_{[1]}h/2, \qquad a_{i,i+n} = -\varepsilon + b_{[2]}h/2, \end{aligned}$$

with obvious modifications near the boundaries that are treated in the usual way and will not be discussed. All other elements are zero. It is immediately clear that we have a K-matrix if and only if

$$(2.8) \qquad \mathrm{Pe}_\alpha \le 2, \qquad \alpha = 1, 2,$$

where Pe_α is the so-called mesh-Péclet number in the x_α-direction given by

$$\mathrm{Pe}_\alpha = \frac{h\,|b_\alpha|}{\varepsilon}.$$

(We assume strict inequality in Definition 2.1 for some i due to boundary conditions.) In the case of variable coefficients, similar reasoning leads to a similar condition on the local Péclet numbers.

In practical applications, notably in computational fluid dynamics, we often have $\varepsilon \ll 1$, $\mathrm{Pe}_\alpha > 2$. In order to handle this case with MG using state-of-the-art methods, one must either increase ε artificially (the *artificial viscosity* method), or replace the central discretization of $(b_\alpha u)_{,\alpha}$ used above by a one-sided discretization (*upwind differencing*). In both cases, the discretization accuracy is lowered, and there is a practical need to increase it again. For this purpose, *defect correction* has been proposed (cf. Böhmer et al. [66] and Hemker [269]). We will not pursue this further.

Before describing upwind differencing, we first introduce the concept of *flux splitting*. We write

$$(b_{[1]}u)|_{(ih,jh)} = f^+_{ij} + f^-_{ij}, \tag{2.9}$$

with f^+, f^- denoting the so-called split fluxes defined below. The term $(b_{[1]}u)_{,[1]}$ is discretized by

$$(b_{[1]}u)_{,[1]} \cong \nabla_1 f^+ + \Delta_1 f^-. \tag{2.10}$$

The advantage of flux splitting is that it ensures that the discretization is conservative, which means that when summing over i we are left with only boundary terms:

$$h \sum_{i=1}^{n} \nabla_1 f^+_{ij} + h \sum_{i=0}^{n-1} \Delta_1 f^-_{ij} = (b_{[1]}u)_{nj} - (b_{[1]}u)_{0j}. \tag{2.11}$$

First order upwind differencing is defined by (2.10) with

$$\begin{aligned} f^+_{ij} &= \tfrac{1}{2}(b_{[1]}(x) + |b_{[1]}(x)|)u(x), \\ f^-_{ij} &= \tfrac{1}{2}(b_{[1]}(x) - |b_{[1]}(x)|)u(x), \end{aligned} \qquad x = (ih, jh). \tag{2.12}$$

The term $(b_{[2]}u)_{,[2]}$ is treated analogously. It is easy to see that now A is a K-matrix for all ε and b_α.

Next, we turn to *finite volume* discretization of (2.1). The computational region (the unit square) is covered by squares of size $h \times h$, $h = 1/n$, $n = 2^q$, where q is a positive integer. These squares are called grid cells, or finite volumes. Their centers are at $((i+\frac{1}{2})h, (j+\frac{1}{2})h)$, $i, j = 0, 1, 2, \cdots, n-1$; their ordering number is $1 + i + jn$. The construction of a finite volume discretization of (2.1) starts with integrating (2.1) over a grid cell and applying Green's formula, which gives

$$\int_{S_{ij}} \{-a_{\alpha\beta} u_{,\alpha} + b_\beta u\} n_\beta \, dS + \iint_{V_{ij}} c u \, dV = \iint_{V_{ij}} s \, dV, \tag{2.13}$$

where V_{ij} is the grid cell with center at $((i+\frac{1}{2})h, (j+\frac{1}{2})h)$, S_{ij} is its boundary, and n_β is the x_β component of the unit normal. We approximate equation

(2.13) numerically by assuming all functions to be constant in V_{ij}. To explain the evaluation of the boundary integral, we describe the treatment of the west face of the cell. For simplicity, assume that $d_{\alpha\beta}=0$, $\alpha \neq \beta$. We approximate the contribution of the diffusion term as follows (temporarily denoting $d_{[1],[1]}$ by d, suppressing subscripts):

$$
(2.14) \quad \int_{jh}^{(j+1)h} (du_{,[1]})\,|_{(ih,x_{[2]})}\, dx_{[2]} \cong h(du_{,[1]})\,|_{(ih,(j+1/2)h)}
$$
$$
\cong 2d_{ij}d_{i-1,j}(u_{ij}-u_{i-1,j})/(d_{ij}+d_{i-1,j}).
$$

The subscript ij refers to the cell with center at $((i+\frac{1}{2})h,\ (j+\frac{1}{2})h)$, and the u_{ij} are the discrete approximations to $u(x)$. The above approximation to $du_{,[1]}$ takes into account the possibility that d_{ij} and $d_{i-1,j}$ differ significantly. The contribution of the convection term is approximated by

$$
(2.15) \quad -\int_{jh}^{(j+1)h} (b_{[1]}u)\,|_{(ih,x_{[2]})}\, dx_{[2]} \cong -h(f^{+}_{i-1,j}+f^{-}_{ij}),
$$

where again the concept of flux splitting is used. If the split fluxes are defined by (2.12), we have upwind flux splitting, and a K-matrix is obtained. The other faces of the cell are treated analogously. The volume integrals in (2.13) are simply approximated by $h^2a_{ij}u_{ij}$ and h^2s_{ij}.

Let us take the example leading to equation (2.7). It is easily seen that finite volume discretization again results in (2.7). What, then, is the distinction between finite difference and finite volume discretization as we have developed them? Apart from the application of MG, there is a distinction in the treatment of the boundary conditions. With finite differences we have grid points on the boundary, whereas with finite volumes the unknown values of u_{ij} are assigned to cell centers located at $((i+\frac{1}{2})h,\ (j+\frac{1}{2})h)$; such points do not lie on the boundary but at a distance $h/2$ from it. For example, assume that the side with $x_1=0$ is a Dirichlet boundary. In the finite difference case, the unknowns u_{0j}, $j=0, 1, 2, \cdots, n-1$ are simply replaced by known values. In the finite volume case, virtual grid cells are introduced with centers at $(-\frac{1}{2}h,\ (j+\frac{1}{2})h)$, $j=0, 1, 2, \cdots, n-1$. For the computation of the diffusive flux along the west face of cell $(0,j)$, a virtual value $u_{-1,j}$ is computed by linear extrapolation:

$$
(2.16) \quad u_{-1,j}=2u_{wj}-u_{1j},
$$

with u_{wj} the prescribed boundary value. This value could also be used for the computation of the convective flux, but then the discretization is no longer conservative. Therefore, it is better to use the approximation

$$
(2.17) \quad f^{+}_{-1,j}=\tfrac{1}{2}(b_{[1]}(x)+|b_{[1]}(x)|)u_{wj},
$$

where $x=(0,\ (j+\frac{1}{2})h)$.

Next, we turn to the discrete approximation of (2.2). Schemes of Lax–Wendroff type (see Richtmyer and Morton [Ri1]) have long been popular and still are widely used. These schemes are explicit and, for time-dependent problems, there is no need for MG: stability and accuracy restrictions on the time step Δt are about equally severe. If the instationary formulation is employed solely as a means to compute a steady state, then we would like to be unrestricted in the choice of Δt and/or use artificial means to get rid of the transient quickly. Ni [428] has proposed a method to do this using multiple grids. This method has been further developed by Johnson [316]. A Runge-Kutta–type discretization combined with artificial dissipation terms and MG has been developed by Jameson [300]. Straightforward nonlinear MG methods have been proposed by Jespersen [310] and Hemker and Spekreijse [274], [275]. (The methods just mentioned are not of linear MG type and are reviewed in Chapter 3, which treats the Euler equations.) Linear MG methods for (2.2) have been proposed by Jespersen [309] and Mulder [423]. These methods use discretizations of flux-splitting type. Jespersen uses the flux splitting proposed by Steger and Warming [St1], whereas Mulder uses the flux splitting proposed by van Leer [Va4]. For a survey of flux splitting, see Harten et al. [Ha1] and van Leer [Va5]. For (2.2), the fluxes $\underline{f}$ and $\underline{g}$ are split by

$$\underline{f} = \underline{f}^{+} + \underline{f}^{-}, \qquad \underline{g} = \underline{g}^{+} + \underline{g}^{-}, \tag{2.18}$$

such that the Jacobians $d\underline{f}^{+}/d\underline{u}$ and $d\underline{g}^{+}/d\underline{u}$ have positive real eigenvalues and $d\underline{f}^{-}/d\underline{u}$ and $d\underline{g}^{-}/d\underline{u}$ have negative real eigenvalues. Then (2.2) is discretized according to the following:

$$h^2[(\underline{\mathbf{u}} - \underline{\tilde{\mathbf{u}}})/\Delta t + \Delta_1 \underline{f}^{-} + \nabla_1 \underline{f}^{+} + \Delta_2 \underline{g}^{-} + \nabla_2 \underline{g}^{+}] = h^2 \underline{s}, \tag{2.19}$$

where $\underline{f}^{\pm} = \underline{f}^{\pm}(\underline{u})$, $\underline{g}^{\pm} = \underline{g}^{\pm}(\underline{u})$, and $\underline{\mathbf{u}}$ ($\underline{\tilde{\mathbf{u}}}$) is the discrete approximation of $\underline{u}$ at time t ($t - \Delta t$). Here we have used an implicit (backward) Euler time discretization. The aim is to rapidly find steady solutions without regard for temporal accuracy. We have multiplied the equation by the cell volume h^2; on a uniform grid this is irrelevant, but on a nonuniform grid it is useful. Equation (2.19) is not solved directly but by way of its Newton-linearization:

$$(h^2/\Delta t I + K)\underline{\mathbf{v}} = \underline{r}, \tag{2.20}$$

where $\underline{\mathbf{v}} = \underline{\mathbf{u}} - \underline{\tilde{\mathbf{u}}}$,

$$K = h \frac{d}{d\underline{u}} [\underline{\tilde{f}}^{+} - \underline{\tilde{f}}^{-} + \underline{\tilde{g}}^{+} - \underline{\tilde{g}}^{-} + T_x \underline{\tilde{f}}^{-} T_x - T_x^{-1} \underline{\tilde{f}}^{+} T_x^{-1} + T_y \underline{\tilde{g}}^{-} T_y - T_y^{-1} \underline{\tilde{g}}^{+} T_y^{-1}], \tag{2.21}$$

and

$$r = h^2[s - \Delta_1\tilde{f}^- - \nabla_1\tilde{f}^+ - \Delta_2\tilde{g}^- - \nabla_2\tilde{g}^+]. \tag{2.22}$$

Here, $\tilde{f}^\pm = \tilde{f}^\pm(\tilde{\mathbf{u}})$, $\tilde{g}^\pm = \tilde{g}^\pm(\tilde{\mathbf{u}})$, and T_x is the forward shift operator in the x-direction defined by

$$(T_x u)_{ij} = u_{i+1,j}, \tag{2.23}$$

and similarly for T_y. The discretization (2.21), (2.22) is to be interpreted as a finite volume discretization.

With Δt large, equation (2.20) begins to look like Newton's method applied directly to the steady state equation. Thus, it can be expected to converge rapidly in the vicinity of a steady state solution, but may diverge far from it when Δt is too large. Therefore, a safe procedure is to start with Δt small and approximately follow the physical time evolution of the solution, increasing Δt as the steady state solution is approached. An algorithmic implementation of this principle is given by Mulder and van Leer [Mu1].

The flux splitting (2.18) is analogous to upwind differencing in the scalar case and increases the main diagonal, which is probably essential for the success of the MG methods of Mulder and Jespersen.

2.3. Prolongation and restriction. The finest grid is Ω^{h_q} or, simply, Ω^q. Coarser grids $\Omega^{q-1}, \Omega^{q-2}, \cdots, \Omega^1$ are generated by successive doubling of the mesh size. The set of grid functions on Ω^k is denoted by H^k, keeping in mind the distinction between finite differences and finite volumes (grid function values are defined at grid nodes and cell centers, respectively). Prolongation and restriction are denoted respectively by

$$I^k_{k-1}: H^{k-1} \to H^k, \qquad I^{k-1}_k: H^k \to H^{k-1}. \tag{3.1}$$

In the finite difference case, I^k_{k-1} is defined by interpolation (see Fig. 2.2). Points 1–4 belong both to the finer grid Ω^k and the coarser grid Ω^{k-1}; points 5–14 belong to Ω^k but not to Ω^{k-1}. Linear interpolation gives

$$(I^k_{k-1}u)_1 = u_1, \qquad (I^k_{k-1}u)_5 = (u_1 + u_2)/2. \tag{3.2}$$

Of course, points 2–4 and 6–8 are treated analogously to points 1 and 5, respectively. For point 9 there are two obvious possibilities, namely,

$$(I^k_{k-1}u)_9 = (u_1 + u_2 + u_3 + u_4)/4 \tag{3.3}$$

and

$$(I^k_{k-1}u)_9 = (u_2 + u_4)/2. \tag{3.4}$$

(Of course, using u_1 and u_3 in (3.4) is analogous.) We will call (3.3) 9-point

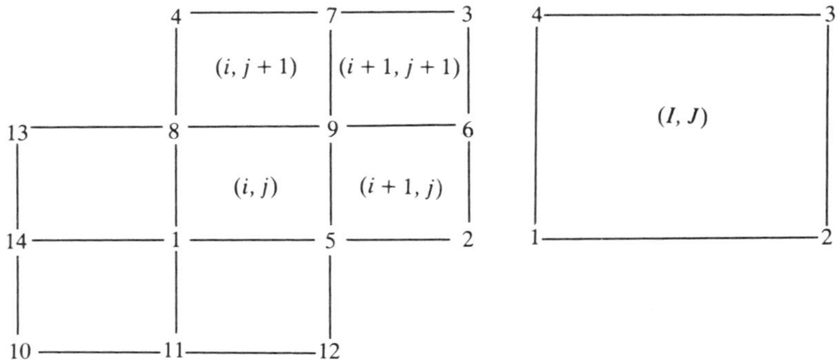

FIG. 2.2. *Coarse and fine grid points.*

and (3.4) 7-point prolongation because, if we define restriction by

$$I_k^{k-1} = (I_{k-1}^k)^T, \tag{3.5}$$

we obtain a weighted average of 9 or 7 grid function values, respectively. Seven-point prolongation leads to somewhat better efficiency if Galerkin coarse grid approximation is used (see below), but has the disadvantage that, for problems in which the solution possesses some symmetry, the numerical approximation may not reproduce this symmetry exactly. Equation (3.5) leads to the restrictions corresponding to prolongations (3.2) and (3.3) given by

$$(I_k^{k-1}u)_1 = u_1 + (u_{11} + u_5 + u_8 + u_{14})/2 + (u_{10} + u_{12} + u_9 + u_{13})/4, \tag{3.6}$$

and to (3.2) and (3.4) given by

$$(I_k^{k-1}u)_1 = u_1 + (u_{11} + u_{12} + u_5 + u_8 + u_{13} + u_{14})/2. \tag{3.7}$$

Other restrictions that occur in the literature are

$$(I_k^{k-1}u)_1 = u_1 \tag{3.8}$$

and

$$(I_1^{k-1}u)_1 = \tfrac{1}{2}u_1 + (u_5 + u_8 + u_{14} + u_{11})/8. \tag{3.9}$$

The restriction defined by equation (3.6) is called *full weighting* or *9-point* restriction; by equation (3.7), *7-point* restriction; by (3.8), *injection*; and by (3.9), *half weighting*. Scaling of restrictions is irrelevant for linear MG, but for nonlinear MG it may be necessary for $I_k^{k-1}u$ to approximate u, so that the sum of the weights should be 1. For further considerations about restrictions, see Hackbusch [244, §§3.5 and 3.8.5].

Prolongation by linear interpolation is inaccurate when u is not locally linear between coarse grid points. This inaccuracy is severe when $d_{\alpha\beta}$ in (2.1) is discontinuous between coarse grid lines. The resulting jump in the slope of the manifold $x_3 = u(x_1, x_2)$ has to be approximately taken into account. Ways to do this are described by Alcouffe et al. [5] and Kettler [328]. It would lead too far to present the resulting prolongations here. For a review, see Hackbusch [244, §10.3].

Finite volume discretization goes hand in hand with a type of prolongation and restriction not covered above. Grid function values represent averages over a cell; these values are assigned more or less arbitrarily to the cell centers. In this context, a "natural" prolongation is piecewise constant interpolation given by (cf. Fig. 2.2)

$$(3.10) \qquad (I_{k-1}^k u)_{ij} = u_{IJ}.$$

The same value u_{IJ} is assigned to cells $(i+1, j)$, $(i, j+1)$, $(i+1, j+1)$. This definition of I_{k-1}^k is also useful on nonuniform grids. If u would instead be interpreted as an integral over a cell, so that the factor h^2 in equation (2.17) would be absorbed in the definition of u, then a spatially varying weight factor would appear in (3.10) for nonuniform grids. This would increase the computational cost.

We now turn to restriction in the finite volume case. Later it will become clear that restriction is applied, not to $\underline{v}$, but to the right-hand side $\underline{r}$ in (2.20), which represents the source strength and the flux balance through the sides of the cells. The motivation for the multiplication of equation (2.8) with the cell volume h^2 is to make this interpretation possible. It follows immediately that when cells coalesce to form a coarse cell, the new source strength and flux balance are simply the sum of the original sources and fluxes. This leads to the following restriction operator:

$$(3.11) \qquad (I_k^{k-1} r)_{IJ} = r_{ij} + r_{i+1,j} + r_{i,j+1} + r_{i+1,j+1}.$$

Without multiplication of (2.17) by the cell volume, spatially varying weight factors would appear in (3.11), increasing the computational cost.

How should we choose prolongation and restriction in practice? Brandt [88] gives the following result. We say that I_{k-1}^k is of order m_p if polynomials of degree $m_p - 1$ are interpolated exactly, and that I_k^{k-1} is of order m_r if $(I_k^{k-1})^T$ interpolates polynomials of degree $m_r - 1$ exactly. If the order of the differential equation is $2m$, then we should have

$$(3.12) \qquad m_p + m_r > 2m.$$

We therefore expect that (3.10) and (3.11) ($m_p = m_r = 1$) will work for (2.2) but not for (2.1). This expectation will be confirmed by a numerical experiment to be described in §6.

2.4. Coarse grid approximation. We consider here two kinds of coarse grid approximations: coarse grid finite difference approximations (CFA) and coarse grid Galerkin approximations (CGA). In CFA, equation (2.1) is discretized on $\Omega^k, k<q$, by (2.6) (or a similar formula, maybe with additional mesh-size–dependent artificial viscosity; cf. de Zeeuw and van Asselt [607]). In CGA the coarse discrete approximation on Ω^{k-1} is defined by

$$A^{k-1} = I_k^{k-1} A^k I_{k-1}^k, \qquad k = q, q-1, \cdots, 2. \tag{4.1}$$

Here A^k is the matrix corresponding to the (linear) discrete approximation on Ω^k. The use of the term "Galerkin approximation" is explained at the end of this section.

Considering first CGA for finite difference discretizations, we have depicted three possible stencil types in Fig. 2.1. Note that the 7-point stencil corresponds to equation (2.6). When we use 9-point prolongation and restriction, CGA leads to 9-point stencils on Ω^k, $k = q-1, q-2, \cdots, 1$, with either a 9-, 7-, or 5-point stencil on Ω^q. With 7-point prolongation and restriction, CGA leads to 7-point stencils on Ω^k, $k = q-1, q-2, \cdots, 1$, with either a 7-point or a 5-point stencil on Ω^q. With finite volume prolongation and restriction (3.10), (3.11), we obtain a 5-point stencil on Ω^k, $k = q-1, q-2, \cdots, 1$, if we have a 5-point stencil on Ω^q.

An algorithm for computing A^k according to (4.1) in the finite difference case is given by Wesseling [592]. The computation of A^{k-1} from A^k with (4.1) requires 64 (7-point A^k) or 47 (5-point A^k) floating point operations (*flops*) per grid point of Ω^k. Whether it costs more or less to compute the coarse grid matrices with CGA than it does with CFA depends on the cost of CFA, which is problem-dependent. An advantage of CGA is that implementation of boundary conditions on coarse grids and homogenization of rapidly varying coefficients are automatic. For comparative experiments with CGA and CFA, see Wesseling [593].

In the finite volume case, computation of A^k with equation (4.1) is especially simple, thanks to the simple form of (3.10) and (3.11). The relation between A^{k-1} and A^k can be derived as follows. Let cell (i, j) be the cell whose center has coordinates $(i-\frac{1}{2})h$, $(j-\frac{1}{2})h$. Let the coarse cell (I, J) be the union of the fine cells (i, j), $(i+1, j)$, $(i, j+1)$, $(i+1, j+1)$, as in Fig. 2.2. Rewrite equation (2.20) on grid Ω^k as

$$A^k \mathbf{v}^k = r^k. \tag{4.2}$$

The coarse grid approximation on Ω^{k-1} becomes

$$I_k^{k-1} A^k I_{k-1}^k \mathbf{v}^{k-1} = I_k^{k-1} r^k. \tag{4.3}$$

Let us temporarily denote the equation in cell (i, j) as

$$\alpha_{ij}^k \mathbf{v}_{i,j-1}^k + \beta_{ij}^k \mathbf{v}_{i-1,j}^k + \gamma^k \mathbf{v}_{ij}^k + \delta^k \mathbf{v}_{i+1,j}^k + \varepsilon^k \mathbf{v}_{i,j+1}^k = r_{ij}^k. \tag{4.4}$$

The multiplication with I_k^{k-1} (defined by (3.11)) amounts to adding together the equations in the cells (i, j), $(i+1, j)$, $(i, j+1)$ and $(i+1, j+1)$. This gives

$$
\begin{aligned}
(4.5)\qquad & \alpha_{ij}^k \mathbf{v}_{i,j-1} + \alpha_{i+1,j}^k \mathbf{v}_{i+1,j-1} + \beta_{ij}^k \mathbf{v}_{i-1,j}^k \\
&+ (\alpha_{i,j+1}^k + \beta_{i+1,j}^k + \gamma_{ij}^k)\mathbf{v}_{ij}^k \\
&+ (\alpha_{i+1,j+1}^k + \gamma_{i+1,j}^k + \delta_{ij}^k)\mathbf{v}_{i+1,j}^k \\
&+ \delta_{i+1,j}^k \mathbf{v}_{i+2,j}^k + \beta_{i,j+1}^k \mathbf{v}_{i-1,j+1}^k \\
&+ (\varepsilon_{ij}^k + \gamma_{i,j+1}^k + \beta_{i+1,j+1}^k)\mathbf{v}_{i,j+1}^k \\
&+ (\varepsilon_{i+1,j}^k + \delta_{i,j+1}^k + \gamma_{i+1,j+1}^k)\mathbf{v}_{i+1,j+1}^k \\
&+ \delta_{i+1,j+1}^k \mathbf{v}_{i+2,j+1}^k + \varepsilon_{i,j+1}^k \mathbf{v}_{i,j+2}^k \\
&+ \varepsilon_{i+1,j+1}^k \mathbf{v}_{i+1,j+2}^k = \cdots,
\end{aligned}
$$

where we do not need the right-hand side. Substitution of $\mathbf{v}^k = I_{k-1}^k \mathbf{v}^{k-1}$, so that $\mathbf{v}_{i,j-1}^k = \mathbf{v}_{I,J-1}^{k-1}$, $\mathbf{v}_{ij}^k = \mathbf{v}_{IJ}^k$, etc., results in

$$(4.6)\qquad \alpha_{IJ}^{k-1}\mathbf{v}_{I,J-1}^{k-1} + \beta_{IJ}^{k-1}\mathbf{v}_{I-1,J}^{k-1} + \gamma_{IJ}^{k-1}\mathbf{v}_{IJ}^{k-1} + \delta_{IJ}^{k-1}\mathbf{v}_{I+1,J}^{k-1} + \varepsilon_{IJ}^{k-1}\mathbf{v}_{I,J+1}^{k-1} = I_k^{k-1} r^k,$$

with

$$
\begin{aligned}
& \alpha_{IJ}^{k-1} = \alpha_{ij}^k + \alpha_{i+1,j}^k, \qquad \beta_{IJ}^{k-1} = \beta_{ij}^k + \beta_{i,j+1}^k, \\
& \gamma_{IJ}^{k-1} = \alpha_{i,j+1}^k + \beta_{i+1,j}^k + \gamma_{ij}^k + \alpha_{i+1,j+1}^k + \gamma_{i+1,j}^k \\
(4.7)\qquad &\qquad + \delta_{ij}^k + \varepsilon_{ij}^k + \gamma_{i,j+1}^k + \beta_{i+1,j+1}^k + \varepsilon_{i+1,j}^k \\
&\qquad + \delta_{i,j+1}^k + \gamma_{i+1,j+1}^k, \\
& \delta_{IJ}^{k-1} = \delta_{i+1,j}^k + \delta_{i+1,j+1}^k, \qquad \varepsilon_{IJ}^{k-1} = \varepsilon_{i,j+1}^k + \varepsilon_{i+1,j+1}^k.
\end{aligned}
$$

This defines A^{k-1}. In the finite volume case, CGA is usually much cheaper than CFA; according to (4.7) the flop count is 15 per coarse grid point.

Having discussed the choice of the coarse grid operators, we now describe how a coarse grid correction is applied. (Here, in the interest of simplicity, we use the so-called *correction scheme* as opposed to the so-called *full approximation scheme*; cf. Brandt [88].) Suppose we have two grids Ω^k and Ω^{k-1}. The fine grid problem to be solved is denoted by

$$(4.8)\qquad A^k \mathbf{w}^k = f^k.$$

Let $w_{(0)}^k$ be the current iterand, and let $w_{(1)}^k$ be the result of applying a coarse grid correction according to the basic two-grid cycling scheme of Chapter 1 so that

$$(4.9)\qquad w_{(1)}^k = w_{(0)}^k + I_{k-1}^k (A^{k-1})^{-1} I_k^{k-1} (f^k - A^k w_{(0)}^k),$$

where we assume for the sake of argument that the coarse grid problem is solved exactly. If A^{k-1} is chosen according to (4.1), then it follows that

$$(4.10)\qquad I_k^{k-1}(f^k - A^k w_{(1)}^k) = 0.$$

In other words,

$$(4.11)\qquad (f^k - A^k w^k_{(1)}, (I^{k-1}_k)^T z^{k-1}) = 0 \quad \text{for all } z^{k-1} \in H^{k-1},$$

which justifies the appellation "Galerkin approximation" for (4.1).

2.5. Smoothing methods. As we have just seen, after exact coarse grid correction we have, for the residue $r = f - Aw$, that

$$(5.1)\qquad r^k_{(1)} \in \ker(I^{k-1}_k),$$

as noted by Hemker [263] and McCormick [388]. Since I^{k-1}_k is a weighted average with positive weights, equation (5.1) implies that $r^k_{(1)}$ has frequent sign changes. In other words, $r^k_{(1)}$ is nonsmooth. In a similar way we can show that the error after coarse grid correction is nonsmooth; see the publications just mentioned. The purpose of smoothing is to efficiently reduce the nonsmooth part of the error.

With few exceptions, the smoothing methods that are used in MG are of Gauss–Seidel (GS) or incomplete factorization (IF) type. Kaczmarz [Ka1] relaxation is equivalent to Gauss–Seidel on AA^T.

GS comes in many variants, depending on the order in which the points of the computational grid are visited and on whether they are updated singly or in blocks. In forward point GS, the points are visited in the order given after equation (2.6). In backward point GS, the order is reversed. In line GS, we solve simultaneously for the unknowns on horizontal and vertical lines; the order is forward (from left to right or from bottom to top) or backward (the reverse). In symmetric GS, forward and backward sweeps alternate.

These methods do not suitably vectorize on a vector computer, because recursion is generally involved. We obtain vectorizable GS variants by suitably ordering the points in patterns (groups of different "colors") and solving simultaneously for a group. For example, one may use red-black (checkerboard) ordering for point GS or zebra ordering for line GS. For 7- or 9-point difference stencils (cf. Fig. 2.1), more than two colors are necessary for vectorization. Another vectorizable GS variant is obtained by forward or backward diagonal ordering. Here the points are visited in the order $n, n-1, 2n, n-2, 2n-1, 3n, \cdots$, etc., or in the reverse order. The resulting GS process vectorizes for the 5- and 7-point stencils of Fig. 2.1, but not for the 9-point stencil. Better smoothing properties may compensate for the fact that with diagonal GS the average vector length is shorter than with pattern GS. This depends on the problem at hand and the type of vector computer that is used.

We consider here two kinds of IF methods: Incomplete LU-factorization (ILU) and Incomplete Block (or line) LU-factorization (IBLU or ILLU). These methods have been developed for a single elliptic equation (2.1), but generalization to systems like (2.2) has yet to be fully explored. For details on IF methods and their application to MG, see Axelsson et al. [Ax1], Hackbusch [244], Hemker [267], Hemker et al. [279], Kettler [328], Kettler and Wesseling [330], Meijerink and van der Vorst [Me1], Sonneveld and Wesseling [523], van der Wees [Va2], van der Wees et al. [Va3], Varga et al. [Va7], Wesseling [592], [593], and Wesseling and Sonneveld [595]. IF methods vectorize somewhat as they stand; full vectorization can be achieved by introducing certain approximations. See Hemker et al. [278] and van der Vorst [Va1].

Smoothing methods may be improved by conjugate gradients, conjugate residuals, or Chebyshev acceleration. This may be helpful in difficult problems where the smoothing method does not perform very well. Bank and Douglas [32] give some analysis of this type of smoothing method improvement. Of course, the same techniques can be used to accelerate the MG method as a whole; practical examples of this approach are given by Kettler [328].

2.6. Smoothing analysis. For general understanding and development of MG methods, the analysis of two-grid cycling schemes can be useful. The general definition of a two-grid method is (cf. Chapter 1)

$$u^k_{(1/3)} = G^k_1(u^k_{(0)}, f^k), \tag{6.1a}$$

$$u^k_{(2/3)} = u^k_{(1/3)} + I^k_{k-1}(A^{k-1})^{-1}I^{k-1}_k r^k_{(1/3)}, \tag{6.1b}$$

$$u^k_{(1)} = G^k_2(u^k_{(2/3)}, f^k). \tag{6.1c}$$

This describes one iteration using the two grids Ω^k and Ω^{k-1}, with starting iterand $u^k_{(0)}$. G^k_1 and G^k_2 represent smoothing on grid Ω^k; because G^k_1 comes before and G^k_2 after coarse grid correction, these processes are called *pre-* and *post-smoothing*, respectively. When one of these smoothings is omitted, we have a *sawtooth cycle*. The iterand after one cycle is $u^k_{(1)}$. The residue r^k is defined by

$$r^k = f^k - A^k u^k. \tag{6.2}$$

Equation (6.1b) implies that we compute an exact solution on Ω^{k-1}, i.e., a coarse grid correction that is carried out exactly. It may be assumed quite generally that the smoothing processes G_i consist of ν_i iterations with iterative methods for the solution of $A^k\mathbf{u}^k = f^k$ of the type

$$u^k \leftarrow u^k + B^k_i(f^k - A^k u^k), \qquad i = 1, 2. \tag{6.3}$$

Hence, we have

$$r^k_{(1/3)} = (I - A^k B^k_1)^{\nu_1} r^k_{(0)}, \tag{6.4a}$$

$$r^k_{(1)} = (I - A^k B^k_2)^{\nu_2} r^k_{(2/3)}. \tag{6.4b}$$

B_1^k, B_2^k and ν_1, ν_2 may or may not be different. In the case of a sawtooth cycle, ν_1 or ν_2 equals 0. Let the exact discrete solution be given by

$$\mathbf{u}^k = (A^k)^{-1} f^k. \tag{6.5}$$

Then the error $e^k = \mathbf{u}^k - u^k$ satisfies

$$e^k_{(1/3)} = (I - B_1^k A^k)^{\nu_1} e^k_{(0)}, \tag{6.6a}$$

$$e^k_{(1)} = (I - B_2^k A^k)^{\nu_2} e^k_{(2/3)}. \tag{6.6b}$$

From (6.1b) it follows that

$$r^k_{(2/3)} = R^k r^k_{(1/3)}, \qquad e^k_{(2/3)} = E^k e^k_{(1/3)}, \tag{6.7}$$

with

$$R^k = I - A^k I^k_{k-1} (A^{k-1})^{-1} I^{k-1}_k$$

and

$$E^k = I - I^k_{k-1} (A^{k-1})^{-1} I^{k-1}_k A^k.$$

Hence,

$$r^k_{(1)} = (I - A^k B_2^k)^{\nu_2} R^k (I - A^k B_1^k)^{\nu_1} r^k_{(0)}, \tag{6.8a}$$

$$e^k_{(1)} = (I - B_2^k A^k)^{\nu_2} E^k (I - B_1^k A^k)^{\nu_1} e^k_{(0)}. \tag{6.8b}$$

If there are no boundaries, and if the coefficients defining A^k and A^{k-1} are constant, then the spectral radii of the operators in (6.8), and hence the rate of convergence of the two-grid method, may be determined by Fourier analysis. We will not pursue two-grid Fourier analysis here; we refer the reader instead to §§ 3 and 5 of Chapter 1 and § 2.8 of Chapter 5. For a full exposition, including some applications, see Stüben and Trottenberg [538].

If A^{k-1} is given by CGA (4.1), the two-grid method has some interesting properties (cf. Hemker [263], McCormick [388], and Sonneveld and Wesseling [523]). According to (6.7),

$$I^{k-1}_k r^k_{(2/3)} = 0. \tag{6.9}$$

As we noted in the beginning of § 2.5, a function satisfying (6.9) has many sign changes and may be called nonsmooth. The functions satisfying (6.9) constitute the subspace $\ker(I^{k-1}_k)$, and its orthogonal complement may be called smooth.

The orthogonal projection operator on $\ker(I^{k-1}_k)$ is given by

$$P_1^k = I - (I^{k-1}_k)^T (I^{k-1}_k (I^{k-1}_k)^T)^{-1} I^{k-1}_k. \tag{6.10}$$

From (6.4b) and (6.9) we deduce that

$$r^k_{(1)} = (I - A^k B_2^k)^{\nu_2} P_1^k r^k_{(2/3)}. \tag{6.11}$$

This leads us to the following definition.

DEFINITION 6.1. The *R-smoothing factor* of the smoothing process (6.1c) is

$$\rho_R = \|(I - A^k B_2^k)^{\nu_2} P_1^k\|.$$

We speak of the R-smoothing factor to emphasize that the kernel of the restriction is involved.

We now take the dual viewpoint of considering the error instead of the residue. Write

$$e^k_{(1/3)} = e_s + e_{ns}, \quad e_s \in R(I^k_{k-1}), \quad e_{ns} \in (R(I^k_{k-1}))^\perp. \tag{6.12}$$

The subscripts s and ns refer to "smooth" and "nonsmooth," respectively. Again, assuming that A^{k-1} is defined by CGA (4.1), we see by writing $e_s = I^k_{k-1}\hat{e}$ for some $\hat{e}$ that

$$e^k_{(2/3)} = E^k e_{ns}. \tag{6.13}$$

Hence, for $e^k_{(2/3)}$ to be small, it is necessary for smoothing to have made e_{ns} small. This motivates the following definition.

DEFINITION 6.2. The *P-smoothing factor* of the smoothing process (6.1a) is

$$\rho_p = \|P_2^k (I - B_1^k A^k)^{\nu_1}\|. \tag{6.14}$$

Here P_2^k is the orthogonal projection operator on $(R(I^k_{k-1}))^\perp$:

$$P_2^k = I - I^k_{k-1}((I^k_{k-1})^T I^k_{k-1})^{-1}(I^k_{k-1})^T. \tag{6.15}$$

We speak of the P-smoothing factor because the prolongation is involved. The quantity ρ_p is defined and used by McCormick [388]. Note that $P_2^k = P_1^k$ if $(I^k_{k-1})^T = I^{k-1}_k$.

The definition of ρ_R and ρ_P is generally valid, but ρ_R and ρ_P are not always easy to compute because the inverse operators that occur are not always easy to evaluate. A more easily computed smoothing factor, originally introduced by Brandt [88], is based on Fourier analysis and is in widespread use. However, it is valid only under ideal assumptions, such as those of constant coefficients and periodic boundaries or infinite regions, so its significance for more general problems is mainly heuristic. Fourier smoothing analysis is explained and applied extensively elsewhere and will not be discussed here. See, for example, Brandt [97], Kettler [328], and Stüben and Trottenberg [538]. The publication by Kettler contains an extensive catalogue of Fourier smoothing analysis results. It has been found that simple smoothing methods, such as point damped Jacobi or point Gauss–Seidel, are not robust (do not work well for many problems of practical interest), but that more complicated smoothing methods, such as block Gauss–Seidel and incomplete factorization, are both robust and efficient.

We now take a closer look at coarse grid approximation. For $i = 1, 2$ let

$$r^k_{(i/3)} = r_{s(i/3)} + r_{ns(i/3)}, \qquad e^k_{(i/3)} = e_{s(i/3)} + e_{ns(i/3)}, \tag{6.16}$$

with $r_{s(i/3)}$, $e_{s(i/3)}$ smooth and $r_{ns(i/3)}$, $e_{ns(i/3)}$ nonsmooth:

$$r_{s(i/3)}, e_{s(i/3)} \in (\ker (I^{k-1}_k))^\perp, \qquad r_{ns(i/3)}, e_{ns(i/3)} \in \ker (I^{k-1}_k). \tag{6.17}$$

From equations (4.1), (6.7), and (6.9), we see that we can write

$$r_{s(2/3)} = 0, \tag{6.18a}$$

$$r_{ns(2/3)} = r_{ns(1/3)} + R^k r_{s(1/3)}. \tag{6.18b}$$

We see once more that coarse grid correction does a proper job of annihilating the smooth part of the residual, but we also see a possibility that the nonsmooth part is amplified. If this amplification is too great, the MG method will not work properly. This can actually happen in practice. To avoid this, prolongation and restriction must satisfy criterion (3.12). We would digress too far by pursuing the matter here, but we will give a practical illustration shortly.

An analogous treatment can be given from a dual viewpoint if we consider the error. Since

$$(\ker (I^{k-1}_k))^\perp = R((I^{k-1}_k)^T), \tag{6.19}$$

there exists a grid function $\hat{e}_{(i/3)} \in H^{k-1}$ such that

$$e_{s(i/3)} = (I^{k-1}_k)^T \hat{e}_{(1/3)}. \tag{6.20}$$

When we assume that $(I^{k-1}_k)^T = I^k_{k-1}$, it follows from equations (4.1), (6.7), and (6.20) that

$$e_{ns(2/3)} = e_{ns(1/3)}, \tag{6.21a}$$

$$e_{s(2/3)} = -I^k_{k-1}(A^{k-1})^{-1} I^{k-1}_k A e_{ns(1/3)}. \tag{6.21b}$$

Again, we see that coarse grid correction does a proper job in the sense that $e_{s(1/3)}$ does not generate a contribution to $e_{(2/3)}$. However, there is a danger that $e_{ns(1/3)}$ generates a new smooth error component $e_{s(2/3)}$ that is too large. We can avoid this by adhering to rule (3.12) for choosing I^k_{k-1} and I^{k-1}_k. The fact that $e_{(2/3)}$ has a smooth part whereas $r_{(2/3)}$ is wholly nonsmooth can be explained by our noting that $r_{(2/3)} = A^k e_{(2/3)}$, and since A^k corresponds to a differential operator, A^k has an "unsmoothing" effect.

The above considerations explain why a finite volume discretization of a partial differential equation of order higher than 1 should not be treated with a MG method using (3.10) and (3.11), although this restriction and prolongation would seem natural for a finite volume formulation. The prolongation (3.10) is not exact for polynomials of degree higher than 0.

The foregoing considerations are illustrated by the following computations. We take equation (2.1) with $d_{[1],[1]} = d_{[2],[2]} = \varepsilon$, $d_{[1],[2]} = d_{[2],[1]} = 0$, $b_{[1]} = \cos\theta$, $b_{[2]} = \sin\theta$, $a = 0$, and the region Ω being the unit square. Discretization is of finite volume type with a $2^q \times 2^q$ finest grid and a 2×2 coarsest grid. Prolongation and restriction are defined by (3.10) and (3.11); coarse grid approximation is defined by (4.1). Smoothing consists of one symmetric point Gauss–Seidel iteration. The MG schedule is the sawtooth cycle (scheme MV of Chapter 1 with $\nu_1 = 0$, $\nu_2 = 1$). We take $\theta = 135°$, which was found to be the worst θ-value for the present smoothing method.

First, ρ_R (Definition 6.1) is determined by using the fact that

$$\rho_R^2 = \rho\{(P_1^k)^T(I - A^k B_2^k)^T(I - A^k B_2^k)P_1^k\}, \tag{6.22}$$

with ρ denoting the spectral radius. We compute ρ by the power method. This is facilitated by the fact that in the present case, as is easily verified,

$$I_k^{k-1}(I_k^{k-1})^T = 4I, \tag{6.23}$$

so that the inverse operator occurring in (6.10) is readily available. Table 2.1 gives results for ρ_R on an $n \times n$ grid. The power method converges rapidly in all cases. ρ_R is bounded independent of n, as a smoothing factor should be for MG to be effective.

TABLE 2.1
R-smoothing factors.

ε	$n=4$	$n=8$	$n=16$	$n=32$	$n=64$
10^7	0.21	0.23	0.23	0.23	0.24
10^{-7}	0.33	0.39	0.43	0.43	0.42

Next, the MG method is applied. Results are given in Table 2.2. The first number of each triplet is the maximum reduction factor of the residue that was observed, the second the average reduction factor, and the third the number of iterations performed. For $\varepsilon = 10^7$ we are solving something very close to the Poisson equation. Clearly, MG does not work: the maximum reduction factor tends to 1 as $n \to \infty$. The cause of failure is not the smoothing process; according to Table 2.1, we have a good smoother. Failure occurs because prolongation and restriction are not accurate enough

TABLE 2.2
Multigrid results.

ε	$n=4$	$n=8$	$n=16$	$n=32$	$n=64$
10^7	0.019, 0.014, 8	0.16, 0.07, 8	0.42, 0.15, 8	0.65, 0.20, 8	0.80, 0.22, 8
10^{-7}	0.03, 10^{-8}, 4	0.10, 10^{-6}, 6	0.14, 0.02, 8	0.20, 0.14, 8	0.29, 0.19, 8

for a second order equation (cf. the end of §2.3). The operator R^k in (6.18b) generates a nonsmooth residue component that is too large. We observed that $\|r_{(2/3)}\|/\|r_{(1/3)}\| > 1$; in fact, this ratio increases with n, with 4.7 being a typical value for $n = 64$. Increasing the number of smoothing steps or using a W-cycle does not help very much.

For $\varepsilon = 10^{-7}$ we are effectively solving a first order equation. According to the discussion in § 2.3, prolongation and restriction should be accurate enough, and indeed we see in Table 2.2 that MG is working well.

2.7. Remarks on software. Since such a large variety of MG algorithms is possible, software development can be approached in various ways, two of which will be examined here.

The first approach is to develop general building blocks and diagnostic tools, which helps users to develop their own software for particular applications without having to start from scratch. Users will therefore need at least a basic knowledge of MG methodology. The software package GRIDPACK (Brandt and Ophir [107]) contains such software tools.

The second approach is to develop *autonomous* (*black box*) subroutines, for which the user has to specify only the problem on the finest grid. (A subroutine may be called autonomous if it operates independently of the user and does not require any additional input from the user beyond problem specification.) The user does not need to know anything about MG methods, and perceives the subroutine as if it were any other linear algebra solution method. This approach is adopted in the AMG method (Chapter 4) and by the MGD codes (MGD1 and MGD5: Hemker and de Zeeuw [277], Hemker et al. [278], [279], Sonneveld and Wesseling [523], Sonneveld et al. [So1], and Wesseling [592]).

An MG code that can be used as an autonomous subroutine is PLTMG (Bank [26] and Bank and Sherman [39], [40], [42]). This widely used subroutine does not accept a given fine grid matrix but rather generates its own with a finite element method. It has a far different scope from that of linear systems solvers: it can generate finite element grids adaptively and is directly applicable to nonlinear problems. However, it is not (meant to be) an efficient linear systems solver and will not be further discussed here.

Of course, it is possible to steer a middle course between the two approaches just outlined. This is done in BOXMG (Dendy [142], [144]) and in the MG00 series of codes developed at GMD (Foerster and Witsch [166], [167] and Stüben et al. [540]).

By sacrificing generality for efficiency, we can obtain very fast MG methods for special problems such as the Poisson or the Helmholtz equation. In MG00 we can do this by setting certain parameters. A very fast MG Poisson code tuned for the CYBER-205 has been developed by Barkai and Brandt [47]. This is probably the fastest 2-D Poisson solver in existence today.

If one wants to emulate a linear algebraic systems solver, with the fine grid matrix and right-hand side supplied by the user, then the use of coarse grid Galerkin approximation (CGA, equation (4.1)) is inevitable. CGA is used in BOXMG and the MGD codes. Instead, the problem may be specified by means of the coefficients in the differential equation. Then the code may generate, by itself, finite difference approximations on all grids (CFA). This approach is followed in MG00. For remarks on the relative advantages and disadvantages of CFA and CGA, see § 2.4 and the references cited therein.

In an autonomous subroutine, the user is not allowed to adapt the method to the problem at hand. Therefore, the method must be robust, i.e., it must work efficiently for a large class of problems.

For linear problems, robustness and efficiency can indeed be realized simultaneously with simple MG algorithms in autonomous codes, as will become clear below. This fact brings out the basic soundness and practical utility of the MG approach. For second order elliptic equations with smooth coefficients, prolongation defined by linear interpolation (equations (3.2)–(3.4)), restriction defined as the transpose of prolongation, and CGA—the coarse grid correction part of the method—are all found to be robust and efficient in practice and are also theoretically well founded. Apart from the theoretical foundation, the same is true when the coefficients are strongly discontinuous (i.e., contain jumps in orders of magnitude) if matrix-dependent prolongation is used. It is found in practice that even the simplest MG schedule (the sawtooth cycle) suffices provided coarse grid correction and smoothing are both effective. Both smoothing analysis (cf. the catalogue of smoothing analysis results in Kettler [328]) and practical experience (cf. the next section) indicate that the simplest smoothing processes, such as Jacobi or point Gauss–Seidel, are not robust, although they are efficient for certain problems, such as those that are close to the Poisson equation. Therefore, block Gauss–Seidel or incomplete factorization must be used.

Nonlinear problems may be solved with linear MG methods in various ways. The problem may be linearized and solved iteratively, for example, by a Newton method. This works well as long as the Jacobian of the nonlinear discrete problem is nonsingular. But it may happen that the given continuous problem has no Fréchet derivative. In that case the condition of the Jacobian deteriorates as the computational grid is refined, and the Newton method does not converge rapidly or even at all. Examples are given in the next section. A systematic way of applying numerical software to problems outside the class of problems to which the software is directly applicable is the defect correction approach. Auzinger and Stetter [19] and Böhmer et al. [66] point out how this ties in with MG methods. On the other hand, codes such as GRIDPACK and PLTMG are directly applicable to nonlinear problems.

We end this section with some remarks on the software packages mentioned above. All are for 2-D second order elliptic partial differential equations.

MG00 is for selfadjoint second order elliptic equations without mixed derivatives on rectangular domains. Finite difference (5-point stencil) approximations are generated on uniform grids. Especially efficient versions are invoked for the Poisson and Helmholtz equations. The user interface is of the ELLPACK type (Rice and Boisvert [464]). The user chooses from various point and block Gauss–Seidel smoothing processes, restriction operators, and MG cycling schedules.

BOXMG uses CGA and can handle strongly discontinuous coefficients because matrix-dependent prolongation is used. It allows mixed derivatives and nonselfadjoint equations. The user provides a finite difference approximation (9-point stencil) on the finest grid, which is uniform and rectangular. More general grids can be handled with boundary-fitted coordinate transformations and/or by adding artificial equations ("padding"). The user has a choice of point and block Gauss–Seidel and Kaczmarz smoothing, and of various MG schedules. BOXMG works well in the selfadjoint case for smooth and strongly discontinuous coefficients. But for equation (8.2) the efficiency degrades badly as $\varepsilon \downarrow 0$; see Dendy [144]. We believe this is due to the use of central differences on the finest grid, so that the matrix is not a K-matrix (Definition 2.1).

MGD1 and MGD5 can handle strongly discontinuous coefficients if discontinuities occur only along parts of lines that belong to the finest grid and to two or more coarser grids as well. They work for mixed derivatives and nonselfadjoint equations, provided there is a K-matrix on the finest grid. The user has to provide a finite difference approximation (7-point stencil) on the finest grid, which is uniform and rectangular. More general grids can be handled in the same way as for BOXMG. The user is not allowed to interfere with the MG algorithm, and perceives these codes as linear algebraic systems solvers; they are completely autonomous. For smoothing, MGD1 and MGD5 use one ILU and one IBLU iteration, respectively. Prolongation and restriction are defined by (3.2), (3.4), and (3.5). The MG schedule is the simplest possible, namely the sawtooth cycle. As we remarked before, the efficiency and robustness attainable with such a simple MG algorithm attest to the basic soundness of the MG approach. These routines are portable and have been designed for auto-vectorization on CRAY-1 and CYBER-205, without much sacrifice of their capacity on scalar machines. More details, including CPU-time measurements on CRAY-1 and CYBER-205, may be found in the publications mentioned above. Users may obtain the MGD software by sending a magnetic tape to the author of this chapter.

2.8. Numerical experiments. The proofs that have been published (see Chapter 5) concerning the h-independent rate of convergence of MG usually assume that damped Jacobi iteration is used for smoothing. Also, most proofs are restricted to selfadjoint problems. However, in practice, damped Jacobi is rarely used, because other smoothing processes such as those mentioned before are found to be more efficient, and because damped Jacobi does not work for important practical problems not covered by the theory. Typical examples of such problems are the following special cases of (2.1):

$$-(\varepsilon a_c^2 + a_s^2)u_{,[1][1]} - 2(\varepsilon - 1)a_s a_c u_{,[1][2]} - (\varepsilon a_s^2 + a_c^2)u_{,[2][2]} = f \tag{8.1}$$

and

$$-\varepsilon u_{,\alpha\alpha} + a_c u_{,[1]} + a_s u_{,[2]} = f, \tag{8.2}$$

with $a_c = \cos\theta$, $a_s = \sin\theta$. Equation (8.1) is called the *rotated anisotropic diffusion equation,* because it is obtained by a rotation of the coordinate axes over an angle θ of the anisotropic diffusion equation

$$-\varepsilon u_{,[1][1]} - u_{,[2][2]} = f. \tag{8.3}$$

Equation (8.2) is the *convection diffusion equation.* It is not selfadjoint. Equation (8.1) models not only physical anisotropic diffusion, but also highly stretched meshes. It also includes mixed derivatives, which arise mostly because of the use of nonorthogonal boundary-fitted coordinate mappings.

MG convergence theory is not uniform in the value of the coefficients in the differential equation, and the theoretical rate of convergence is not bounded away from 1 as $\varepsilon \downarrow 0$. In the absence of theoretical justification, we must resort to numerical experiments to validate a method, and equations (8.1) and (8.2) constitute a discriminating test bed. The fact that the coefficients in (8.1) and (8.2) are constant does not make these equations easier to solve with MG. It may happen that the method does not work well for certain values of ε and θ. This difficulty is more noticeable when ε and θ have these values throughout the domain than when they occur only locally.

Of course, numerical experiments cannot guarantee good performance of a method for problems that have not yet been attempted, but can be used only to rule out methods that fail. However, we can build up confidence by choosing test problems carefully, trying to make them representative for large classes of problems, and keeping the field of application in mind.

The practical utility of MG was demonstrated at an early stage by the numerical experiments of Brandt [88], Hackbusch [215], [224], Nicolaides [434], and Wesseling and Sonneveld [595]. Here we will concentrate on experiments with software that is now available.

Constant coefficient test problems are representative of equations with smooth coefficients. For equations with strongly discontinuous coefficients, such as those that occur in petroleum reservoir engineering and in neutron diffusion problems, other test problems are needed. Suitable ones have been proposed by Kershaw [Ke1] and Stone [St2]; a definition of these test problems may also be found in Kettler [328].

The four test problems mentioned above and similar ones are rapidly gaining acceptance among MG practitioners as the standard. Sonneveld and Wesseling [523] give results for (8.1) and (8.2) with the MGD1, MGD5, MGHZ, and MGAZ codes. MGHZ and MGAZ differ from MGD1 and MGD5 (described in the preceding section) in one respect: smoothing is done with one block Gauss–Seidel iteration, with horizontal lines in a zebra pattern, and with combined horizontal/vertical lines in a zebra pattern, respectively. Table 2.3 gives some results for these methods. In all cases, $\varepsilon = 10^{-8}$ and the computational grid has size 65×65. In Table 2.3, w is the estimated number of operations per iteration and per grid point and l is the number of iterations required for one decimal digit. The θ-values were samples with an interval of 15°; Table 2.3 presents the worst case for each method. The block Gauss–Seidel methods would work better for equation (8.2) if a successive ordering (both backward and forward) instead of a zebra pattern were used. However, the only way to preserve vectorizability would be to adopt the vectorizable point GS variant referred to in §2.5. For further details, see Sonneveld and Wesseling [523].

For some comparative experiments with AMG, BOXMG, MG00, and MGD1, see Ruge and Stüben [486]. In selfadjoint cases, MG00 is significantly faster when the coefficients are constant. With variable coefficients, the methods are about equally fast, with AMG needing a large setup time, which is inherent in the nature of the method. (The statement in the aforementioned publication that MGD1 is not applicable to equations in divergence form is not correct.) AMG and BOXMG are most robust, since they can handle discontinuous coefficients. For the convection-diffusion

TABLE 2.3
Numerical results for equations (8.1) *and* (8.2).

		Equation (8.1)		Equation (8.2)	
	w	θ	l	θ	l
MGD1	30	30°	4.9	165°	3.5
MGD5	54	135°	3.5	ANY	0.1
MGHZ	22	45°	4.9	MANY	>10
MGAZ	40	45°	4.5	MANY	>10

equation, AMG and MGD1 are compared. Due to the smoothing properties of ILU (see Kettler [328]), MGD1 performance depends on ε and θ (see Sonneveld and Wesseling [523]). But in all cases tested, MGD1 has produced eight or more decimal digits within the setup time needed by AMG, with one exception, where AMG overtakes MGD1 after four digits. On the basis of Table 2.3, MGD5 is expected to be much faster in all convection-diffusion cases. Also, if we disregard setup time, MGD1 is (much) faster than AMG in most examples. This is due to the greater generality and complexity of AMG. However, AMG is the method of choice for problems with very complex structure, where the matrix does not have the regular 5-, 7-, or 9-diagonal structure assumed by MG00, MGD1/5, and BOXMG, respectively.

MG tests on equations with discontinuous coefficients are described by Alcouffe et al. [5], Behie and Forsyth [60], Kettler [328], Ruge and Stüben [486], and Sonneveld and Wesseling [523]. MG works well for these problems provided matrix-dependent prolongation and CGA are used. Here we will give no further discussion of this type of problem.

In practice, linear MG (or other) solvers will often be used to iteratively solve nonlinear problems. For purposes of illustration we mention two examples.

Nowak and Wesseling [439] apply MGD4 (an extension of MGD1 to 9-point stencils) to the full potential equation for the compressible flow around an airfoil with lift. The discrete approximation on a 128×41 grid is linearized with Newton's method, and the linear system is solved with MGD4, which is found to converge rapidly in all cases. For subsonic flow, the Newton method also converges rapidly and quadratically. For transonic flow with a fairly strong shock, the Newton method converges slowly and erratically because the Fréchet derivative of the global system is ill conditioned. This trouble can probably be avoided in nonlinear MG methods that use Newton-linearization only locally in a nonlinear smoothing process. This implies that, for the given problem, the development of a special MG method would be more effective than a standard (black box) linear MG subroutine. Another remedy is to modify the global Newton process. This approach is followed by Boerstoel and Kassies [Bo1]. It turns out that the underlying cause of the ill-conditioning of the global Fréchet derivative is erratic change of the shock position, in mutual interference with the circulation (lift). Therefore, Boerstoel and Kassies freeze the shock position and circulation within their linear MG computation, and update the shock position and circulation in a special way outside the MG part of the algorithm. This results in very satisfactory convergence rates. It is to be noted that they use a special purpose linear MG solver.

An example of the application of a linear MG solver to a nonlinear hyperbolic system is given by Mulder [423]. The equation is given by (2.2), its discretization by (2.19), and the linearized version by (2.20). Prolongation and restriction are given by (3.10) and (3.11), and CGA is used. Smoothing is done with forward/backward, horizontal/vertical line Gauss–Seidel. The V-cycle is used, with two pre- and two post-smoothing operations. The rate of convergence is almost independent of grid size, and important reductions in computer time are realized as compared with the corrresponding single grid method. Equation (2.19) is first order accurate. A second order scheme is solved also by the same method with defect correction. The rate of convergence then is weakly dependent on grid size, but the acceleration compared with single grid iteration is still significant. See also Jespersen [309] for a linear MG approach to the same type of problem.

2.9. Final remarks. We have reviewed the state of the art of the solution of discretized elliptic and hyperbolic partial differential equations by means of linear MG. The material presented indicates that dependable and very rapid methods are evolving.

The effort required for the programming of an MG method is appreciable. Preconditioned conjugate gradient (PCG) methods provide an alternative to MG for which the programming required may be less and simpler. MG and PCG can be viewed as acceleration techniques, and have been compared from this point of view by Sonneveld and Wesseling [523] and Sonneveld et al. [So1]. They have in common the iterative method (called smoother in MG and preconditioner in PCG) that is to be accelerated. Experience indicates that a good smoother is usually also a good preconditioner and vice versa. Numerical experiments comparing MG and PCG are reported by Behie and Forsyth [60], Dendy and Hyman [147], Sonneveld and Wesseling [523], and Wesseling and Sonneveld [595]. Because the computational complexity of PCG is a (weakly) growing function of the number of unknowns n ($O(n^{5/4})$ has been proven for the Poisson equation; see Gustafsson [Gu1]), for n large enough MG is faster since its computational complexity is $O(n)$. But the experiments in the publications cited above indicate that for medium-sized problems, PCG and MG are fairly comparable.

REFERENCES

[Ax1] O. Axelsson, S. Brinkkemper and V. P. Il'in, *On some versions of incomplete block-matrix factorization iterative methods*, Linear Algebra Appl., 58 (1984), pp. 3–15.

[Bo1] J. W. Boerstoel and A. Kassies, *On the integration of multi-grid relaxation into a robust fast-solver for transonic potential flows around lifting airfoils*, in Colloquium Topics in Applied Numerical Analysis, J. G. Verwer, ed., CWI Syllabus, Centre for Mathematics and Computer Science, Amsterdam, 1984, pp. 107–148.

[Co1] R. COURANT, AND K. O. FRIEDRICHS, *Supersonic Flow and Shock Waves,* Springer–Verlag, New York, 1949.

[Gu1] I. GUSTAFSSON, *A class of first order factorization methods,* BIT, 18 (1978), pp. 142–156.

[Ha1] A. HARTEN, P. D. LAX AND B. VAN LEER, *On upstream differencing and Godunov-type schemes for hyperbolic conservation laws,* SIAM Rev., 25 (1983), pp. 35–61.

[Ka1] S. KACZMARZ, *Angenäherte Auflösung von Systemen Gleichungen,* Bull. Int. Acad. Polon. Sci. Cl.A, 35 (1937), pp. 355–357.

[Ke1] D. S. KERSHAW, *The incomplete Choleski-conjugate gradient method for the iterative solution of systems of linear equations,* J. Comput. Phys., 26 (1978), pp. 43–65.

[Me1] J. A. MEIJERINK AND H. A. VAN DER VORST, *An iterative solution method for linear systems of which the coefficient matrix is a symmetric M-matrix,* Math. Comp., 31 (1977), pp. 148–162.

[Mu1] W. A. MULDER AND B. VAN LEER, *Experiments with upwind methods for the Euler equations,* J. Comput. Phys., 59 (1985), pp. 232–246.

[Ri1] R. D. RICHTMYER, AND K. W. MORTON. *Difference Methods for Initial Value Problems,* John Wiley, New York, 1967.

[So1] P. SONNEVELD, P. WESSELING AND P. M. DE ZEEUW, *Multigrid and conjugate gradient acceleration of basic iterative methods,* in Numerical Methods for Fluid Dynamics II, K. W. Morton and M. J. Baines, eds., Clarendon Press, Oxford, 1986, pp. 347–368.

[St1] J. STEGER AND R. WARMING, *Flux vector splitting of the inviscid gas dynamic equations with applications to finite difference methods,* J. Comput. Phys., 40 (1981), pp. 263–293.

[St2] H. L. STONE, *Iterative solution of implicit approximations of multidimensional partial differential equations,* SIAM J. Numer. Anal., 5 (1968), pp. 530–558.

[Va1] H. A. VAN DER VORST, *A vectorizable variant of some ICCG methods,* SIAM J. Sci. Statist. Comput., 3 (1982), pp. 350–356.

[Va2] A. J. VAN DER WEES, *Robust calculation of 3D potential flow based on the non-linear FAS multi-grid method and a mixed ILU/SIP algorithm,* in Colloquium Topics in Applied Numerical Analysis, J. G. Verwer, ed., CWI Syllabus, Centre for Mathematics and Computer Science, Amsterdam, 1984, pp. 419–459.

[Va3] A. J. VAN DER WEES, J. VAN DER VOOREN AND J. H. MEELKER, *Robust Calculation of 2D Transonic Potential Flow Based on the Non-Linear FAS Multi-Grid Method and Incomplete LU Decomposition,* AIAA Paper 83–1950, 1983.

[Va4] B. VAN LEER, *Flux-vector splitting for the Euler equations,* in Proc. Eighth International Conference on Numerical Methods in Fluid Dynamics, E. Krause, ed., Aachen, 1982, pp. 507–512. Lecture Notes in Physics 170, Springer-Verlag, Berlin, 1982.

[Va5] ———, *On the relation between the upwind-differencing schemes of Godunov, Engquist-Osher and Roe,* SIAM J. Sci. Statist. Comput., 5 (1984), pp. 1–20.

[Va6] R. S. VARGA, *Matrix Iterative Analysis,* Prentice-Hall, Englewood Cliffs, NJ, 1962.

[Va7] R. S. VARGA, E. B. SAFF AND V. MEHRMANN, *Incomplete factorizations of matrices and connections with H-matrices,* SIAM J. Numer. Anal., 17 (1980), pp. 787–793.

CHAPTER 3

Multigrid Approaches to the Euler Equations

P. W. HEMKER AND G. M. JOHNSON

3.1. Introduction. In this chapter we give a survey of the present state of the art for multigrid solution of the Euler equations for inviscid compressible flow. This is an example of a branch of multigrid research in which a thorough mathematical basis is still missing. What does exist to guide applications is an abundance of heuristic arguments and analogues from areas with better theoretical foundations, yet there is a scarcity of solid theory to account for the convergence speed and efficiency shown in practice. Such theory as there is lags well behind both practical development and the excellent results that have already been obtained with multigrid Euler solvers applied to large scale problems.

We chose to restrict our discussion in this chapter to the Euler equations because there are a few visible lines of development that can easily be treated within the scope of this chapter. Much interesting work has also been done in the general field of compressible and incompressible Navier–Stokes equations (cf. the pioneering work by Brandt [92], [95], [99], [105]), but the state of the art in this area is changing too rapidly to be suitable for discussion here. We refer the reader instead to the literature for other fluid flow applications. In particular, the KWIC index to the Multigrid Bibliography included in Appendix 2 of this book lists a collection of papers on the compressible and incompressible Navier equations, potential flow, and the Stokes equations.

Because even the multigrid Euler-solver discipline is continually changing, in this chapter we adopt the perspective of an overview rather than one of prescription and detailed guidance. We hope that this overview and the cited references will prepare the reader for further studies in this advancing field.

The efficient solution of flow problems was one of the early aims in the applications of multigrid (MG) methods [86]. However, in recent years most of the progress in the development of MG has been made in the field of elliptic partial differential equations and other fields where a solid mathematical theory exists (e.g., integral equations). For the inherently more complex equations that describe flow problems, the theoretical development of MG did not proceed at the same pace. Early numerical work was done by Brandt [99], [105] and South and Brandt [524], where, for example, the Stokes equations and the incompressible and compressible Navier–Stokes equations were considered.

On the other hand, triggered by practical interest from the engineering sciences, several attempts have been made to apply MG ideas to improving the efficiency of flow computations. If the flow is assumed to be irrotational, then it can best be described by the potential equation, which—in the interesting case of transonic flow—is of mixed hyperbolic and elliptic type. By the use of MG, substantial improvements were made in the procedures for solving these equations [70], [123], [297], [382], [439], [524]. When the assumption of irrotational flow is dropped, an exact description of inviscid flow is given by the Euler equations. When the physical effects of viscosity and heat conduction are also included, these equations extend to the Navier–Stokes equations. Models of turbulence can also be included in the Navier–Stokes equations.

In this chapter we will treat several multiple grid approaches that are used for the solution of the equations of compressible flow. We restrict ourselves to problems in 2 space dimensions. Almost all techniques discussed here can be applied in 3-D as well, but the burden of 3-D notation makes the description unattractive. Also, in practice, most codes are written for 2-D problems because the complexity of 3-D computations and the computational requirements for their implementation are at the limit of present-day computer capabilities. The advent of more powerful computers will certainly change this situation in the near future.

Although practical problems that arise in the aircraft and turbomachinery industries are often described by the compressible Navier–Stokes equations, we shall consider mainly the Euler equations of inviscid flow. The reason for this is the assumption that a good method for the solution of the Euler equations may be extended to those situations where viscosity plays a significant role.

In those cases where the solution of the Euler equations can be used as a first approximation to the solution of the full Navier–Stokes equations, it may be a convenient approach to compute (an approximation to) this Euler flow first. This approximation can then be corrected for viscous effects. Most simply, this is done by a defect correction approach [66], where the solution of the Navier–Stokes equations is found by an iterative process in which only Euler-type equations are (approximately) solved and the heat conduction and viscous Navier–Stokes terms are taken care of by adding the corresponding corrections as forcing terms. In practice, a simple method to realize such an iterative process for the solution of the Navier–Stokes equation is to neglect the extra Navier–Stokes terms at particular stages of the solution process.

3.1.1. The equations. The 2-D *Navier–Stokes equations,* describing the physical laws of conservation of mass, momentum and energy, can be written in conservation form as

$$\frac{\partial}{\partial t}\mathbf{q}+\frac{\partial}{\partial x}F(\mathbf{q})+\frac{\partial}{\partial y}G(\mathbf{q})=0, \tag{1.1a}$$

where

$$F(q)=f(q)-\mathrm{Re}^{-1}r(q),\qquad G(q)=g(q)-\mathrm{Re}^{-1}s(q), \tag{1.1b}$$

and

$$q=\begin{pmatrix}\rho\\ \rho u\\ \rho v\\ \rho e\end{pmatrix},\quad f=\begin{pmatrix}\rho u\\ \rho u^2+p\\ \rho uv\\ \rho uH\end{pmatrix},\quad g=\begin{pmatrix}\rho v\\ \rho vu\\ \rho v^2+p\\ \rho vH\end{pmatrix},$$

$$r=\begin{pmatrix}0\\ \tau_{xx}\\ \tau_{xy}\\ \kappa\,\mathrm{Pr}^{-1}(\gamma-1)^{-1}(c^2)_x+u\tau_{xx}+v\tau_{xy}\end{pmatrix},$$

$$s=\begin{pmatrix}0\\ \tau_{xy}\\ \tau_{yy}\\ \kappa\,\mathrm{Pr}^{-1}(\gamma-1)^{-1}(c^2)_y+u\tau_{xy}+v\tau_{yy}\end{pmatrix}.$$

Here ρ, u, v, e and p, respectively, represent density, velocity in x- and y-direction, specific energy and pressure; $H=e+p/\rho$ is the specific enthalpy. The pressure is obtained from the equation of state, which for a perfect gas reads

$$p=(\gamma-1)\rho(e-\tfrac{1}{2}(u^2+v^2));$$

γ is the ratio of specific heats. $q(t,x,y)$ describes the state of the gas as a function of time and space and f and g are the convective fluxes in the x- and y-direction, respectively. Re and Pr denote the Reynolds and Prandtl numbers; thermal conductivity is given by κ; $c=\sqrt{\gamma p/\rho}$ is the local speed of sound; and

$$\tau_{xx}=(\lambda+2\mu)u_x+\lambda v_y,\quad \tau_{xy}=\mu(u_y+v_x),\quad \tau_{yy}=(\lambda+2\mu)v_y+\lambda u_x,$$

where λ and μ are viscosity coefficients. Stokes assumption of zero bulk viscosity may reduce the number of coefficients by one: $3\lambda+2\mu=0$.

We denote the open domain of definition of (1.1) by Ω^*.

The *Euler equations* are obtained from (1.1a) by neglecting viscous and heat conduction effects:

$$F(q)=f(q), \qquad G(q)=g(q). \tag{1.1c}$$

The time-dependent Euler equations form a hyperbolic system: written in the quasi-linear form

$$\frac{\partial \mathbf{q}}{\partial t}+\frac{\partial f}{\partial q}\cdot\frac{\partial \mathbf{q}}{\partial x}+\frac{\partial g}{\partial q}\cdot\frac{\partial \mathbf{q}}{\partial y}=0,$$

the matrix

$$k_1A+k_2B=k_1\frac{\partial f}{\partial q}+k_2\frac{\partial g}{\partial q} \tag{1.2}$$

has real eigenvalues for all directions (k_1, k_2). These eigenvalues are $(k_1u+k_2v)\pm c$ and (k_1u+k_2v) (a double eigenvalue). The sign of the eigenvalues determines the direction in which the information about the solution is carried along the line (k_1, k_2) as time develops (i.e., it determines the direction of flow of characteristic information). It locates the direction of the domain of dependence.

It is well known that, because of the nonlinearity, solutions of the Euler equations may develop discontinuities, even if the initial flow $(t=t_0)$ is smooth. To allow discontinuous solutions, (1.1) is rewritten in its integral form

$$\frac{\partial}{\partial t}\iint_{\Omega}\mathbf{q}\,dx\,dy+\int_{\partial\Omega}(f\cdot n_x+g\cdot n_y)\,ds=0 \quad \text{for all } \Omega\subset\Omega^*; \tag{1.3}$$

$\partial\Omega$ is the boundary of Ω and (n_x, n_y) is the outward normal vector at the wall $\partial\Omega$.

The form (1.3) of equation (1.1) shows clearly the character of the system of conservation laws: the increase of q in Ω can be caused only by the inflow of q over $\partial\Omega$. In symbolic form we write (1.3) as

$$\mathbf{q}_t+N(\mathbf{q})=0. \tag{1.4}$$

The solution of the weak form (1.3) of (1.1a, c) is known to be nonunique, and a physically realistic solution (which is the limit of a flow with vanishing viscosity) is known to satisfy the additional entropy condition (cf. [La1], [La2]). The entropy condition implies that characteristics do not emerge at a discontinuity in the flow.

The *steady state* equations are obtained by the assumption $\partial\mathbf{q}/\partial t=0$. Guided by the defect correction principle and knowing how the viscous effects change the governing equations, for the Navier–Stokes equations with large Reynolds number we can concentrate on the solution methods for the stationary Euler equations

$$N(\mathbf{q})=0. \tag{1.5}$$

3.1.2. The discretizations. For the discretization of (1.1) or (1.3), two different approaches can be taken. First, the time and space discretizations can be made at once. This leads, for example, to discretization schemes of Lax–Wendroff type. An initial state of the fluid $q^h_{(n)}$, defined on a discrete grid, is advanced over one time step. Using a second order approximation in time yields

$$q^h_{(n+1)}=q^h_{(n)}+\Delta t(q^h)_t+\tfrac{1}{2}(\Delta t)^2(q^h)_{tt}. \tag{1.6}$$

With the equation (1.1a, c), we arrive at

$$q^{(n+1)}_{ij}=q^{(n)}_{ij}-\Delta t(f_x+g_y)_{ij}+\tfrac{1}{2}(\Delta t)^2\Big\{[A(f_x+g_y)]_x+[B(f_x+g_y)]_y\Big\}_{ij},$$

where A and B are defined by (1.2). Using various difference approximations of the bracketed terms in the right-hand side, we may obtain different Lax–Wendroff type discretizations.

This type of discretization is usually made on a rectangular grid. If the domain Ω^* is not rectangular, a 1-1-mapping $(x, y)\longleftrightarrow(\xi, \eta)$ between the physical domain and a rectangular computational domain can be constructed. Then the differential equation and the boundary conditions are reformulated on this computational domain.

A property of most of these Lax–Wendroff discretizations is that, when by time stepping a stationary state is obtained such that $q^h_{(n+1)}=q^h_{(n)}$, the discrete stationary state still depends on Δt. This is caused by the fact that the discrete term with $(\Delta t)^2$ in (1.6) in general does not vanish.

A second approach is to distinguish clearly between the time and the space discretization by the method of lines. First, a space discretization is made for the partial differential equation (1.4) by which it is reduced to the large system of ordinary differential equations

$$\frac{\partial}{\partial t}\mathbf{q}^h=N^h(\mathbf{q}^h). \tag{1.7}$$

Now, to find an approximation of the time-dependent solution of (1.4), any method can be used for the integration of this system of ordinary differential equations. The solution of the steady state can be computed by solving (1.7) until the transients have died out. Alternatively, we can avoid the ordinary differential equations (1.7) and solve the nonlinear system

$$N^h(\mathbf{q}^h)=0 \tag{1.8}$$

by other (more direct) means. In both cases (1.7) and (1.8), we find a steady approximate solution $\mathbf{q}^h$ independent of the choice of a time step.

For the construction of the semi-discrete system (1.7) or (1.8) on a nonrectangular domain Ω^*, a mapping $(x, y) \longleftrightarrow (\xi, \eta)$ can again be introduced and finite difference approximations (of arbitrarily high order) can be used to construct a space discretization of the transformed steady equation

$$[y_\eta F(\mathbf{q}) - x_\eta G(\mathbf{q})]_\xi + [-y_\xi F(\mathbf{q}) + x_\xi G(\mathbf{q})]_\eta = 0.$$

Another way to construct system (1.7) on a nonrectangular grid is by a *finite volume* technique. Here, the starting point for the discretization is (1.3). Without an a priori transformation, the domain Ω^* is divided into a set of disjoint quadrilateral cells Ω_{ij}. The discrete representation $\mathbf{q}^h$ of $\mathbf{q}$ is given by the values $\mathbf{q}_{ij}$, the (mean) values of $\mathbf{q}$ in the cell Ω_{ij}. Using different approximations for the computation of fluxes between the cells Ω_{ij}, we obtain various finite volume discretizations. We can easily obtain a conservative scheme by computing a unique approximation for each flux over the boundary between two neighboring cells.

In order to define a proper sequence of discretizations as $h \to 0$ for a nonrectangular grid, a formal relation between the vertices of cells Ω_{ij} and a regular grid can be given, again by a mapping $(x, y) \longleftrightarrow (\xi, \eta)$. If this mapping is smooth enough, it can be proved that, for refinements $h \to 0$ corresponding to regular refinements in (ξ, η), space discretizations up to second order can be obtained. An advantage of the finite volume technique is that the untransformed equations can be used, even for a complex region. Boundary condition information is also usually simpler for finite volume methods.

With the finite volume technique, both central difference and upwind type finite volume schemes are used. They differ by the computation of the flux between neighboring cells Ω_{ij}:

(1) For a central difference type, the flux over a cell wall Γ_{LR} between two cells with states q_L and q_R is computed as $\frac{1}{2}f^*(q_L) + \frac{1}{2}f^*(q_R)$, where $f^* = k_1 f + k_2 g$ is the flux normal to Γ_{LR}. On a Cartesian grid this scheme reduces to the usual central difference scheme. In order to stabilize this scheme, and to prevent the uncoupling of odd and even cells in the grid, it is necessary to supplement the scheme with some kind of artificial dissipation (artificial viscosity).

(2) For upwind difference type discretizations, numerical flux functions $f^*(q_L, q_R)$ are introduced to compute the flux over Γ_{LR}. Several functions f^* are possible. They solve approximately the Riemann problem of gas dynamics: they approximate the flux between two (initially) uniform states q_L and q_R. Approximate Riemann solvers have been proposed by Steger and Warming [St1], van Leer [Va1], Roe [Ro1], and Osher [Os1], [Os2]. A description of these upwind schemes and their properties can be found in the cited literature. For a consistent scheme, $f^*(q, q) = f^*(q)$, i.e., the

numerical flux function with equal arguments conforms with the genuine flux function in (1.1c). All these upwind flux functions are purely one-sided if all characteristics point in the same direction, i.e., $f^*(q_L, q_R) = f^*(q_L)$ if the flow of all information is from left to right.

3.1.3. The multiple grid methods. When a multiple grid technique is used to solve the system of nonlinear (differential) equations (1.7) or (1.8), we assume the existence of a nested set of grids. Usually this nesting is such that a set of 2×2 cells in a fine mesh forms a single cell in the next coarser one. (No staggered grids!) The coarser grids are used to effect the acceleration of a basic iterative (time marching or relaxation) procedure on the finest grid.

Slightly generalizing equations (1.7) and (1.8) to

$$\frac{\partial}{\partial t}\mathbf{q}^h = N^h(\mathbf{q}^h) - r^h \tag{1.9}$$

and

$$N^h(\mathbf{q}^h) = r^h, \tag{1.10}$$

where r^h denotes a possible correction or forcing term, we can write the basic iterative procedure as

$$q^h \leftarrow J^h(q^h, r^h). \tag{1.11}$$

Generally, for a nonlinear equation this will be a nonlinear operation (e.g., a nonlinear Gauss–Seidel relaxation scheme).

The usual coarse grid acceleration algorithm is as follows: with an approximation $q^h_{(k)}$ on the finest mesh, and some approximation $q^{2h}_{(0)}$ on the next coarser one (e.g., $q^{2h}_{(0)} = I^{2h}_h q^h_{(k)}$), first an approximate solution $q^{2h}_{(1)}$ is found for the coarse grid problem

$$N^{2h}(\mathbf{q}^{2h}) = N^{2h}(q^{2h}_{(0)}) - \hat{I}^{2h}_h(N^h(q^h_{(k)}) - r^h), \tag{1.12}$$

and then the value $q^h_{(k)}$ is updated by

$$q^h_{(k+1)} = q^h_{(k)} + I^h_{2h}(q^{2h}_{(1)} - q^{2h}_{(0)}). \tag{1.13}$$

Notice that $\hat{I}^{2h}_h$ is a restriction operator similar to I^{2h}_h; the difference is that I^{2h}_h works on approximate solutions q^h (the state of the flow), whereas $\hat{I}^{2h}_h$ works on residuals (rates of change of the flow). The difference is not only formal: in the simplest case I^{2h}_h takes the mean value of states in a set of cells, but $\hat{I}^{2h}_h$ performs a summation of rates of change over a set of cells.

The combination of (1.12) and (1.13) is a *coarse grid correction* (CGC). The solution $\mathbf{q}^{2h}$ of (1.12) can be approximated, e.g., by an (accelerated) iteration process on the $2h$-grid again. As for linear problems, by the recursive application of this idea we can form V-cycles or μ-cycles.

We will see in §3.2 that the coarser grids sometimes play a role in the acceleration process that is different than the one we have just described [316], [474].

The *nonlinear multigrid cycle* (also called the *FAS-cycle*)

$$q^h \leftarrow FAS^h_\mu(q^h, r^h)$$

for the solution of (1.10) now consists of the following steps:

Step 0: Start with an approximate solution q^h.
Step 1: Improve q^h by application of ν_1 nonlinear (pre-) relaxation iterations (1.11) to $N^h(\mathbf{q}^h) = r^h$.
Step 2: If the present grid is the coarsest, go to Step 4. Otherwise improve q^h by application of a coarse grid correction, where the approximation of (1.12) is effected by μ FAS-cycles to this coarser grid problem; that is, compute

$$r^{2h} \leftarrow N^{2h}(q^{2h}_{(0)}) - \hat{I}^{2h}_h(N^h(q^h) - r^h),$$

and perform μ times

$$q^{2h} \leftarrow FAS^{2h}_\mu(q^{2h}, r^{2h}).$$

Step 3: $q^h \leftarrow q^h + I^h_{2h}(q^{2h} - q^{2h}_{(0)})$.
Step 4: Improve q^h by application of ν_2 nonlinear (post-) relaxation iterations to $N^h(\mathbf{q}^h) = r^h$.

Again, the case with $\mu = 1$ is called a V-cycle; $\mu = 2$ yields a W-cycle. A V-cycle with $\nu_1 + \nu_2 = 1$ is called a *sawtooth cycle.*

3.2. Methods based on Lax–Wendroff type time stepping. Ni [428] was among the first to apply an MG acceleration to the (isenthalpic) Euler equations. He uses the following time stepping procedure as a basic iteration. Starting with an initial state $q^h_{(n)}$, where the values $q^{(n)}_{ij}$ are given at the grid points, he first computes the following quantities by means of a control volume centered integration method with fluxes interpolated from corner values:

$$\Delta q_{i+1/2,j+1/2} = -\frac{1}{2}\frac{\Delta t}{\Delta x}[(F_{i+1,j} - F_{i,j}) + (F_{i+1,j+1} - F_{i,j+1})]$$

(2.1)
$$-\frac{1}{2}\frac{\Delta t}{\Delta y}[(G_{i,j+1} - G_{i,j}) + (G_{i+1,j} - G_{i+1,j+1})],$$

$$F_{i,j} = F(q^{(n)}_{ij}), \quad \text{etc.}$$

These increments are then distributed over the mesh points using direction-weighted means (cell increments are distributed over mesh point values):

$$\Delta q_{ij} = \frac{1}{4} \sum_{l=\pm 1} \sum_{k=\pm 1} \left[I - k \frac{\Delta t}{\Delta x} A_{i+k/2,j+l/2} - l \frac{\Delta t}{\Delta y} B_{i+k/2,j+l/2} \right] \Delta q_{i+k/2,j+l/2}, \tag{2.2}$$

$$q_{ij}^{(n+1)} = q_{ij}^{(n)} + \Delta q_{ij}.$$

By way of the Jacobian matrices A and B, this distribution formula has an upwinding effect, but for transonic or supersonic cases an artificial damping is still necessary.

Symbolically, this time stepping process (2.1)–(2.2) is described as follows:

(2.3a) compute Δq_{cell}^h,

with cell values $\Delta q_{i+1/2,j+1/2} \approx -\Delta t \left(\int_{\partial\Omega_{i+1/2,j+1/2}} (f \cdot n_x + g \cdot n_y)\, ds \right) / (\Delta x \cdot \Delta y)$;

(2.3b) $q_{(n+1)}^h \leftarrow q_{(n)}^h + D^h \, \Delta q_{\text{cell}}^h$.

The operator D^h is the distribution operator that transfers the cell centered corrections to the grid points by means of (2.2).

The coarse grid acceleration as introduced in [428] by Ni deviates from the canonical coarse grid scheme (1.12), (1.13). In [428] the coarse grid correction is obtained by first computing corrections at coarser cells, $\Delta q_{\text{cell}}^{2h}$. This can be done by restriction of Δq^h to the $2h$-grid. Then the corrections $\Delta q_{\text{cell}}^{2h}$ are distributed to the coarser meshpoints as in (2.2), and the coarse grid correction is interpolated to the fine grid. Thus, here the coarse grid correction reads

(2.4a) $\Delta q_{\text{cell}}^{2h} \leftarrow I_h^{2h} \, \Delta q_{\text{cell}}^h$,

(2.4b) $q_{(n+1)}^h \leftarrow q_{(n)}^h + I_{2h}^h D^{2h} \, \Delta q_{\text{cell}}^{2h}$,

where I_{2h}^h is a (bi-) linear interpolation operator. Since the coarse grid corrections are based on fine grid residuals, it is obvious that the possible convergence to a steady state yields a solution of the system (1.8).

In the same way the correction procedure can be repeated on progressively coarser grids. Therefore, in (2.4), $2h$ should be replaced by $2^k h$. We notice that the corrections on the different levels may be made independently of each other. This makes it possible to compute all coarse grid corrections, $k = 1, \cdots, m$, in parallel and to form the correction

$$q_{(n+1)}^h = q_{(n)}^h + \sum_{k=1}^{m} I_{2^k h}^h D^{2^k h} \, \Delta q_{\text{cell}}^{2^k h}$$

at once [541]. When optimal use of modern multi-processor computers is made, it is also possible to perform both computations (2.3a) and (2.4) in parallel [320], [541].

We see that it is still possible to form different variants of the Ni-type multigrid Euler solver. First, any other Lax–Wendroff type time-marching procedure can be used for (2.3a). In [134], [314], [318] Johnson applies the popular MacCormack scheme. Further, in (2.4a) various restrictions, I_h^{2h}, can be used. Equation (2.4a) transfers the values of the fine grid corrections to a single value for each control volume in the coarser grid. Injection of the correction to the corresponding point of the coarse grid cell is often used [316], but weighted averages are also an obvious choice.

Heuristically, the coarse grid corrections in (2.4) have an accelerating effect because they may move disturbances of the steady state over the distance of many mesh cells in one time step. Apparently, the Lax–Wendroff schemes used in combination with this coarse grid correction must be sufficiently dissipative to reduce the high frequency disturbances present in the initial approximation or introduced by linear interpolation. One way to do this is to make a careful choice of Δt. Until now, no complete mathematical theory has been developed to explain or quantify the amount of acceleration clearly found in the use of this approach.

As an alternative to (2.2), where Jacobians are used to form the correction, Johnson [315] introduced a correction that is based on extrapolation (in time) of the computed fluxes.

3.3. Methods based on semidiscretization and time stepping. When only the solution of the steady state is to be computed, the time-accurate integration of the system of ordinary differential equations is wasteful. The convergence of (1.4) to steady state is slow. However, there may be several reasons to prefer time stepping methods, such as the desire to have a procedure that solves transient as well as steady state problems, coding convenience, or the restrictions imposed by the optimal use of vector computers. When no time accuracy is desired, many devices are known to accelerate the integration process (cf. [305]). For the solution of the Euler equations, these devices include: (i) local time stepping, which means that the step size in the integration process may differ over different parts of the domain Ω^*; (ii) enthalpy damping, where a priori knowledge about the behavior of the enthalpy over Ω^* is used (e.g., H constant over Ω^*); (iii) residual smoothing; and (iv) implicit residual averaging, which uses the fact that instability effects appear first for high frequencies, so that larger time steps are possible when the residual is smooth.

For all explicit integration methods, stability requirements set a limit on the size of the possible time steps (CFL limits). Implicit integration procedures can be unconditionally stable, but they require the solution of a nonlinear system at each individual time step.

An important code based on a time stepping method has been developed by Jameson, Schmidt and Turkel [305]. They use an explicit time stepping method of Runge–Kutta type. This *multistage time stepping procedure* is a specially adapted Runge–Kutta method, where the hyperbolic (convective) and the parabolic (dissipative) parts of $N^h(q^h)$ are treated separately. The Runge–Kutta coefficients in the k-stage Runge–Kutta schemes ($k=3, 4$) are selected not only for their large stability bounds, but also with the aim of improving the damping of the high frequency modes. In the k stages of the Runge–Kutta process, the updating of the dissipative part is frozen at the first stage. This saves a substantial part of the computational effort.

The multigrid scheme used by Jameson [300] is an FAS sawtooth cycle with $\nu_1 = 1$. The restriction $I_h^{2h}(\hat{I}_h^{2h})$ is defined by volume-weighted averaging of the states (summation of changes of states, respectively). The prolongation I_{2h}^h is defined by bilinear interpolation. The basic smoothing procedure is the "multistage time stepping scheme." On the coarser grids the stability bounds for the time step, which are $O(h)$, allow larger time steps. On each grid the time step is varied locally to yield a fixed Courant number, and the same Courant number is used on all grids, so that progressively larger time steps are used after each transfer to a coarser grid. As for Ni's method, the reasoning is that disturbances from the steady state will be more rapidly expelled from the domain Ω^* by the larger time steps. The interpolation of corrections back to the fine grid introduces high frequency errors, which cannot be rapidly expelled. These errors should be locally damped. Hence, to obtain a fast rate of convergence, the time stepping process should rapidly damp the high frequency errors.

In [311] Jespersen announced an interesting theorem on the use of the MG process in combination with a time stepping procedure. This theorem asserts the following. Let I^h (resp. $\hat{I}^h$) be defined as a restriction operator from the continuous state space (resp. space of rates of change) to its discrete equivalent on Ω^h, and let I_h be a prolongation operator that interpolates states on Ω^h to states on Ω. Let $N^h(\mathbf{q}^h) = 0$ be a space discretization of $N(\mathbf{q}) = 0$ which is consistent, i.e.,

$$N^h(I^h(\mathbf{q})) - \hat{I}^h N(\mathbf{q}) = O(h),$$

and let the time stepping procedure be consistent in time, i.e.,

$$q_{(n+1)}^h = q_{(n)}^h + \Delta t_{(n)}[N^h(q_{(n)}^h) - r^h] + O((\Delta t_{(n)})^2).$$

If we consider the sawtooth algorithm, with $\nu_1 = 1$, $\nu_2 = 0$, $\mu = 1$, and if I_h and I^h satisfy an approximation property (i.e., for a smooth function q the prolongation and restriction in the state space are such that $I_h I^n q - q = O(h)$), then the MG algorithm on m grids is a consistent, first order in time, discretization of (1.4) with time step $\Delta t_{\text{tot}} = \sum_{j=1,\cdots,m} \Delta t_j$.

In a sense this theorem formalizes the heuristic reasoning that on coarser grids the deviations from steady state can be expelled faster by the use of larger time steps. This may suggest that more, say $\nu > 1$, steps on the coarser grids would further improve the convergence. However, the theorem addresses consistency; stability is not considered. Hence, in the same paper [311] Jespersen shows by an example that convergence is lost when a large number of relaxations is made on the coarse grid. In fact, a strong stability condition of the form $\Delta t/\Delta x \leq O(\nu^{-1})$ seems to appear.

3.4. Fully implicit methods. Most methods considered so far are based on the concept of integrating the equations (1.4) in time until a steady state is reached. If we are only interested in a possible solution of the steady state equation (1.5) and assume that this solution is unique, we may disregard the time-dependence completely. Further, assuming that a suitable space discretization takes into account the proper directions of dependence in Ω^*, we can restrict ourselves simply to the solution of the nonlinear system (1.8) or

$$N^h(\mathbf{q}^h) = r^h. \tag{1.10}$$

Also, if the time-dependent system (1.9) is solved by means of an implicit time stepping method in order to circumvent the stability bounds on Δt, we have to solve systems (1.10) at each time step. Using these implicit solution methods and giving up time accuracy for (1.10) means that there is little or no difference between these time stepping procedures and (nonlinear) relaxation methods for (1.10).

If we start with the nonlinear system (1.10), two direct MG approaches can be used. We can either apply the nonlinear multiple grid algorithm (FAS) directly to the system (1.10), or we may apply linearization (Newton's method) and use the linear version of multiple grid (CS) for the solution of the resulting linear systems. Jespersen [310] gives an extensive recital of the (dis)advantages of both approaches. Both have been used with success for the Euler equations.

Linearization and CS have been used by Jespersen [309] and Mulder [423]; the nonlinear FAS procedure is used by Steger [528], Jespersen [309], and Hemker and Spekreijse [274], [275].

In all of these papers upwind discretizations have been used. In [309], [528] the Steger–Warming scheme is used; [423] uses the differentiable van Leer flux splitting method; [274], [275] use Osher's flux difference splitting. In [150] Dick also considers Roe's flux difference splitting for the 1-D Euler equations.

When Newton's method is applied for linearization, it may be difficult to start in the domain of contraction of the iteration. Therefore, Mulder [423] introduces the so-called Switched Evolution Relaxation (SER) scheme, which is a chimera of a forward Euler time stepping and a Newton method:

$$\left[\frac{1}{\Delta t} I - \frac{\partial}{\partial q} N^h(q^h_{(n)})\right](q^h_{(n+1)} - q^h_{(n)}) = N^h(q^h_{(n)}). \tag{4.1}$$

For $\Delta t \to 0$, this gives the simple time stepping procedure; for $\Delta t \to \infty$, (3.1) is equivalent to Newton's method. In the actual computation Δt varies, depending on the size of the residual, such that (3.1) is initially a time stepping procedure and becomes Newton's method in the final stages of the solution process.

In an FAS procedure, a natural way to obtain an initial estimate is of course to use full multigrid (FMG) [97]. The initial estimate is obtained by interpolation from the approximate solution on the coarser grid(s). For many problems this process gives very good results, even if one starts with rough approximations on every coarse grid.

3.4.1. A nested sequence of Galerkin discretizations. When (1.3) is discretized by a finite volume method, and if a conservative first order upwind (or a central difference) discretization is used as described in §3.1, it can be shown [275] that with a particularly simple restriction $\hat{I}_h^{2h}$ and prolongation I_{2h}^h, the coarse discrete operator N^{2h} is a Galerkin approximation to the fine grid discretization N^h. With I_{2h}^h the piecewise constant interpolation over cells, and $\hat{I}_h^{2h}$ the summation of the residual over fine mesh cells to form a residual on the corresponding coarse cell, the following relation holds:

$$N^{2h}(q^{2h}) = \hat{I}_h^{2h} N^h(I_{2h}^h q^{2h}). \tag{4.2}$$

This formula has an interesting implication for a coarse grid correction that is constructed by means of these operators. If the coarse grid correction (1.12), (1.13) transforms the approximation q_h into $\tilde{q}^h$, the residual of $\tilde{q}^h$ satisfies

$$\hat{I}_h^{2h}[r^h - N^h(\tilde{q}^h)] = \hat{I}_h^{2h}[(N^h q^h - N^h I_{2h}^h \hat{I}_h^{2h} q^h) - (N^h \tilde{q}^h - N^h I_{2h}^h \hat{I}_h^{2h} \tilde{q}^h)]. \tag{4.3}$$

For a smooth operator N^h, this implies

$$\hat{I}_h^{2h}[r^h - N^h(\tilde{q}^h)] = O(\|q^h - \tilde{q}^h\|^2).$$

This means that the restriction of the residual mainly contains high frequency components. As is the case with common elliptic problems, it is the task of the relaxation method to efficiently damp these highly oscillating residuals.

3.4.2. Relaxation methods. Clearly, whether a sequence of Galerkin approximations is used or not, the important feature of a relaxation method in a multiple grid context (both CS and FAS) is its capability to damp the high frequency components in the error (or in the residual). Therefore, the difference scheme should be sufficiently dissipative as first order upwind schemes usually are. An advantage of these schemes over central differences is that this numerical dissipation is well defined and independent of an artificial parameter for the added dissipation necessary for the central difference schemes. The lack of differentiability of the numerical flux function may create a problem, but some differentiable flux functions are now available [528], [Os1], [Os2], [Va1].

Both in the linearized (CS) and in the nonlinear (FAS) application, well-known and simple relaxation procedures such as Gauss–Seidel (GS), symmetric Gauss–Seidel (SGS) and line Gauss–Seidel (LGS) are reported to work well when applied to the discrete Euler equations. (All of these relaxation methods are used in their "collective" version, i.e., the 3 or 4 variables corresponding to a single point or cell are relaxed simultaneously.) The smoothing behavior of these relaxations can be analyzed by local mode analysis. Here we should notice that the *smoothing factor,* as used for common elliptic problems, has no significant meaning for the Euler equation because we have to take into account characteristic (unstable) modes. A local mode analysis should follow more along the lines used for singularly perturbed elliptic problems (cf. e.g. [328]). Jespersen [309] has published some results in this regard. He shows that for a subsonic and supersonic case, SGS has a reasonably good smoothing behavior when applied to a first order scheme. Of course, the nonsymmetric GS relaxation is only effective if the direction of the relaxation sufficiently conforms with the direction of the characteristics. If we study plots of reduction factors of Fourier components (spectral radii, or norms for the error or residual amplification operator), e.g., when SGS is applied to the Euler equations, we see that two SGS sweeps are usually sufficient for a significant reduction of the high frequencies (Hemker, unpublished results). For second order schemes the smoothing rates are not satisfactory.

Van Leer and Mulder published a study [Va2] where several relaxation schemes (GS, LGS, ZEBRA, point Jacobi, line Jacobi, ADI, AF) were compared when applied to the linearized isenthalpic Euler equations.

3.4.3. Higher order schemes. When both first and second order upwind schemes are studied, the best MG performance is found for the first order discretizations. This can be explained by the fact that first order upwind schemes are more dissipative and hence more able to damp high frequencies. As first order schemes may not be accurate enough for practical computations and, moreover, have the unpleasant property of

TABLE 3.1
Fully implicit multiple grid approaches.

	Discretization scheme	MG	Relaxation
Steger (1981)	Steger–Warming Finite differences	FAS	AF
Jespersen (1983)	Steger–Warming Finite differences	FAS/CS	SGS, GS
Mulder (1984)	van Leer Finite differences	CS	SGS
Hemker and Spekreijse (1985)	Osher Finite volumes	FAS Nested Galerkin	SGS, Damped Jacobi

smearing out skew discontinuities, second order schemes are highly desirable.

Beside the possibility of applying the MG acceleration directly to the second order scheme—with the unwanted effect of slowing the convergence rate—another possibility exists. Starting with a first approximation, we can improve the accuracy by the defect correction iteration [66], [271], [525]

$$N_1^h(q_{(n+1)}^h) = N_1^h(q_{(n)}^h) - N_2^h(q_{(n)}^h). \tag{4.4}$$

Here N_p^h, $p = 1, 2$, denotes the pth order discretization. A theorem [225] has shown that for smooth solutions a single correction step (3.4) is sufficient to obtain the higher order of accuracy. Also, for solutions with discontinuities (where the formal order of convergence has no practical meaning), it is shown in [271] that one or a few steps (3.4) improve the accuracy of the solution significantly.

In Table 3.1 we summarize the several attempts to solve the steady Euler equations by an MG method with implicit relaxation. It is our opinion that the recent methods of this class are the most robust and efficient ones for solving the steady Euler equations. The development in the last few years has led to a significant improvement of the algorithms. However, the fully implicit methods have a rather complex structure and are not directly suited for vector computers. Furthermore, at the moment there is much less practical experience with these methods than, e.g., with Jameson's multistage time stepping procedure or the commonly used Beam–Warming [Be1] algorithm.

Acknowledgments. We are grateful to B. van Leer, whose comments on a draft of this paper led to several improvements. The investigations were supported in part by the Netherlands Technology Foundation.

REFERENCES

[Be1] R. M. BEAM AND R. F. WARMING, *An implicit finite-difference algorithm for hyperbolic systems in conservation-law form,* J. Comput. Phys., 22 (1976), pp. 87–110.

[La1] P. D. LAX, *Shock waves and entropy,* in Contributions to Non-linear Functional Analysis, E. H. Zarantonello, ed., Academic Press, New York, 1971.

[La2] ———, *Hyperbolic Systems of Conservation Laws and the Mathematical Theory of Shock Waves,* CBMS-NSF Regional Conference Series in Applied Mathematics 11, Society for Industrial and Applied Mathematics, Philadelphia, PA, 1973.

[Os1] S. OSHER, *Numerical solution of singular perturbation problems and hyperbolic systems of conservation laws,* in Analytical and Numerical Approaches to Asymptotic Problems in Analysis, O. Axelsson, L. S. Frank and A. van der Sluis, eds., Mathematics Studies 47, North-Holland, Amsterdam, 1981.

[Os2] S. OSHER AND F. SOLOMON, *Upwind difference schemes for hyperbolic systems of conservation laws,* Math. Comp., 38 (1982), pp. 339–374.

[Ro1] P. L. ROE, *Approximate Riemann solvers, parameter vectors and difference schemes,* J. Comput. Phys., 43 (1981), pp. 357–372.

[St1] J. L. STEGER AND R. F. WARMING, *Flux vector splitting of the inviscid gasdynamics equations with application to finite difference methods,* J. Comput. Phys., 40 (1981), pp. 263–293.

[Va1] B. VAN LEER, *Flux-vector splitting for the Euler equations,* Proc. 8th Internat. Conf. on Numerical Methods in Fluid Dynamics, Aachen, June 1982, Lecture Notes in Physics 170, Springer-Verlag, New York–Berlin–Heidelberg.

[Va2] B. VAN LEER AND W. A. MULDER, *Relaxation methods for hyperbolic conservation laws,* in Numerical Methods for Euler Equations of Fluid Dynamics, F. Angrand, A. Dervieux, J. A. Desideri and R. Glowinski, eds., Society for Industrial and Applied Mathematics, Philadelphia, PA, 1984.

CHAPTER 4

Algebraic Multigrid

J. W. RUGE AND K. STÜBEN

4.1. Introduction. The focus in the application of standard multigrid methods is on the continuous problem to be solved. With the geometry of the problem known, the user discretizes the corresponding operators on a sequence of increasingly finer grids, each grid generally being a uniform refinement of the previous one, with transfer operators between the grids. The coarsest grid is sufficiently coarse to make the cost of solving the (residual) problem there negligible, while the finest is chosen to provide some desired degree of accuracy. The solution process, which involves relaxation, transfer of residuals from fine to coarse grids, and interpolation of corrections from coarse to fine levels, is a very efficient solver for the problem on the finest grid, provided the above "multigrid components" are properly chosen.

Roughly, the efficiency of proper multigrid methods is due to the fact that error only slightly affected by relaxation (*smooth* error) can be easily approximated on a coarser grid by solving the residual equation there, where it is cheaper to compute. This error approximation is interpolated to the fine grid and used to correct the solution. Generally, uniform coarsening and linear interpolation are used, so the key to constructing an efficient multigrid algorithm is to pick the relaxation process that quickly reduces error not in the range of interpolation.

The algebraic multigrid (AMG) approach is developed to solve *matrix equations* using the principles of usual multigrid methods. In contrast to "geometric" multigrid methods, the relaxation used in AMG is *fixed.* The coarsening process (picking the coarse "grid" and defining interpolation) is performed *automatically* in a way that ensures the range of interpolation approximates those errors not efficiently reduced by relaxation. From a theoretical point of view, the process is best understood in the context of symmetric M-matrices, although, in practice, its use is not restricted to such cases. The underlying idea of the coarsening process is to exploit the fact that the form of the error after relaxation can be approximately expressed using the equations themselves, so that the coarse grid can be chosen and interpolation defined if the equations are used directly. This makes AMG attractive as a "black box" solver. In addition, AMG can be used for many kinds of problems, described below, where the application of standard multigrid methods is difficult or impossible.

(1) The first kind of problem that presents difficulties for geometric multigrid methods is one in which the domain of the problem is complex enough so that any sensible discretization is too fine to serve as the coarsest grid. Even if the finest grid can be coarsened sufficiently, the necessary pattern of coarsening may be quite complex, and the work required to write the needed interpolation routines would be prohibitive. AMG can be used in such situations, since coarsening is automatic. As long as solution of the problem by relaxation remains inefficient, a coarser grid that will help the process can be chosen, without regard to the underlying geometry.

(2) A second set of problems, related to the first, consists of those for which the discretization on the finest grid does not allow uniform coarsening at all. For example, finite element discretizations using irregular triangulations result in such problems. These often arise when finite element applications packages produce meshes adapted to particular domains or structures. These packages were not written with multigrid solution methods in mind, and to change them would be impractical. Again, AMG deals effectively with such problems, since uniform grids are not at all necessary.

(3) A third case consists of problems caused by the operator itself, not the domain. In such cases, when uniform coarsening is used with linear interpolation, it may be difficult to find a relaxation process that smoothes the error sufficiently to admit a good coarse grid correction. At times this may not even be possible; for example, when we are solving diffusion problems with discontinuous coefficients, it is necessary to change the definition of interpolation across the discontinuities (cf. [5]). Another example is the skewed Laplace operator, for which no smoothing in the usual sense is possible. In effect, the discretization splits the grid into two sets (in a checkerboard fashion) that are not connected at all by the matrix, and interpolation between the two sets is useless. AMG, however, is not affected in such cases. Interpolation is defined only from points that are "strongly connected" as defined by the matrix entries, and weights are chosen that reflect the relative strength of each connection.

(4) Finally, some problems are purely discrete. It is even possible (though unlikely) that such a problem has *no* geometric background. Examples of discrete problems arise in geodesy (cf. [Po1]), structural mechanics (trusses and frames, cf. [Ch1]), and economics. Often simplified models of such problems resemble those arising from elliptic partial differential equations, so multigrid processes should provide efficient solutions, but a straightforward application of usual multigrid methods is impossible. In such cases, AMG can be used, since it does not rely on an underlying continuous problem in order to obtain the coarse grid operators.

Section 4.2 of this chapter introduces the general AMG components and contains some introductory remarks. Sections 4.3–4.5 cover theoretical aspects of AMG, while §§ 4.6–4.8 concentrate on practical considerations. These theoretical and practical parts can be read independently; references between them are provided when necessary.

Sections 4.3–4.5 discuss in detail the interaction between the relaxation process and the coarse grid correction necessary for proper behavior of the solution process, adapting and developing theoretical results of Brandt, Mandel, McCormick and others to suit our purposes. Section 4.3 gives sufficient conditions on relaxation and interpolation for the convergence of the V-cycle. Section 4.4 focuses on the relaxation used in AMG, what smoothing means in an algebraic setting, and how it relates to the existing theory. Section 4.5 discusses the coarse grid correction and the definition of interpolation. Some properties of the coarse grid operators are discussed, and results on the convergence of two-level and multi-level convergence are given.

In § 4.6, we give details of an algorithm particularly suited for problems obtained by discretizing a single elliptic, second order partial differential equation (PDE), as well as problems of a similar nature, which we call "scalar" problems. We also give some information on the work and storage requirements. Results of experiments with such problems using both finite difference and finite element discretizations are presented. In addition, we discuss several modifications to the method, including the explicit use of geometric information.

Although the convergence behavior of AMG is insensitive to complex domains and discretizations, the algorithm described in § 4.6 can fail when more than one function is being approximated, as in a discretized system of PDEs ("system" problems). In § 4.7, we discuss modifications of the algorithm that account for this fact, and give results for several examples, including some discrete problems and some problems from elasticity and structural mechanics.

The last section contains concluding remarks on the method and directions of present and future research. These directions include the use of AMG for solving chains of linear problems efficiently. Such chains arise in the solution of time-dependent and nonlinear problems. The approach is to update the coarser grids and grid transfer operators only when necessary, and only locally.

4.2. Terminology, assumptions and notation. Formally, an AMG cycle can be described in the same way as any other multigrid cycle. Differences are mainly due to the different meaning of the terms *grid, grid points,* etc. In this section we will give a formal description, specify certain assumptions and introduce some notation. In particular, in § 4.2.3 we introduce three inner products, important mainly because their related norms behave differently if applied to "smooth" error vectors. This makes it possible to identify smooth errors by simply comparing different corresponding norm values. Also, and for the same reason, these norms will be used in convergence theorems to formulate reasonable assumptions on the smoothing operators and the coarse grid correction operators *separately*.

4.2.1. AMG components in general. In order to solve a (sparse) linear system of equations

$$Au = b \quad \text{or} \quad \sum_{j=1}^{n} a_{ij}u_j = b_i \quad (i = 1, \cdots, n) \tag{2.1}$$

by means of a "multigrid-like" cycling process, we first have to generate a sequence of smaller and smaller systems of equations:

$$A^m u^m = b^m \quad \text{or} \quad \sum_{j=1}^{n_m} a_{ij}^m u_j^m = b_i^m \quad (i = 1, \cdots, n_m) \tag{2.2}$$

($m = 1, 2, \cdots, q$ with $n = n_1 > n_2 > \cdots > n_q$) which, for $m > 1$, will formally play the same role as coarse grid equations in geometric multigrid methods.[1] In particular, b^m and u^m will be residuals and corrections, respectively. For $m = 1$, (2.2) is identical to (2.1). Thus, in the context of AMG, a *grid* can simply be regarded as a set of unknowns. Corresponding *grid functions* u^m and b^m are just vectors in $\mathbf{R}^{n_m}$ and the *coarse grid operators* A^m are matrices mapping $\mathbf{R}^{n_m}$ into itself.

Furthermore, we need (linear) transfer operators

$$I_{m+1}^m : \mathbf{R}^{n_{m+1}} \to \mathbf{R}^{n_m} \quad \text{and} \quad I_m^{m+1} : \mathbf{R}^{n_m} \to \mathbf{R}^{n_{m+1}}$$

called *interpolation* and *restriction*, respectively, and smoothing processes of the form

$$u_{\text{new}}^m = G^m u_{\text{old}}^m + (I^m - G^m)(A^m)^{-1} b^m,$$

with corresponding (linear) *smoothing operators* $G^m : \mathbf{R}^{n_m} \to \mathbf{R}^{n_m}$.

Once the above components are known, we can set up a multigrid cycle in the usual way (cf. [538,§ 4.4.2]). Our main concern in §§ 4.3–4.5 will be to get some theoretical insight and to derive some general rules on how to proceed in constructing these components. In § 4.6 we will give an explicit algorithm that has turned out to be quite efficient for many applications.

[1] Compared to geometric multigrid descriptions, the indexing is "backwards" in AMG. Here it is more convenient to use *increasing* numbers m for denoting coarser and coarser "levels," since, starting from a given finest level, the number of coarser levels to be used is not known a priori.

4.2.2. Assumptions and remarks. A striking difference between a geometric and an algebraic multigrid solver is the fact that the latter includes a separate (A-dependent) preparation phase in which the *complete* coarsening process is *fully automatic*. No geometric grid structure is needed. In fact, in the ideal case, an AMG algorithm uses only information explicitly contained in the given matrix A (e.g., in terms of the size of the matrix entries). The coarsening performed by AMG is required to be such that the resulting cycle will become efficient, i.e., it should exhibit typical multigrid convergence speed and should need $O(n)$ operations only. (Of course, the preparation phase itself should also not need more than $O(n)$ operations.)

There is no general algorithm which allows for this without any further assumptions on the matrix A. For our theoretical discussion we will always assume A to be *symmetric* and *positive definite.*[2] (These terms, as well as the definition of the transpose of a matrix, always refer to the *Euclidean inner product,* if there is no explicit statement to the contrary.) Concerning the AMG components, we generally assume the interpolation operators I^m_{m+1} to have *full rank* and the restriction and coarse grid operators to be defined by

$$I^{m+1}_m = (I^m_{m+1})^T \quad \text{and} \quad A^{m+1} = I^{m+1}_m A^m I^m_{m+1}, \tag{2.3}$$

respectively. These special choices are very convenient, as the symmetry and positive definiteness of A imply that of A^m for all m and all intermediate $(m, m+1)$ *coarse grid correction operators* (cf. [538, § 2.3]),

$$T^m = I^m - I^m_{m+1}(A^{m+1})^{-1} I^{m+1}_m A^m, \tag{2.4}$$

become orthogonal projectors (see § 4.3.1). Multigrid methods using (2.3) are often referred to as *Galerkin type* methods due to their origin in finite element formulations, where these equalities hold naturally in an implicit way. While in geometric multigrid methods (applied to finite difference equations) there are several different possible ways to choose the restriction and coarse grid operators, in a purely algebraic setting there is hardly an alternative to the above choice.

As a smoothing process we usually take the simplest one, namely *Gauss–Seidel relaxation.* Summarizing, we see that in the preparation phase of AMG *only the interpolation operators remain to be defined recursively.*

4.2.3. Some further notation. For any square matrix A we write $A > 0$ ($A \geq 0$) if A is symmetric and positive definite (positive semi-definite). Correspondingly, $A > B$ ($A \geq B$) stands for $A - B > 0$ ($A - B \geq 0$).

[2] We could also allow for positive semi-definite row sum zero matrices. For simplicity, however, we exclude this case from our theoretical discussion. The concrete AMG algorithm which will be presented in § 4.6 does not require A to be necessarily symmetric and positive definite. Limitations of this algorithm will be discussed.

Important parts of our theoretical discussion refer to the "model class" of *symmetric M-matrices,* where a symmetric matrix A is defined to be an M-matrix if it is positive definite and off-diagonally nonpositive. For vectors, $u > 0$ and $u \geq 0$ mean that the corresponding inequalities hold componentwise.

We will usually use the letter u to denote *solution* quantities and the letters v or e to denote *correction* or *error* quantities. The *range* of some matrix operator L is denoted by $R(L)$ and its *spectral radius* by $\rho(L)$. In addition to the *Euclidean inner product* $\langle \cdot, \cdot \rangle$, we will need three different inner products

$$\langle u, v \rangle_0 = \langle Du, v \rangle, \quad \langle u, v \rangle_1 = \langle Au, v \rangle, \quad \langle u, v \rangle_2 = \langle D^{-1}Au, Av \rangle,$$

along with their corresponding norms $\|\cdot\|_i$ $(i = 0, 1, 2)$. Here $A > 0$ is assumed and $D = \operatorname{diag}(A)$. The Euclidean norm is denoted by $\|\cdot\|$. Although these definitions will be used on different levels in the multigrid hierarchy (with A^m instead of A), we will, for simplicity, always use the same symbols.[3]

Usually, when no confusion can arise, we simplify notation. For instance, when we are considering one multigrid level at a time, we omit the indices $m, m+1, \cdots$. On the other hand, when we are considering two consecutive levels only, we use indices h and H instead of m and $m+1$, respectively, to distinguish fine and coarse level quantities. For instance, $A^h \mathbf{u}^h = b^h$ denotes the current fine grid problem, and the corresponding coarse grid operator is denoted by $A^H = I_h^H A^h I_H^h$ with $I_h^H = (I_H^h)^T$ (cf. (2.3)). These particular indices have been chosen only to have a formal similarity to known geometric two-level descriptions; generally, h and H have nothing to do with any kind of discretization parameter.

4.3. General results on convergence. In this section we give some general convergence theorems. This will be in terms of estimates of the V-cycle convergence factor, assuming a fixed A and that corresponding multigrid components are given. Clearly, we are not interested in having

---

[3] Using the energy inner product $\langle \cdot, \cdot \rangle_1$, we find that these inner products actually belong to a *scale of inner products* defined by

$$\langle u, v \rangle_\alpha := \langle (D^{-1}A)^{\alpha-1} u, v \rangle_1.$$

A more general scale would be given by using some matrix $B > 0$ instead of D. (Note that $B^{-1}A$ is always symmetric and positive definite with respect to the energy inner product $\langle \cdot, \cdot \rangle_1$.) While, in particular, the corresponding norm $\|\cdot\|_\alpha$ for $\alpha = 2$ plays an important role in geometric multigrid (V-cycle) theory for regular elliptic partial differential problems, the norms for $\alpha < 2$ allow for a more general (W-cycle) theory in less regular cases [375]. However, we will not consider "fractional norms" in this chapter, nor will we consider any B different from D.

convergence for one particular A only, but rather in having *uniform* convergence if A ranges over some reasonable *class* of matrices. Corresponding to the fact that in geometric multigrid the convergence factor of a cycle does not depend on a discretization parameter, AMG interpolation should be such that, in particular, convergence speed does not depend on the size of A. This has to be kept in mind in interpreting the theorems of the next sections which are usually formulated assuming a fixed A.

The results below are contained in [375], [395], but they have been included here to serve as a starting point for a theoretical discussion. Also, they have been reformulated to better suit our purposes and our terminology.

4.3.1. General estimates of the V-cycle convergence factor. In the symmetric and positive definite cases (and assuming (2.3)), it is easiest to investigate convergence with respect to the *energy norm* $\|\cdot\|_1$. This is because all $(m, m+1)$ coarse grid correction operators (2.4) are orthogonal projectors with respect to the energy inner product $\langle\cdot,\cdot\rangle_1$ with $R(T^m)$ being orthogonal to $R(I^m_{m+1})$. In particular, $\|T^m\|_1 = 1$ and the energy norm of errors after an $(m, m+1)$ coarse grid correction step is minimum (with respect to changes in $R(I^m_{m+1})$). The following theorem gives a quantitative estimate for the V-cycle convergence factor.

THEOREM 3.1. *Let $A > 0$. Assume that the interpolation operators I^m_{m+1} have full rank and that the restriction and coarse grid operators are defined by* (2.3) *Furthermore, suppose that, for all e^m,*

$$\|G^m e^m\|_1^2 \le \|e^m\|_1^2 - \delta\,\|T^m e^m\|_1^2 \tag{3.1a}$$

holds with some $\delta > 0$ independently of e^m and m. Then $\delta \le 1$, and—provided that the coarsest grid equation is solved and that at least one smoothing step is performed after each coarse grid correction step—the V-cycle to solve (2.1) *has a convergence factor* (*with respect to the energy norm*) *bounded above by $\sqrt{1-\delta}$.*

Proof. The proof is recursive. Let us therefore consider any two consecutive levels of the multigrid hierarchy described in § 4.2.1 (denoted by indices h and H instead of m and $m+1$, respectively), and let us suppose the convergence factor of the V-cycle on the H-level to be $0 \le \eta_H < 1$. In order to derive a bound η_h for the convergence factor of the V-cycle on the h-level as a function of η_H, we first note that (with $\tilde{T}^h$ denoting the corresponding coarse grid correction operator) the error after a coarse grid correction step can be written as

$$\tilde{T}^h e^h = e^h - I^h_H \tilde{v}^H = e^h - I^h_H v^H + I^h_H(v^H - \tilde{v}^H) = T^h e^h + I^h_H(v^H - \tilde{v}^H).$$

Here $\tilde{v}^H$ and v^H denote the *V-cycle* correction and the corresponding (h, H)-*two-level* correction (i.e., the exact solution of $A^H v^H = I^H_h A^h e^h$), respectively, and T^h is the (h, H)-two-level correction operator (2.4). Due to our above assumption on the V-cycle convergence on the H-level, we can estimate the following:

$$\|I_H^h(v^H - \tilde{v}^H)\|_1 = \|v^H - \tilde{v}^H\|_1 \le \eta_H \|v^H\|_1 = \eta_H \|I_H^h v^H\|_1.$$

(Note that because of (2.3), $\|w^H\|_1 = \|I_H^h w^H\|_1$ for every w^H.) As $R(T^h)$ is orthogonal to $R(I_H^h)$ with respect to $\langle \cdot, \cdot \rangle_1$, we obtain

$$\begin{aligned} \|\tilde{T}^h e^h\|_1^2 &= \|T^h e^h\|_1^2 + \|I_H^h(v^H - \tilde{v}^H)\|_1^2 \\ &\le \|T^h e^h\|_1^2 + \eta_H^2 \|I_H^h v^H\|_1^2 \\ &= \|T^h e^h\|_1^2 + \eta_H^2(\|e^h\|_1^2 - \|T^h e^h\|_1^2). \end{aligned} \tag{3.2}$$

Using assumption (3.1a), and because $T^h \tilde{T}^h = T^h$ and $\|T^h\|_1 = 1$, we find that the following estimation now shows $\eta_h \le \max\{\eta_H, \sqrt{1-\delta}\}$:

$$\begin{aligned} \|G^h \tilde{T}^h e^h\|_1^2 &\le \|\tilde{T}^h e^h\|_1^2 - \delta \|T^h \tilde{T}^h e^h\|_1^2 = \|\tilde{T}^h e^h\|_1^2 - \delta \|T^h e^h\|_1^2 \\ &\le (1 - \delta - \eta_H^2) \|T^h e^h\|_1^2 + \eta_H^2 \|e^h\|_1^2 \le \|e^h\|_1^2 \max\{\eta_H^2, 1-\delta\}. \end{aligned}$$

A recursive application of this result proves the theorem.

While (3.1a) is a natural requirement if smoothing is performed *after* the coarse grid correction steps (cf. the next section), a slightly different condition is natural if smoothing is done *before* the coarse grid correction steps:

Remark 3.2 If, instead of (3.1a), we have

$$\|G^m e^m\|_1^2 \le \|e^m\|_1^2 - \delta \|T^m G^m e^m\|_1^2, \tag{3.1b}$$

the V-cycle convergence factor is bounded above by $1/\sqrt{1+\delta}$ if at least one smoothing step is performed before each coarse grid correction step.

Proof. The proof is similar to the one of the previous theorem. From (3.2) (applied to $G^h e^h$ instead of e^h), and from our assumption (3.1b), we obtain the following two inequalities:

$$\|\tilde{T}^h G^h e^h\|_1^2 \le (\xi + \eta_H^2(1-\xi)) \|G^h e^h\|_1^2, \qquad (1 + \delta\xi) \|G^h e^h\|_1^2 \le \|e^h\|_1^2,$$

respectively, with $\xi = \xi(e^h) = \|T^h G^h e^h\|_1^2 / \|G^h e^h\|_1^2$. By combining these inequalities, we see that $\|\tilde{T}^h G^h e^h\|_1 \le \eta_h \|e^h\|_1$ with

$$\eta_h^2 = \max_{0 \le \xi \le 1} \frac{\xi + \eta_H^2(1-\xi)}{1+\delta\xi} = \max\left\{\eta_H^2, \frac{1}{1+\delta}\right\},$$

which proves the remark.

In practice, we usually perform one smoothing step both before *and* after the coarse grid correction. Then, clearly, either one of the conditions (3.1a) and (3.1b) can be used. Moreover, if *both* conditions hold with constants δ_1 and δ_2, respectively, it is easy to see that the convergence factor of the

V-cycle is bounded above by $\sqrt{(1-\delta_1)/(1+\delta_2)}$. The proof is the same as that of Theorem 3.1 (with $G^h e^h$ instead of e^h) with the only exception that, just before taking the maximum, we apply (3.1b) in a similar way as in the proof of Remark 3.2.

4.3.2. Interpretation and sufficient conditions. The conditions (3.1), which reflect the basic interplay between the two essential multigrid processes, smoothing and coarse grid correction, have a simple interpretation. For instance, (3.1a) means that error components e^m that cannot efficiently be reduced by T^m(i.e., for which $\|T^m e^m\|_1 \approx \|e^m\|_1$) have to be effectively and uniformly reducible by G^m. On the other hand, for components that *can* be efficiently reduced by T^m (i.e., for which $\|T^m e^m\|_1 \ll \|e^m\|_1$, or, in other words, that are approximately in $R(I^m_{m+1})$), the smoothing operator G^m is allowed to be ineffective.

Errors for which smoothing is ineffective (i.e., for which $G^m e^m \approx e^m$) are called *smooth*. Thus, the essence of the above remark is, roughly, that *smooth errors have to be approximately in* $R(I^m_{m+1})$. Achieving this will be one main objective in the explicit construction of interpolation operators in AMG. Note that this is the most basic conceptual difference between a usual geometric multigrid solver and an AMG solver. In geometric solvers, all coarser level components are predefined, and the smoothing process has to be *selected* so that the above objective is satisfied. On the other hand, in AMG the smoothing process is *fixed*, and, by exploiting properties of corresponding smooth errors, we explicitly construct suitable (operator-dependent) interpolation operators. Of course, we still need a handy algebraic characterization of smooth errors. For this see § 4.4.3.

In applications, the assumption (3.1a) (and similarly (3.1b)) is usually replaced by two separate inequalities for G^m and T^m which are often called the *smoothing assumption* and the *approximation assumption*, respectively. For this, observe that $\|T^m e^m\|_1$ can be regarded as a (semi-) norm for e^m which, as we mentioned above, necessarily has to be $\ll |e^m\|_1$ for smooth errors. We now replace this norm by a simpler one which does not depend on T^m, namely by $\|e^m\|_2$, but which will turn out to still have a similar property. It is obvious that the following two conditions imply (3.1a) with $\delta = \alpha/\beta$:

$$\text{(3.3a)} \qquad \begin{aligned} &\|G^m e^m\|_1^2 \le \|e^m\|_1^2 - \alpha\, \|e^m\|_2^2, \\ &\|T^m e^m\|_1^2 \le \beta\, \|e^m\|_2^2 \quad (\alpha, \beta > 0). \end{aligned}$$

Similarly,

$$\text{(3.3b)} \qquad \begin{aligned} &\|G^m e^m\|_1^2 \le \|e^m\|_1^2 - \alpha\, \|G^m e^m\|_2^2, \\ &\|T^m e^m\|_1^2 \le \beta\, \|e^m\|_2^2 \quad (\alpha, \beta > 0) \end{aligned}$$

are sufficient for (3.1b) to hold.

The first inequalities in (3.3a) and (3.3b) hold under quite general circumstances, as the next section shows. It is the corresponding second inequalities that present the most problems (cf. § 4.5.5). In § 4.5.3 we will consider a weaker condition that can easily be satisfied under reasonable assumptions on A and still allow for a *two-level* convergence theory.

4.4. The smoothing property. In this section, we will consider smoothing only. Omitting the index m, we will see that, for typical relaxation operators G like that of Gauss–Seidel relaxation, the inequalities (cf. (3.3))

$$\|Ge\|_1^2 \le \|e\|_1^2 - \alpha \|e\|_2^2, \tag{4.1a}$$

$$\|Ge\|_1^2 \le \|e\|_1^2 - \alpha \|Ge\|_2^2 \tag{4.1b}$$

hold for all e with some constant $\alpha > 0$ (which is usually different for the two different inequalities). Moreover, under reasonable assumptions on A, they also hold *uniformly* with respect to A.

4.4.1. An auxiliary result. Note first that $\alpha \|e\|_2^2 \le \|e\|_1^2$ and $\alpha \|Ge\|_2^2 \le \|e\|_1^2$, respectively, are *necessary* for the inequalities in (4.1) to hold. In order to have smoothing in the sense of (4.1a), for instance, uniformly for A ranging over a certain class $\mathcal{A}$ of matrices, this means that the 2-norm has to be weaker than the 1-norm, uniformly for $A \in \mathcal{A}$. Because $\rho(D^{-1}A) = \sup \{\|e\|_2^2/\|e\|_1^2\}$, this is equivalent to the uniform boundedness of $\rho(D^{-1}A)$ which is satisfied for all important cases under consideration (cf. Examples 4.3).

To be more specific, let G be of the general form

$$G = I - Q^{-1}A, \tag{4.2}$$

with some nonsingular matrix Q. We then have [377] the following lemma.

LEMMA 4.1. *Given any $A > 0$, the two inequalities in* (4.1) *are equivalent to*

$$\alpha Q^T D^{-1} Q \le Q + Q^T - A$$

and

$$\alpha (A - Q^T) D^{-1} (A - Q) \le Q + Q^T - A,$$

respectively.

Proof. Using the definition of G, we obtain

$$\|Ge\|_1^2 = \|e\|_1^2 - \langle (Q + Q^T - A) Q^{-1} Ae, Q^{-1} Ae \rangle.$$

Thus, the two inequalities in (4.1) are equivalent to the following:

$$\alpha \|e\|_2^2 \le \langle (Q + Q^T - A) Q^{-1} Ae, Q^{-1} Ae \rangle,$$
$$\alpha \|Ge\|_2^2 \le \langle (Q + Q^T - A) Q^{-1} Ae, Q^{-1} Ae \rangle,$$

which in turn are equivalent to

$$\alpha\langle D^{-1}Qe, Qe\rangle \le \langle (Q+Q^T-A)e, e\rangle,$$
$$\alpha\langle D^{-1}(Q-A)e, (Q-A)e\rangle \le \langle (Q+Q^T-A)e, e\rangle,$$

respectively. This proves the lemma.

4.4.2. Specific cases. We first look at standard *Gauss–Seidel relaxation* (cf. [103]). In this case, Q is just the lower triangular part of A (including the diagonal).

THEOREM 4.2. *Let $A>0$ and define, with any positive vector $w=(w_i)$,*

$$\gamma_- = \max_{1\le i\le n}\left\{\frac{1}{w_i a_{ii}}\sum_{j<i} w_j\,|a_{ij}|\right\}, \qquad \gamma_+ = \max_{1\le i\le n}\left\{\frac{1}{w_i a_{ii}}\sum_{j>i} w_j\,|a_{ij}|\right\}.$$

Then Gauss–Seidel relaxation satisfies (4.1a) *and* (4.1b) *if* $\alpha \le 1/(1+\gamma_-)(1+\gamma_+)$ *and* $\alpha \le 1/\gamma_-\gamma_+$, *respectively.*

Proof. We first observe that $Q+Q^T-A=D$. Thus, because of Lemma 4.1, the two inequalities (4.1) are equivalent to

$$\alpha \le 1/\rho(D^{-1}QD^{-1}Q^T) \quad \text{and} \quad \alpha \le 1/\rho(D^{-1}(A-Q)D^{-1}(A-Q^T)),$$

respectively, which in turn are implied by

$$\alpha \le 1/|D^{-1}Q|\,|D^{-1}Q^T| \quad \text{and} \quad \alpha \le 1/|D^{-1}(A-Q)|\,|D^{-1}(A-Q^T)|,$$

respectively. Here $|\cdot|$ stands for any arbitrary matrix norm induced by a vector norm. The special choice

$$|L| = |L|_w = \max_{1\le i\le n}\left\{\frac{1}{w_i}\sum_{j=1}^{n} w_j\,|l_{ij}|\right\} \tag{4.3}$$

proves the theorem.

Examples 4.3. The inequalities (4.1) hold uniformly for essentially all matrices that are of interest here. For instance, for any sparse $A>0$ with no more than a *fixed* number l of nonvanishing off-diagonal entries per row, the choice $w_i = 1/\sqrt{a_{ii}}$ gives $\gamma_- < l$ and $\gamma_+ < l$ because $a_{ij}^2 < a_{ii}a_{jj}$ ($i\ne j$). Thus, for any such A, (4.1a) and (4.1b) are satisfied with $\alpha=(1+l)^{-2}$ and $\alpha = l^{-2}$, respectively. From a practical point of view, however, these values are far too pessimistic in typical situations. In practice, we usually have $\sum_{j\ne i}|a_{ij}| \approx a_{ii}$, which means that by selecting $w_i=1$ we can expect γ_- and γ_+ to be close to or even less than 1. An important class of matrices in which we have uniform smoothing is the class of *symmetric M-matrices*. To see this, note that for any such matrix A there exists a vector $z>0$ with $Az>0$ (cf. [Sc1]). By choosing $w=z$ in Theorem 4.2, we obtain

$$\gamma_- = \max_{1\le i\le n}\left\{\frac{1}{z_i a_{ii}}\sum_{j<i} z_j\,|a_{ij}|\right\} = \max_{1\le i\le n}\left\{1-\frac{1}{z_i a_{ii}}\sum_{j\le i} z_j a_{ij}\right\} < 1.$$

Similarly, we obtain $\gamma_+ < 1$. Thus, (4.1a) and (4.1b) are satisfied with $\alpha = \frac{1}{4}$ and $\alpha = 1$, respectively.

Gauss–Seidel relaxation is the scheme we usually use for smoothing in AMG. With respect to vector and parallel computers, Jacobi-type relaxations are also of interest. In particular, standard *Jacobi relaxation* has properties similar to Gauss–Seidel relaxation, if some relaxation parameter ω is used, i.e., if $Q = D/\omega$.

THEOREM 4.4. *Let $A > 0$ and $\gamma_0 \geq \rho(D^{-1}A)$. Then Jacobi relaxation with relaxation parameter $0 < \omega < 2/\gamma_0$ satisfies* (4.1a) *and* (4.1b) *if $\alpha \leq \omega(2 - \omega\gamma_0)$ and $\alpha \leq \omega \min\{2, (2-\omega\gamma_0)/(1-\omega\gamma_0)^2\}$, respectively.*

Proof. Using Lemma 4.1, we see that the two inequalities (4.1) are equivalent to

$$\rho\left(D^{-1}A + \frac{\alpha}{\omega^2} I\right) \leq \frac{2}{\omega} \quad \text{and} \quad \rho\left(D^{-1}A + \alpha\left(D^{-1}A - \frac{1}{\omega} I\right)^2\right) \leq \frac{2}{\omega},$$

which in turn are implied by

$$\gamma_0 + \frac{\alpha}{\omega^2} \leq \frac{2}{\omega} \quad \text{and} \quad \max_{0 \leq \lambda \leq \gamma_0} \left\{\lambda + \alpha\left(\lambda - \frac{1}{\omega}\right)^2\right\} \leq \frac{2}{\omega},$$

respectively. From this, our statement follows by a simple calculation.

As a bound for $\rho(D^{-1}A)$ we can use, for instance, $\gamma_0 = |D^{-1}A|_w$ with any vector $w > 0$ (see (4.3)). For scalar PDE applications we typically have $\gamma_0 \approx 2$. Also, for symmetric M-matrices we can choose $\gamma_0 = 2$ (cf. Examples 4.3). In terms of γ_0, the optimal values for ω (which give the largest values of α) are easily computed to be $\omega^* = 1/\gamma_0$ in the case of (4.1a) and $\omega^* = 3/2\gamma_0$ in the case of (4.1b). For these optimal parameters, the smoothing inequalities are satisfied with $\alpha = 1/\gamma_0$ and $\alpha = 3/\gamma_0$, respectively, giving $\alpha = \frac{1}{2}$ and $\alpha = \frac{3}{2}$ if $\gamma_0 = 2$.

4.4.3. Interpretation of algebraic smoothness. In AMG we *define* an error e to be smooth if it is slow to converge with respect to G, i.e., if $\|Ge\|_1 \approx \|e\|_1$. Note that, in this generality, the term "smooth" is not always related to what is called smooth in a geometric environment. Actually, e being smooth does not mean much more here than that e has to be taken care of by the coarse grid correction in order to obtain an efficient multigrid method. From an algebraic point of view, this is the important point in distinguishing smooth and nonsmooth error. Geometrically, however, the term smooth is used in a more restrictive (viz. the "natural") way. In fact, without any further assumptions on A (besides $A > 0$), an algebraically smooth error may well be highly oscillating geometrically, e.g., in certain cases of off-diagonally nonnegative matrices. Also, there may not be any algebraically smooth error at all, e.g., if A is a strongly diagonally dominant

M-matrix. Such cases, however, are generally not those that typically arise from interesting PDE applications or those that require a particular fast solution method.

For common relaxation schemes, a smooth error e is algebraically characterized by the fact that the residual $r = Ae$ is small in some sense compared to the error itself. This can be seen either directly from their definition (4.2) or from the properties (4.1). The latter, for instance, show that a smooth error necessarily has to satisfy $\|e\|_2 \ll \|e\|_1$, or, more explicitly, $\sum_i r_i^2/a_{ii} \ll \sum_i r_i e_i$. Therefore, on the average, we can expect $|r_i| \ll a_{ii}\,|e_i|$ for all i. Consequently, we obtain a quite good approximation for e_i as a function of its "neighboring" error values e_j ($j \in N_i$) by evaluating

$$(4.4) \qquad (r_i =)\; a_{ii}e_i + \sum_{j \in N_i} a_{ij}e_j = 0,$$

where $N_i = \{j \neq i : a_{ij} \neq 0\}$ denotes the *neighborhood* of i. This simple fact about smooth errors gives the basic information needed for the construction of interpolation, one goal of which is to guarantee that smooth error is approximately in the range of interpolation (cf. § 4.3.2). However, as (4.4) is usually not "local enough," it cannot be used directly in order to obtain an efficient interpolation formula for AMG (see Example 5.1 below), but rather has to be "truncated." In § 4.5 we will discuss some possibilities.

In the important case of symmetric M-matrices one easily obtains some more information on what smooth error looks like (cf. [103]). For this, note first that

$$(4.5) \qquad \|e\|_1^2 \leq \|e\|_0\,\|e\|_2,$$

which follows from $\langle Ae, e\rangle = \langle D^{-1/2}Ae, D^{1/2}e\rangle$ by applying Schwarz' inequality. This shows that $\|e\|_2 \ll \|e\|_1$ implies $\|e\|_1 \ll \|e\|_0$, or, more explicitly,

$$\langle Ae, e\rangle = \frac{1}{2}\sum_{i,j} (-a_{ij})(e_i - e_j)^2 + \sum_i \Big(\sum_j a_{ij}\Big) e_i^2 \ll \sum_i a_{ii}e_i^2.$$

For the most important case that $\sum_{j\neq i} |a_{ij}| \approx a_{ii}$, this means that, on the average for each i,

$$\sum_{j \neq i} \frac{|a_{ij}|}{a_{ii}} \frac{(e_i - e_j)^2}{e_i^2} \ll 1.$$

In other words, smooth error generally *varies slowly in the direction of strong connections*, i.e., from e_i to e_j if $|a_{ij}|/a_{ii}$ is relatively large. This observation is important in connection with the question of the truncation mentioned above (see §§ 4.5.4 and 4.6.1).

4.5. The coarse grid approximation and convergence. The main topic of this section is how the coarsening enters the question of smoothing, which was not considered in the previous section. As the corresponding discussion will refer to only two consecutive levels at a time, we will, according to our convention, use indices h and H to distinguish fine and coarse level quantities.

All of the following considerations could be applied in the framework of vectors, matrices, etc. The transfer of information between different levels, however, becomes (at least formally) more transparent and easier to describe if we use some kind of grid terminology as introduced in § 4.5.1. Also, an AMG method then becomes formally completely analogous to a usual geometric multigrid method.

We have seen in § 4.3 that the approximation property (second inequality in (3.3)) is sufficient for a V-cycle convergence theory. However, even if we know that coarsening strategies with this property exist (e.g., in regular elliptic scalar PDE problems), it is quite difficult to determine one in an automatic AMG process that uses *only algebraic information.* This will be discussed in § 4.5.5. In § 4.5.3, we first consider a weaker condition, due to Brandt [103], which will turn out to be sufficient for a *two-level* convergence theory. As an application, we will treat symmetric M-matrices in § 4.5.4.

Some major results given in §§ 4.5.3 and 4.5.4 are also contained in [103]. However, we consider some additional results which are relevant for the practical development of an AMG algorithm and we also have a different point of view here in that we stress connections to the general convergence theorems of § 4.3.

4.5.1. Grid terminology. As was mentioned earlier, we can, for instance, introduce grids as sets of unknowns. By accepting some loss in generality, however, we can go one step further, namely, if we regard the sets of coarse level variables as being "subsets" of the finer level variables: in terms of any two consecutive levels, h and H, we assume that each coarse level variable u_k^H is used to directly correct some uniquely defined finer level variable $u_{i(k)}^h$.[4] Thus, the set of variables on level h can be split into two disjoint subsets: the first one contains the variables also represented in the coarser level (*C-variables*), and the second one is just the complementary set (*F-variables*). The corresponding sets of indices are denoted by C^h and F^h, respectively. By renumbering the coarse level unknowns (and all related quantities) such that they have the same index as their finer grid analogues, we can now, for instance, write the H-level equations in the form

$$\sum_{j \in C^h} a_{ij}^H \mathbf{u}_j^H = b_i^H \qquad (i \in C^h). \tag{5.1}$$

[4] This assumption is "natural" for all cases we have in mind in this paper, and it corresponds formally to the geometric multigrid case if coarser grids are defined to be subsets of the finer ones.

By identifying each $i \in \Omega^h = C^h \cup F^h$ with some fictitious *point* in the plane, we interpret the equations $A^h \mathbf{u}^h = b^h$ and $A^H \mathbf{u}^H = b^H$ as *grid equations* on the fictitious *grid* Ω^h and the subgrid $\Omega^H = C^h$, respectively. In this sense, referring to a point $i \in \Omega^h$ means nothing else than referring to the variable u_i^h.

According to the above terminology, the actual h to H coarsening proceeds in three steps. First we have to decide which variables (points) are to become C- and which F-variables (-points), i.e., we have to define a splitting $\Omega^h = C^h \cup F^h$. Then, with $\Omega^H = C^h$, the weights $w_{ik} = w_{ik}^h$ of the interpolation

$$(5.2) \qquad (I_H^h e^H)_i = \sum_{k \in C^h} w_{ik} e_k^H \quad (i \in \Omega^h) \quad \text{with } w_{ik} = \delta_{ik} \text{ if } i \in C^h$$

have to be defined (δ_{ik} denotes the Kronecker symbol). Coarsening is concluded by the computation of $I_h^H = (I_H^h)^T$ and $A^H = I_h^H A^h I_H^h$. Clearly, for reasons of efficiency, interpolation should be "local," i.e., for each $i \in F^h$, we must have $w_{ik} = 0$ unless $k \in C_i^h$, where $C_i^h \subset C^h$ is some reasonable small set of interpolation points. For reasons of easy referencing, we call an interpolation of the above form *standard interpolation*. Note that each standard interpolation has full rank. In practice, the selection of a coarser grid and the definition of a corresponding interpolation formula are closely coupled processes. Whenever we talk about "construction of interpolation," we actually mean both processes.

When we are setting up more than two levels, all of the above applies recursively in a straightforward manner starting from the given finest level.

Finally, we define *connections* between grid points in a natural way in the sense of the directed graph which is associated with any matrix. That is, for any level h, we define a point $i \in \Omega^h$ to be (directly) *connected* to a point $j \in \Omega^h$ if $a_{ij}^h \neq 0$. Correspondingly, we define the (direct) *neighborhood* of a point $i \in \Omega^h$ by

$$N_i^h = \{j \in \Omega^h : j \neq i, a_{ij}^h \neq 0\}.$$

4.5.2. Convergence versus numerical work. Before we investigate the question of coarsening, we want to point out that, in the context of AMG, it is not enough to merely look at convergence: convergence does not mean anything if *numerical work* is not taken into account. This point is much more crucial here than it is in standard geometric multigrid applications where the numerical work (per cycle, say) is approximately known a priori. In an algebraic multigrid process, in general, nothing definite is known a priori about the numerical work. Consequently, it is very important to find conditions (for the practical construction of interpolation) that not only give good convergence, but also allow for a reasonable control of the numerical work by the algorithm (cf. § 4.6.3). The following example stresses the importance of this remark.

Example 5.1. It is easy to develop an AMG solver that has a convergence factor of 0, i.e., that solves (2.1) exactly in one cycle only. Assuming any $A > 0$, we find that the preparation phase of such an algorithm runs as follows. First, the set of all n variables is subdivided somehow into two nonempty sets of C- and F-variables (cf. § 4.5.1) in such a way that each F-variable has *no connection to any other F-variable.* (Note that this is always possible unless $n = 1$.) The interpolation (5.2) is defined by $w_{ik} = -a_{ik}/a_{ii}$ ($i \in F$, $k \in C$). After the computation of (2.3), this process is recursively repeated until only very few unknowns are left. As there are no F-to-F connections on any level, it is easy to see that $R(G^m) \subseteq R(I^m_{m+1})$ holds for all m, if only the order of (Gauss–Seidel) relaxation is arranged such that *F-variables are always relaxed last.* Because $T^m e^m = 0$ if $e^m \in R(I^m_{m+1})$, we therefore have $T^m G^m e^m \equiv 0$. As a consequence, (3.1b) is satisfied for any arbitrarily large δ, which proves our statement. However, a closer examination immediately shows that such an algorithm is completely useless for practical purposes: with increasing m, not only does the reduction of the number of unknowns usually become extremely slow from one level to the next, but there is also a tremendous fill-in in the coarse grid matrices. The major drawback of the above coarsening process is that it is actually defined just to give $R(G^m) \subseteq R(I^m_{m+1})$. *Smoothing effects are not really exploited,* and this condition is too strong to be practical.

4.5.3. Two-level convergence. In this section we will consider two-level methods only. Without essential loss of generality, we therefore restrict our investigation to the case where smoothing is performed *after* each coarse grid correction step. Assuming that the corresponding smoothing property (cf. (3.3a))

$$\|G^h e^h\|_1^2 \le \|e^h\|_1^2 - \alpha \|e^h\|_2^2 \qquad (\alpha > 0) \tag{5.3}$$

holds, we will give a general condition on I^h_H that allows for an estimation of the two-level convergence factor $\|G^h T^h\|_1$. As a motivation recall that, due to (5.3), error reduction by smoothing becomes inefficient if $\|e^h\|_2 \ll \|e^h\|_1$. Therefore, in order to obtain reasonable two-level convergence, the least we have to require is that the coarse grid correction T^h raises the error norm $\|e^h\|_2$ relative to $\|e^h\|_1$. More precisely, we need an inequality of the form

$$\|T^h e^h\|_1^2 \le \beta \|T^h e^h\|_2^2 \qquad (\beta > 0). \tag{5.4}$$

Note that this is just the approximation property in (3.3), required for $e^h \in R(T^h)$ only. Instead of (5.4), a more practical, sufficient condition is used in the following theorem.

THEOREM 5.2. *Let $A^h > 0$ and let G^h satisfy* (5.3). *Suppose that the interpolation I_H^h has full rank and that, for each e^h,*

$$\min_{e^H} \|e^h - I_H^h e^H\|_0^2 \le \beta \|e^h\|_1^2, \tag{5.5}$$

with $\beta > 0$ independent of e^h. Then $\beta \ge \alpha$, and the (h, H)-two-level convergence factor satisfies $\|G^h T^h\|_1 \le \sqrt{1-\alpha/\beta}$.

Proof. We first show that assumption (5.5) implies (5.4). As $R(T^h)$ is orthogonal to $R(I_H^h)$ with respect to $\langle \cdot . \cdot \rangle_1$, we have, for all $e^h \in R(T^h)$ and for all e^H,

$$\|e^h\|_1^2 = \langle A^h e^h, e^h - I_H^h e^H \rangle.$$

A simple calculation using Schwarz' inequality (writing A and D instead of A^h and D^h, respectively, and observing that $D^{-1}A > 0$ with respect to $\langle \cdot, \cdot \rangle_1$) yields

$$\begin{aligned}\|e^h\|_1^2 &= \langle A^{1/2}(D^{-1}A)^{1/2} e^h, A^{1/2}(D^{-1}A)^{-1/2}(e^h - I_H^h e^H)\rangle \\ &\le \|A^{1/2}(D^{-1}A)^{1/2} e^h\| \, \|A^{1/2}(D^{-1}A)^{-1/2}(e^h - I_H^h e^H)\| \\ &= \|e^h\|_2 \, \|e^h - I_H^h e^H\|_0.\end{aligned}$$

Because of (5.5), we therefore obtain $\|e^h\|_1^2 \le \beta \|e^h\|_2^2$ for all $e^h \in R(T^h)$, which proves (5.4). The convergence estimate of the theorem is now an immediate consequence of (5.4) and (5.3):

$$\begin{aligned}\|G^h T^h e^h\|_1^2 &\le \|T^h e^h\|_1^2 - \alpha \|T^h e^h\|_2^2 \le \|T^h e^h\|_1^2 - \frac{\alpha}{\beta} \|T^h e^h\|_1^2 \\ &= \left(1 - \frac{\alpha}{\beta}\right) \|T^h e^h\|_1^2 \le \left(1 - \frac{\alpha}{\beta}\right) \|e^h\|_1^2.\end{aligned}$$

Assume the interpolation to be of the standard type (5.2); the following theorem gives a condition sufficient for (5.5) and relating the interpolation weights w_{ik} to the matrix entries a_{ij}^h. Here and in the following we use the notation $s_i = \sum_{k \in C} w_{ik}$ for the ith row sum of interpolation.

THEOREM 5.3. *Let $A^h > 0$ and assume, for any given set $C = C^h$ of C-points, that I_H^h is of the form* (5.2) *with $w_{ik} \ge 0$ and $s_i \le 1$. Then property* (5.5) *is satisfied if the following two inequalities hold with $\beta > 0$ independently of $e = e^h$:*

$$\begin{aligned}&\sum_{i \in F} \sum_{k \in C} a_{ii}^h w_{ik} (e_i - e_k)^2 \le \frac{\beta}{2} \sum_{i,j} (-a_{ij}^h)(e_i - e_j)^2, \\ &\sum_{i \in F} a_{ii}^h (1 - s_i) e_i^2 \le \beta \sum_i \left(\sum_j a_{ij}^h\right) e_i^2.\end{aligned} \tag{5.6}$$

Proof. Let e^h be arbitrary and define $e_k^H = e_k^h (k \in C)$. With this particular e^H and omitting the index h, inequality (5.5) reads

$$\sum_{i\in F} a_{ii}\Big(e_i - \sum_{k\in C} w_{ik}e_k\Big)^2 \le \beta\Big(\frac{1}{2}\sum_{i,j}(-a_{ij})(e_i-e_j)^2 + \sum_i\Big(\sum_j a_{ij}\Big)e_i^2\Big).$$

Estimating the left-hand side by

$$\sum_{i\in F} a_{ii}\Big(e_i - \sum_{k\in C} w_{ik}e_k\Big)^2 = \sum_{i\in F} a_{ii}\Big(\sum_{k\in C} w_{ik}(e_i-e_k) + (1-s_i)e_i\Big)^2$$

$$\le \sum_{i\in F} a_{ii}\Big(\sum_{k\in C} w_{ik}(e_i-e_k)^2 + (1-s_i)e_i^2\Big),$$

we see that (5.6) is sufficient for (5.5).

4.5.4. Application to M-matrices. In this section, we will discuss interpolation in some detail for symmetric M-matrices A^h. Recall that, in this class of matrices, we have uniform smoothing (cf. § 4.4.2). Under the additional assumption of *weak diagonal dominance,* i.e., $\sum_j a_{ij}^h \ge 0$ for all i,[5] we will consider ways to interpolate such that (5.5) also holds uniformly. Theorem 5.2 then guarantees uniform two-level convergence. The following considerations will be mainly theoretical; for an efficient algorithm and practical applications we refer to §§ 4.6.1, 4.6.2 and 4.6.4.

As we already mentioned in § 4.4.3, the general idea in deriving an interpolation formula for some F-point i is to use the ith residual equation (4.4) of A^h. In particular, for each $i \in F$ with $N_i \subseteq C$, a good formula is obtained by defining the set of interpolation points by $C_i = N_i$ with corresponding weights $w_{ik} = |a_{ik}^h|/a_{ii}^h$ $(k \in C_i)$ (cf. Example 5.1). For reasons of localness, however, we want C_i to be small. Therefore, we generally allow $D_i = N_i - C_i \neq \emptyset$ and we have to make a decision about how to "distribute" the noninterpolatory connections a_{ij}^h $(j \in D_i)$ in interpolating to i. This distribution is the most crucial point in the development of a good interpolation: the goal of obtaining local interpolation has to be achieved without losing too much precision compared to (4.4).

The interpolation formula introduced in § 4.5.4.1 below is quite simple in that all a_{ij}^h $(j \in D_i)$ are added to the diagonal element a_{ii}^h. In terms of (4.4), this means that we replace e_j by e_i for all $j \in D_i$. If $|a_{ij}^h|$ is small, this replacement should not cause too much harm. If, on the other hand, $|a_{ij}^h|$ is large, we generally can expect this replacement to be reasonable due to smoothness arguments (cf. § 4.4.3). Although this approach leads to a quite acceptable AMG algorithm (cf. [536], [486]), it can be improved considerably by interpolating in a different way which is described in § 4.5.4.2. The main idea here is to distribute the noninterpolatory connections a_{ij}^h

[5] Such matrices are called *positive type* matrices in [103].

($j \in D_i$) in a weighted average way to the interpolation points in C_i instead of simply adding them to the diagonal.

A combination of the above two approaches is used in actual practice. This is, in general, considerably better for geometrically posed problems than the simple approach of § 4.5.4.1. The reasons, however, are not apparent from merely algebraic estimates as given below; rather, we must take certain geometric arguments into account (see § 4.6.1).

4.5.4.1. Interpolation along direct connections. In this section we consider interpolation along *direct* connections. That is, we consider only interpolation points C_i with $C_i \subseteq N_i \cap C$ and corresponding weights of the form

$$w_{ik} = \eta_i \, |a_{ik}^h| \qquad (i \in F,\, k \in C_i), \tag{5.7}$$

with $0 \le \eta_i \le 1/\sum_{l \in C_i} |a_{il}^h|$, which just ensures $s_i \le 1$. In order for property (5.5) to hold, it is, due to Theorem 5.3, obviously sufficient to require for every $i \in F,\ k \in C_i$

$$0 \le a_{ii}^h w_{ik} \le \beta \, |a_{ik}^h|, \qquad 0 \le a_{ii}^h(1 - s_i) \le \beta \sum_j a_{ij}^h. \tag{5.8}$$

From these simple inequalities, we can easily derive more practical conditions, which can be used to effectively develop an automatic coarsening algorithm that has β as input parameter and chooses interpolation with weights (5.7) satisfying (5.8). An example which has been tested in practical applications is given in the following theorem.

THEOREM 5.4. *Let $\beta \ge 1$ be fixed. Assume for any symmetric, weakly diagonally dominant M-matrix A^h the C-points are picked such that, for each $i \in F$, there is a nonempty set $C_i \subseteq N_i \cap C$ with*

$$\beta \sum_{j \notin C_i} a_{ij}^h \ge a_{ii}^h \tag{5.9}$$

and define the interpolation weights (5.7) *by $\eta_i = 1/\sum_{j \notin C_i} a_{ij}^h$. Then property* (5.5) *is satisfied.*

For fixed β, there are usually many possible ways to choose C and C_i so that they satisfy (5.9). Aside from this purely algebraic condition, there are, however, others that should be met (at least approximately) to ensure efficiency in practice. For instance (further requirements are discussed in § 4.6.1), we want C to be as small as possible. Consequently, we should try to satisfy (5.9) with sets C_i as small as possible. This, in turn, requires arranging a concrete algorithm such that each F-point i is *strongly* connected to its interpolation points, i.e., we actually want each $|a_{ij}|$ ($j \in C_i$) to be comparable in size to the largest of the $|a_{ik}|$ ($k \in N_i$).

We want to make a few remarks concerning the choice of the parameter β. Clearly, the larger β, the weaker the assumption (5.9) is. In particular, for large β, (5.9) allows for rapid coarsening, but, due to Theorem 5.2, the two-level convergence speed will be very slow. On the other hand, the choice $\beta = 1$ gives best convergence, but will lead to an extremely expensive method (cf. Example 5.1). For real computations, experience shows that, on the average, a reasonable compromise is given by $\beta = 2$, which means that about 50% of the total strength of connections of every F-point will be represented on the coarser level (and used for interpolation). If applied recursively in a real multi-level cycle (cf. § 4.5.4.4), this still limits the efficiency of coarsening so that, in practice, a corresponding AMG algorithm is not fully satisfactory (in particular, if the occurring matrices have many row entries of similar size (see [486])).

4.5.4.2. More general interpolation. Instead of (5.7), we now consider weights of the more general form

$$w_{ik} = \eta_i\Big(|a_{ik}^h| + \sum_{j\in D_i} \eta_{ik}^{(j)}\,|a_{ij}^h|\Big) \qquad (i \in F,\ k \in C_i), \tag{5.10}$$

with nonnegative η_i and $\eta_{ik}^{(j)}$. This form is obtained if we perform, for every F-point i, a *weighted distribution* of all noninterpolatory connections a_{ij}^h ($j \in D_i$) to the interpolation points $k \in C_i$ using the distribution weights $\eta_{ik}^{(j)}$. Here we do not require C_i to be a subset of N_i, which means that we allow *long-range interpolation,* i.e., interpolation from points to which there is no *direct* connection.

Again, there are several reasonable ways to specify η_i and the distribution weights $\eta_{ik}^{(j)}$. In the following theorem, we choose $\eta_{ik}^{(j)}$ to be proportional to a_{jk}^h, i.e., we distribute only along direct connections. The essential condition for this to be reasonable is that, for every F-point i, the *total* connection of each noninterpolatory point j to the set of interpolation points C_i is strong enough, i.e., $\sum_{l\in C_i}|a_{jl}^h|$ is sufficiently large. The following theorem expresses this more quantitatively.

THEOREM 5.5. *Let $\zeta > 0$ be a fixed number. Assume, for any symmetric, weakly diagonally dominant M-matrix A^h, that the C-points and $C_i \subseteq C$ ($i \in F$) are picked such that, for each j and $k \in N_j \cap C$, we have*

$$\zeta \sum_{l\in C_i} |a_{jl}^h| \geq \sum_{v\in F_{jk}} |a_{jv}^h| \qquad (i \in F_{jk}), \tag{5.11}$$

where $F_{jk} = \{i \in F : k \in C_i$ and $j \in D_i\}$. Then, if the weights (5.10) *of interpolation are defined by $\eta_i = 1/a_{ii}^h$ and $\eta_{ik}^{(j)} = |a_{jk}^h|/\sum_{l\in C_i}|a_{jl}^h|$, property* (5.5) *is satisfied with some β that depends only on ζ but not on A^h.*

Proof. For the proof, we apply Theorem 5.3. We first observe that the second inequality in (5.6) is satisfied for every $\beta \geq 1$ because $a_{ii}(1 - s_i) =$

$\sum_j a_{ij}$ $(i \in F)$. The first inequality in (5.6) follows from the estimate

$$a_{ii}w_{ik}(e_i - e_k)^2 = |a_{ik}|\,(e_i - e_k)^2 + \sum_{j \in D_i} \eta_{ik}^{(j)}\,|a_{ij}|\,(e_i - e_k)^2$$

$$\leq |a_{ik}|\,(e_i - e_k)^2 + 2\sum_{j \in D_i} \eta_{ik}^{(j)}\,|a_{ij}|\,((e_i - e_j)^2 + (e_j - e_k)^2)$$

by observing that

$$\sum_{i \in F}\sum_{k \in C_i}\sum_{j \in D_i} \eta_{ik}^{(j)}\,|a_{ij}|\,(e_i - e_j)^2 = \sum_{i \in F}\sum_{j \in D_i}\Big(\sum_{k \in C_i}\eta_{ik}^{(j)}\Big)\,|a_{ij}|\,(e_i - e_j)^2$$

$$= \sum_{i \in F}\sum_{j \in D_i} |a_{ij}|\,(e_i - e_j)^2$$

and

$$\sum_{i \in F}\sum_{k \in C_i}\sum_{j \in D_i} \eta_{ik}^{(j)}\,|a_{ij}|\,(e_j - e_k)^2 = \sum_{j}\sum_{k \in C}\Big(\sum_{i \in F_{jk}} \frac{|a_{ji}|}{\sum_{l \in C_i}|a_{jl}|}\Big)\,|a_{jk}|\,(e_j - e_k)^2$$

$$\leq \zeta \sum_{j}\sum_{k \in C}\Big(\sum_{i \in F_{jk}} \frac{|a_{ji}|}{\sum_{v \in F_{jk}}|a_{jv}|}\Big)\,|a_{jk}|\,(e_j - e_k)^2$$

$$= \zeta \sum_{j}\sum_{k \in C} |a_{jk}|\,(e_j - e_k)^2.$$

4.5.4.3. Extensions and further remarks. In constructing interpolation formulas in the previous two sections, we used Theorem 5.3 as our theoretical tool. From the basic inequalities (5.6), it can be seen that the essential ideas carry over to matrices that also contain some (small) *positive* off-diagonal elements. For instance, for the construction of interpolation in § 4.5.4.1, it is sufficient to require that A^h satisfies $a_{ij}^h < 0$ $(i \in F,\ j \in C_i)$ and, for all e^h,

$$\sum_{i,j} (-a_{ij}^h)(e_i^h - e_j^h)^2 \geq \zeta \sum_{i \in F}\sum_{j \in C_i} (-a_{ij}^h)(e_i^h - e_j^h)^2$$

with $\zeta > 0$.[6] For cases involving higher order difference approximations to second order elliptic problems, or for certain problems involving mixed derivatives, such relations can usually be derived by using simple estimates like

$$\frac{\phi\psi}{\phi + \psi}(\alpha + \beta)^2 \leq \phi\alpha^2 + \psi\beta^2 \qquad (\phi,\ \psi > 0).$$

A last remark refers to the assumption of *weak diagonal dominance*. While uniform smoothing is guaranteed in the class of *all* symmetric M-matrices A^h, (5.5) cannot be expected to hold in this class with β being independent

[6] Such matrices are called *essentially positive type* in [103].

of A^h. This becomes obvious if, e.g., we consider the subclass of matrices A^h_μ ($\mu < \lambda_0$) defined by discretizing the Helmholtz operator $-\Delta - \mu I$ on the unit square with fixed mesh size h and Dirichlet boundary conditions. Here, λ_0 is the first eigenvalue of A^h_0. If we select $e^h = e^h_0$ as the corresponding (normalized) eigenfunction, we have for each $A^h = A^h_\mu$ that $\|e^h\|_1^2 = \lambda_0 - \mu$. Thus, for (5.5) to hold independently of μ if $\mu \to \lambda_0$, its left side necessarily has to approach zero, i.e., I^h_H has to be such that this particular eigenfunction e^h is arbitrarily close to $R(I^h_H)$, which is generally not true unless special techniques are used (see § 4.6.6).

Note, however, that A^h can be allowed to have *zero row sums*. This is because the first eigenvector, which *is* in the range of interpolation by definition, is then constant. In fact, as is easily seen, constants are always interpolated to constants and the zero row sum property is preserved on all coarser grids.

4.5.4.4. Recursive application in multi-level cycles. In order to recursively set up a multi-level cycle, we need, first of all, that the finest grid operator's essential properties, which have been exploited in our considerations, carry over to the coarser grids. The following theorem shows that this can be achieved for M-matrices.

THEOREM 5.6. *Let A^h be any symmetric, weakly diagonally dominant M-matrix and let the interpolation weights satisfy* (5.8) *with some $\beta \le 2$. Then A^H is also a symmetric, weakly diagonally dominant M-matrix.*

Proof. First, because of the Galerkin formulation, A^H is symmetric and positive definite. In order to show that A^H is off-diagonally nonpositive, we write its entries in their explicit form:

$$a^H_{kl} = \sum_{i,j} w_{ik} a^h_{ij} w_{jl} \qquad (k, l \in C).$$

Because $w_{ik} = \delta_{ik}$ (if $i, k \in C$) and for reasons of symmetry, we can rewrite this in the following way:

$$\begin{aligned} a^H_{kl} &= a^h_{kl} + \sum_{i \in F} (w_{ik} a^h_{il} + w_{il} a^h_{ik}) + \sum_{i \in F} \sum_{j \in F} w_{ik} a^h_{ij} w_{jl} \\ &= a^h_{kl} + \sum_{i \in F} w_{ik} \left(a^h_{il} + \frac{1}{2} \sum_{j \in F} w_{jl} a^h_{ij} \right) + \sum_{i \in F} w_{il} \left(a^h_{ik} + \frac{1}{2} \sum_{j \in F} w_{jk} a^h_{ij} \right). \end{aligned}$$

Due to our assumption that $w_{ik} a^h_{ii} \le 2\,|a^h_{ik}|$ ($i \in F$, $k \in C$), we have

$$a^h_{ik} + \frac{1}{2} \sum_{j \in F} w_{jk} a^h_{ij} \le a^h_{ik} + \frac{1}{2} w_{ik} a^h_{ii} \le 0 \qquad (i \in F, k \in C),$$

which implies $a^H_{kl} \le 0$ ($k, l \in C$, $k \ne l$). To show that A^H is weakly diagonally dominant, we compute $\sum_{l\in C} a_{kl}^H$ by adding up the above equations for a_{kl}^H:

$$\sum_{l\in C} a_{kl}^H = \sigma_k^C + \sum_{i\in F} w_{ik}\left(\sigma_i^C + \frac{1}{2}\sum_{j\in F} s_j a_{ij}^h\right) + \sum_{i\in F} s_i\left(a_{ik}^h + \frac{1}{2}\sum_{j\in F} w_{jk}a_{ij}^h\right)$$

$$= \sigma_k + \sum_{i\in F} w_{ik}\left(\sigma_i + \frac{1}{2}\sum_{j\in F}(s_j - 1)a_{ij}^h\right) + \sum_{i\in F}(s_i - 1)\left(a_{ik}^h + \frac{1}{2}\sum_{j\in F} w_{jk}a_{ij}^h\right).$$

Here we used the abbreviations $\sigma_i = \sum_j a_{ij}^h$ and $\sigma_i^C := \sum_{l\in C} a_{il}^h$. Using (5.8), we obtain

$$\sigma_i + \frac{1}{2}\sum_{j\in F}(s_j - 1)a_{ij}^h \geq \sigma_i + \frac{1}{2}(s_i - 1)a_{ii}^h \geq 0,$$

which implies the diagonal dominance.

The situation is more involved if the interpolation strategy of § 4.5.4.2 is used. Instead of going into detail, we just mention that the coarsening process, if arranged as described in § 4.6, will then generate coarse grid matrices which are generally close to being weakly diagonally dominant M-matrices. In any case, as long as the remarks of the previous section apply, there will be no essential deterioration in practice.

4.5.5. Remarks on multi-level convergence. So far, we have given conditions which, e.g., in our model class of symmetric, weakly diagonally dominant M-matrices or certain perturbations thereof, guarantee uniform two-level convergence. We have also seen that, in these cases, a recursive application in a real multi-level cycle is formally possible in the sense that the important properties of the finest level operator carry over to the coarser level ones. In this section we now discuss the question of multi-level convergence.

We first note that the two-level theory does not directly carry over to a multi-level theory. That is, requiring (5.5) to hold on each of the intermediate levels with the same β is not sufficient to guarantee that the V-cycle convergence factor is independent of q, the number of levels. This can be seen from the following simple counterexample (cf. [103]).

Example 5.7. Let A^h be derived from discretizing $-u''$ on the unit interval with mesh size h, i.e., the rows of A^h correspond to the difference stencil $1/h^2[-1\ \ 2\ \ -1]_h$. (For our purpose we may ignore the boundary in the following.) One possibility for satisfying (5.5) with $\beta = 2$ is to coarsen the grid by doubling the mesh size, and to define *strictly one-sided* interpolation to each F-point (using only the respective neighboring C-point to the right, say, with the interpolation weight being 1). The corresponding coarse grid operator A^{2h} is then easily computed to correspond to the difference stencil $1/(2h)^2[-4\ \ 8\ \ -4]_{2h}$, which, after proper scaling of the restriction operator by $\frac{1}{2}$, is seen to be off by a factor of 2 compared to the "natural" $2h$-discretization of $-u''$. Due to this, for a *very smooth* error frequency, $\sin(\pi x)$ say, we obtain $T^h e^h \approx \frac{1}{2} e^h$. Consequently, as smoothing hardly affects this frequency, we cannot expect the asymptotic two-level convergence factor to be better than $\frac{1}{2}$. If the same coarsening strategy is now used recursively to introduce coarser and coarser levels, the above arguments carry over to each of the intermediate levels. In particular, each coarser grid operator is off by a factor of 2 compared to the previous one. A simple recursive argument, applied to the same frequency as above, shows that errors are accumulated from grid to grid, and that the asymptotic (V-cycle) convergence factor cannot be expected to be better than $1 - 1/2^{q-1}$.

Clearly, one standard way to overcome q-dependent convergence factors is to use "better" cycles such as F- or W-cycles (cf. [101]). Apart from the fact that such cycles are more expensive (expense may be considerable in AMG, depending on the actual coarsening), they will have at best the same convergence factor as the corresponding two-level method. This method, in turn and for the same reasons that may lead to q-dependent convergence, may not be satisfactory. Therefore, a sufficient understanding of the reasons for q-dependent convergence is necessary.

As seen from the above example, a basic deficiency of the essential condition (5.5) is that it, by itself, does not guarantee a sufficiently good interpolation for smooth errors. A stronger condition that has been shown to be suitable for a V-cycle convergence theory is the approximation property given in § 4.3.2, namely

$$\|T^h e^h\|_1^2 \le \beta \, \|e^h\|_2^2. \tag{5.12}$$

Its essential relation to condition (5.5) is seen from the following lemma.

LEMMA 5.8. *If* (5.12) *holds, then we also have*

$$\min_{e^H} \|e^h - I_H^h e^H\|_0^2 \le \beta^2 \, \|e^h\|_2^2. \tag{5.13}$$

Proof. We first observe that (5.12) is equivalent to

$$\|T^h e^h\|_0^2 \le \beta \, \|e^h\|_1^2, \tag{5.14}$$

which, omitting the index h, follows from

$$\sup_e \frac{\|Te\|_1^2}{\|e\|_2^2} = \rho(E^{-1/2} T E^{-1/2}) = \rho(T E^{-1} T) = \sup_e \frac{\|Te\|_0^2}{\|e\|_1^2},$$

with $E = D^{-1} A$. Applying (5.14) to $T^h e^h$ and using (5.12), we have

$$\|T^h e^h\|_0^2 \le \beta \, \|T^h e^h\|_1^2 \le \beta^2 \, \|e^h\|_2^2,$$

which, in particular, implies (5.13).

For smooth errors e^h, (5.13) is much stronger than (5.5) because $\|e^h\|_2 \ll \|e^h\|_1$. For instance, in the case of second order scalar PDE problems and smooth error frequencies on a regular grid of mesh size h, we typically have that $\|e^h\|_2 = O(h)\, \|e^h\|_1$. In such situations, (5.13) *increases the order of interpolation* compared to (5.5).

In summary, it becomes clear that, in order to obtain robust and efficient V-cycle convergence, we have to increase the accuracy of interpolation for smooth errors. Theoretically, we could achieve this by constructing interpolation that satisfies (5.13) rather than (5.5), a sufficient condition being that the inequality in (5.13) must hold for each e^h with the special choice $e_k^H = e_k^h$ $(k \in C)$:

$$\sum_{i \in F} a_{ii}^h \Big(e_i^h - \sum_{k \in C} w_{ik} e_k^h \Big)^2 \le \beta^2 \sum_i a_{ii}^h \Big(\frac{r_i^h}{a_{ii}^h} \Big)^2,$$

where $r^h = A^h e^h$. Unfortunately, it is hardly possible to construct the interpolation so that we satisfy this inequality *exactly* by using *only algebraic information* (such as the matrix entries), unless we define it in the inefficient way described in Example 5.1.

In practice, however, it turns out that it is not really necessary to satisfy this inequality exactly. For geometrically posed problems, a sufficient improvement of the accuracy of interpolation is generally obtained if we add certain objectives to the interpolation requirements that result from § 4.5.4. These objectives (see criteria (C1) and (C2) in § 4.6.1), although partly motivated by geometric arguments, do not explicitly use geometric information. In particular, extremely one-sided interpolation like the one in Example 5.7 will be avoided.

Of course, if there is really no geometric background to a problem, or if the underlying structure of the connections is far from being local, there is no a priori guarantee of highest efficiency. However, a large number of practical tests, including complicated problems with random matrices and quite irregular finite element triangulations, show that obeying the objectives mentioned above results in a very robust and efficient method. For more information and, in particular, a collection of examples, see the following numerical sections.

For solving problems with underlying geometry, there are ways to improve interpolation further. One possibility is to supply AMG with some minimum geometric information, e.g., point locations. Along with other possibilities, this will also be discussed in the following section. All these "more sophisticated" techniques are not really needed in connection with the problems we had in mind up to now: the extra work does not pay. They become, however, quite important for certain "systems" problems.

4.6. AMG for scalar problems. AMG, as we stated previously, is designed to use the principles of geometric multigrid methods without relying on the geometry of the problem being solved. In the developmental phase, design choices were often based on experiences with discretized second order elliptic PDEs, but an effort was made to keep the explicit dependence on the underlying geometry to a minimum. As a result, the method is applicable to a wide class of problems including, but not limited to, those involving symmetric M-matrices. The method developed for such problems, called the *scalar algorithm,* only makes explicit use of the information contained in the matrix itself and does not rely on the geometry of the problem.

The application of AMG to a matrix equation is a two-part process. The first part, called the *setup phase,* consists of choosing the coarser grids and defining the grid transfer and coarse grid operators. The second, called the *solution phase,* uses the components defined in order to perform standard multigrid cycling until a desired level of tolerance is reached.

The purpose of this section is to describe the setup phase in some detail, to give the work and storage requirements, and to give results for a number of sample problems. First, in § 4.6.1, an overview of the principles and the reasons for our choice of the algorithm will be given. Section 4.6.2 covers the technical details of the basic setup algorithm. This will be of interest primarily to the reader wishing to program AMG. Section 4.6.3 contains a discussion of the work and storage required for the AMG algorithm. Numerical results for a number of sample problems are given in § 4.6.4. Section 4.6.5 contains a simple method for computing a posteriori error estimates for problems involving symmetric M-matrices. Finally, § 4.6.6 outlines ways to improve the performance of the algorithm by, in some cases, explicitly using geometric information, which is usually readily available.

In the remainder of the paper, we assume that the reader is familiar with the notation and terminology introduced in §§ 4.2 and 4.5.1.

4.6.1. The setup phase: a general discussion. In any multigrid approach, relaxation and coarse grid correction are used in conjunction to eliminate the error. This requires that any error which cannot be corrected by appealing to a coarser grid problem (i.e., any error not approximately represented in the range of interpolation) must be effectively reduced by relaxation. In geometric multigrid methods, linear interpolation is normally used; then relaxation must be chosen that smoothes the error in the usual geometric sense. The approach in AMG is just the opposite. The relaxation method is fixed, and the coarse grid and interpolation operator are chosen so that the range of interpolation is forced to (approximately) contain those functions unaffected by relaxation. The other multigrid components needed are then defined as in § 4.2.2.

Consider the application of AMG to a problem $A\mathbf{u} = b$. An outline for the setup algorithm, provided $A^1 = A$, is as follows:

A1. *The setup phase:*
Step 1. Set $m = 1$.
Step 2. Choose the coarse grid Ω^{m+1} and define the interpolation operator I^m_{m+1}.
Step 3. Set $I^{m+1}_m = (I^m_{m+1})^T$ and $A^{m+1} = I^{m+1}_m A^m I^m_{m+1}$.
Step 4. If Ω^{m+1} is small enough to allow for inexpensive inversion of A^{m+1}, set $q = m + 1$ and stop. Otherwise, set $m = m + 1$ and go to Step 2.

Step 2, called the *coarsening step,* is the only process not fully specified. In this section, we define the interpolation formula and determine the properties that the coarse grid should have. This will be done for any fixed level, denoted by the index h, since the method is the same for all levels. The corresponding coarser level to be constructed will be denoted by the index H. All sets of points used below (C, F, S, D and N) actually depend on h, but the index will be omitted for convenience.

Now, to determine what is required of the coarsening procedure, we consider the matrix equation

$$A^h \mathbf{u}^h = b^h, \tag{6.1}$$

where, for ease of motivation, we assume that A^h is a symmetric M-matrix. In this case, the theoretical results of the previous section apply. (The algorithm itself, however, does not require this to be true, and it can formally be applied in more general situations. Some examples of such situations are presented in § 4.6.4.) In particular, we have shown in § 4.4.3 that Gauss–Seidel relaxation converges slowly if and only if the (properly scaled) residuals are small compared to the error, e^h, between the current approximation and the exact solution. Thus, error that is slow to converge is characterized by

$$a^h_{ii} e^h_i \approx -\sum_{j \in N_i} a^h_{ij} e^h_j \qquad (i \in \Omega^h). \tag{6.2}$$

Such error is called *algebraically smooth.* Functions with this property should be well represented by the range of interpolation.

We assume that the interpolation operator is constructed as in § 4.5.1. That is, the grid Ω^h is partitioned into two sets, C and F, called C- and F-points, respectively. C is identified with the coarse grid Ω^H. Then, for any function e^H,

$$(I^h_H e^H)_i = \begin{cases} e^H_i & \text{if } i \in C, \\ \sum_{k \in C_i} w_{ik} e^H_k & \text{if } i \in F, \end{cases} \tag{6.3}$$

with some small sets of interpolation points $C_i \subset C$. Formula (6.2) is used to obtain the interpolation weights $w_{ik} = w_{ik}^h$ for the points in F. For instance, if C and F are chosen so that, for each $i \in F$, $N_i \subseteq C$ and we set $C_i = N_i$, then (6.2) can be used directly to define the weights (i.e., set $w_{ik} = -a_{ik}^h / a_{ii}^h$ for $i \in F$, $k \in C$). However, as noted in Example 5.1, this is seldom practical, and we would generally like to interpolate to $i \in F$ from as small a set C_i as possible. Then, in order to obtain an interpolation formula from (6.2), for each $j \in D_i = N_i - C_i$, e_j^h must be approximated in terms of e^h at points in C_i.

Formula (6.2) states that, at each point i, e_i^h is essentially determined by those e_j^h for which $-a_{ij}^h$ is large. For this reason, we introduce the following definition. A point i is *strongly connected to j*, or *strongly depends on j*, if $-a_{ij}^h \geq \theta \max_{l \neq i} \{-a_{il}^h\}$ with some $0 < \theta \leq 1$ (usually taken to be .25 in practice). With S_i denoting the set of all strong connections of point i, we will take $C_i = C \cap S_i$.[7] (Here, C is assumed to be given. Criteria for how to actually choose C will be derived below.) Then let $D_i^s = D_i \cap S_i$ and $D_i^w = D_i - S_i$. These are called the *strong* and *weak noninterpolatory connections*, respectively. Now, for points $j \in D_i^w$, we can replace e_j^h by e_i^h in (6.2). For points $j \in D_i^s$, however, this is generally not sufficient. Since the value of e_j^h is also determined by e^h at the points to which j is strongly connected, we make the approximation

$$e_j^h \approx \left(\sum_{k \in C_i} a_{jk}^h e_k^h \right) \Big/ \left(\sum_{k \in C_i} a_{jk}^h \right). \tag{6.4}$$

These approximations are substituted into (6.2), which is then solved for e_i^h, yielding the interpolation weights for point i. The more strongly connected j is to the points in C_i, the more accurate we can expect (6.4) to be (cf. Theorem 5.5). In practice, we only require at least one such strong connection. This requirement is considerably weaker than the corresponding one of Theorem 5.5, but it turned out to be sufficient in all our applications and gave highest efficiency in terms of overall work. Thus, the following criterion is suggested for choosing the sets C and F:

(C1) For each $i \in F$, each point $j \in S_i$ should either be in C, or should be strongly connected to at least one point in C_i $(= C \cap S_i)$.

A primary concern in the coarsening process is to produce multigrid components that result in an efficient solution process. The efficiency is

[7] Note that if, in contrast to our assumption, A^h is not an M-matrix, but rather contains *few positive off-diagonal entries* per row, the corresponding connections are not defined to be strong. Correspondingly, an F-point i does not interpolate from any point j with $a_{ij}^h > 0$. Unless such coefficients become substantial, AMG performance will not seriously deteriorate.

determined by the convergence factors and the amount of work needed per cycle. As a general rule, the more points Ω^H has, the better h-level convergence can be. On the other hand, the amount of work necessary for relaxation on the H-level increases with the size of A^H, so it is advantageous to limit the number of C-points while still enforcing (C1). We do this by introducing the following additional criterion:

(C2) C should be a maximal subset of all points with the property that no two C-points are strongly connected to each other.

It is easy to construct examples for which it is impossible to strictly satisfy both (C1) and (C2). However, (C2) is used as a guideline for constructing C while ensuring that (C1) holds. The details of the C-point choice, and how it relates to these criteria, are discussed in the next section.

Remark 6.1. The coarse grids chosen according to (C1) and (C2) tend to produce an especially efficient solution process in problems with a geometric background, such as discretized PDEs. Since the work of relaxation, which dominates the cycling time, is proportional to the number of nonzero entries in the matrices, it is important to keep the "stencil size" as small as possible on all levels. This is generally achieved by the criteria used. Another consequence of the coarsening produced is that the grids chosen generally avoid "one-sided" interpolation, such as that discussed in Example 5.7. Lemma 5.8 essentially states that, in solving a regular, second order elliptic partial differential equation, second order interpolation is required for proper V-cycle convergence. Although this cannot be guaranteed without the explicit use of geometric information, grids chosen according to (C1) and (C2) typically give balanced sets of interpolation points, allowing for interpolation that is sufficiently accurate for the problems tested.

4.6.2. Technical details of the coarsening process. This section is not essential for the remainder of the paper and can be skipped by the casual reader. Its purpose is to describe the coarsening process (Step 2) of the setup phase A1 in the previous section in detail. In addition to the notation used there, we define the set of points strongly connected *to* i by $S_i^T = \{j : i \in S_j\}$. Furthermore, for a set P, $|P|$ denotes the number of elements in P.

In order to keep the work involved in the choice of the coarse grid small, a two-part process is used. First, a relatively quick C-point choice is performed; this attempts to enforce the criterion (C2) by distributing C-points uniformly over the grid. In the second part, which is combined with the computation of interpolation weights, the tentative F-points resulting from the first part are tested to ensure that (C1) holds, with new C-points added as necessary. The first part is given below.

A2. *Preliminary C-point choice:*

Step 1. Set $C = \emptyset$, $F = \emptyset$, $U = \Omega^h$, and $\lambda_i = |S_i^T|$ for all i.

Step 2. Pick an $i \in U$ with maximal λ_i. Set $C = C \cup \{i\}$ and $U = U - \{i\}$.

Step 3. For all $j \in S_i^T \cap U$, perform Steps 4 and 5.

Step 4. Set $F = F \cup \{j\}$ and $U = U - \{j\}$.

Step 5. For all $l \in S_j \cap U$, set $\lambda_l = \lambda_l + 1$.

Step 6. For all $j \in S_i \cap U$ set $\lambda_j = \lambda_j - 1$.

Step 7. If $U = \emptyset$, stop. Otherwise, go to Step 2.

Roughly speaking, the tendency is to "build" the grid starting from one point and continuing outwards until the domain is covered. In Step 2, λ_i, which can be written as $|S_i^T \cap U| + 2|S_i^T \cap F|$, measures the value of the point $i \in U$ as a C-point given the current status of C and F, since a point that can be used for interpolation to a large number of F-points should help to satisfy (C1) with smaller C. Initially, points with many others strongly connected to them become C-points, while later the tendency is to pick as C-points those on which many F-points depend. This tends to produce a grid with a maximal number of C-points subject to the restriction that there are few, if any, strong C-C connections. In addition, this guarantees that each F-point produced has at least one strong connection to a C-point.

The algorithm for performing the final C-point choice and defining the interpolation weights is as follows:

A3. *Final C-point choice and definition of interpolation weights:*

Step 1. Set $T = \emptyset$.

Step 2. If $T \supseteq F$, stop. Otherwise, pick $i \in F - T$ and $T = T \cup \{i\}$.

Step 3. Set $C_i = S_i \cap C$, $D_i^s = S_i - C_i$, $D_i^w = N_i - S_i$ and $\tilde{C}_i = \emptyset$.

Step 4. Set $d_i = a_{ii}^h + \sum_{j \in D_i^w} a_{ij}^h$ and for $k \in C_i$, set $d_k = a_{ik}^h$.

Step 5. For each $j \in D_i^s$ do Steps 6 through 8.

Step 6. If $S_j \cap C_i \neq \emptyset$, then go to Step 8.

Step 7. If $\tilde{C}_i \neq \emptyset$, set $C = C \cup \{i\}$, $F = F - \{i\}$, and go to Step 2. Otherwise, set $\tilde{C}_i = \{j\}$, $C_i = C_i \cup \{j\}$, $D_i^s = D_i^s - \{j\}$, and go to Step 4.

Step 8. Set $d_k = d_k + a_{ij}^h a_{jk}^h / \sum_{l \in C_i} a_{jl}^h$ for $k \in C_i$.

Step 9. Set $C = C \cup \tilde{C}_i$, $F = F - \tilde{C}_i$, and $w_{ik} = -d_k/d_i$ for each $k \in C_i$, and go to Step 2.

The main idea of this algorithm is to sequentially test each F-point i in order to ensure that each point in D_i^s has a strong connection to at least one point in C_i. The set of F-points tested is denoted by T, and when all F-points have been tested, the algorithm terminates. When an F-point i is found to have a noninterpolatory strong connection $j \in D_i^s$ that does *not* strongly depend on C_i, then j is tentatively made into a C-point (it is put in the sets $\tilde{C}_i$ and C_i), and testing of the points in D_i^s (which no longer contains j) starts again. If all those points now depend strongly on C_i, then interpolation is defined for i (Step 8) and j is put into C (Step 9). However, if the algorithm finds another point in D_i^s that does not depend strongly on C_i, then the point i itself is put into C. This is done in an effort to minimize the number of additional C-points introduced. From this viewpoint, it is better to make i itself a C-point than to force more than one of its strong connections to be C-points.

In practice, this two-part coarsening process performs well. The preliminary choice is fast, and the number of C-points added in the final choice is normally small.

4.6.3. Work and storage requirements. It is important to have some idea of the work (in terms of floating point operations) and the storage necessary to solve a given problem with AMG. Since the process is fully automatic, this cannot be predicted exactly. However, the total work necessary for both setup and cycling can be estimated in terms of several basic quantities. Ideally, this work should be $O(n)$, and the constant should depend only on the nature of the problem, not the size.

The quantities used in the estimates are: κ^A, the average "stencil size" over all grids; κ^I, the average number of interpolation points per F-point, σ^Ω, the ratio of the total number of points on all grids to that on the fine grid (called the *grid complexity*); and σ^A, the ratio of the total number of nonzero entries in all the matrices to that in the fine grid matrix (called the *operator complexity*). The goal of the AMG setup phase is to keep these values as small as possible in order to ensure efficient cycling. Typical values are given in § 4.6.4 for a number of problems.

In usual geometric multigrid, these quantities are known in advance. The grids are generally coarsened by a factor of 2 in all directions, so the grid complexity σ^Ω is $1/(1-2^{-d})$, where d is the dimension of the domain. If the same type of approximation to the continuous operator is used on all grids, κ^A is equal to the stencil size of the fine grid problem, and the operator complexity σ^A is also $1/(1-2^{-d})$. For linear interpolation, we have $\kappa^I = d/(1-2^{-d})$. Thus, for example, a PDE defined on a region of the plane discretized with a 5-point stencil will give the values $\sigma^A = \sigma^\Omega = 4/3$, $\kappa^A = 5$, and $\kappa^I = 8/3$.

Numerical work in the setup phase. The floating point arithmetic needed for the setup phase represents only a small amount of the total time required, since the dominant part involves sorting, keeping counters, looping, maintaining linked lists, dynamic storage manipulation, garbage collection, etc. For this reason, we do not go into detail, but only present the final estimates for the total number of floating point operations needed to define the interpolation weights and the coarse grid operators. These are denoted by ω^I and ω^A, respectively, and are as follows[8]:

$$\omega^I = n\kappa^I(3(\kappa^A - \kappa^I) - 2), \qquad \omega^A = n\kappa^I(2\kappa^I(\kappa^A - \kappa^I) + 3\kappa^I + \kappa^A). \tag{6.5}$$

Numerical work per cycle. The approximate numbers of floating point operations on level m required for one relaxation sweep, residual transfer, and interpolation are

$$2n_m^A, \quad 2n_m^A + 2\kappa^I n_m^F, \quad n_m^C + 2\kappa^I n_m^F,$$

respectively. Here, n_m^A is the number of nonzero entries of A^m, and n_m^C and n_m^F denote the numbers of C- and F-points on Ω^m. The actual work per cycle, of course, depends on its type. The most common one is the (ν_1, ν_2) V-cycle, in which ν_1 and ν_2 are the numbers of relaxation sweeps performed before and after each coarse grid correction. Then, for example, the total amount of work per V-cycle, if we let $\nu = \nu_1 + \nu_2$ (and note that $\sum_m n_m^F \approx n$), is approximately

$$n(2(\nu + 1)\kappa^A \sigma^\Omega + 4\kappa^I + \sigma^\Omega - 1). \tag{6.6}$$

Remark 6.2. The number of operations required for relaxation dominates the work per V-cycle and is proportional to the number of nonzero entries in all the matrices. Thus, the operator complexity, σ^A, which does not occur in the above estimate, is actually a convenient approximation of the ratio of the total work per V-cycle to the relaxation work on the finest grid.

Remark 6.3. Some cycling schemes used in usual multigrid methods may not be appropriate at times for AMG, because the reduction of the problem size from one level to the next is not fast enough. For example, for a 2-D anisotropic problem, this reduction is given by a factor of around .5 (cf. Table 4.2). In this case, the total work per point in a W-cycle is known to depend on the size of the fine grid problem (cf. [538, § 4.4]).

Storage requirements. The most convenient storage scheme for AMG is that used in the Yale Sparse Matrix Package (cf. [Ei1]). This form allows the use of general matrices and does not require storage of the zero entries. Three one-dimensional arrays are used to specify a matrix: A, IA, and JA. We assume that the rows of the matrix are ordered so that the ith equation corresponds to the ith variable. The diagonal entry a_{ii} is stored in the real vector A at the position indicated by the ith entry of IA, with the nonzero off-diagonals following in any order. The rows are stored consecutively. The vector JA, which has the same length as A, contains the column number of the corresponding entries of A. The interpolation operators are stored in a

[8] This estimate for ω^I assumes that *all* connections are strong, which is the worst case. In addition, there has been no effort to take advantage of the symmetry in defining the coarse grid operator. An algorithm so designed could cut the work for the operator construction by close to a factor of 2.

similar form. The restriction operator does not need to be stored, since it is simply the transpose of interpolation. In addition to the matrices and the interpolation operators, we need storage for the solution approximation and the right-hand side, and pointers giving the correspondence between points on different grids. We need a total of $(\kappa^A + 2)\sigma^\Omega + \kappa^I$ real and $(\kappa^A + 3)\sigma^\Omega + \kappa^I + 1$ integer storage locations per point on the finest grid.[9]

Remark 6.4. The values of σ^A, σ^Ω, κ^A, and κ^I for geometric multigrid applied to a given problem can generally be regarded as "best-case" values for AMG applied to the same problem. Consequently, using the values given before in the above formulas, we can expect the solution of a PDE discretized on a uniform grid with a 5-point stencil AMG to require at least 12 real storage locations and 14 integer locations per point on the finest grid. In practice, values may be higher, but will seldom exceed about 19 real and 22 integer locations.

Remark 6.5. During the setup phase some work space is necessary. However, at any stage of the process, there is available storage, which will be needed for matrices and pointers for levels not yet determined; this usually suffices.

4.6.4. Numerical results. In this section we present results of experiments with AMG on a number of discretized partial differential equations. They demonstrate the robustness and efficiency of the method in a variety of cases, including irregular domains and discretizations, anisotropic and discontinuous coefficients and nonsymmetric operators.

In all cases, the parameter θ defining strong connections and described in § 4.6.2 was taken as 0.25. (1, 1) V-cycles with Gauss–Seidel relaxation and C/F-ordering of points were used in all tests, and the convergence factors were computed by a von Mises vector iteration. The values given in Tables 4.1–4.7 are as follows (cf. § 4.6.3).[10]

ρ	Asymptotic V-cycle convergence factor,
t_V	Time required for one V-cycle,
t_{prep}	Time required for the setup phase,
σ^A	Operator complexity,
σ^Ω	Grid complexity,
κ^A	Average stencil size,
κ^I	Average number of interpolation points per F-point.

[9] Again, there can be some savings if we take advantage of symmetry. The setup phase is most efficient when both the upper and lower off-diagonal nonzero matrix entries are stored, but it is possible to keep all nonzeros stored only for the current working level, and to discard the lower (or upper) off-diagonal elements before proceeding.

[10] All the time measurements reported in this section were obtained on an IBM 3083 computer at the Gesellschaft für Mathematik und Datenverarbeitung, St. Augustin, West Germany.

The Laplace operator. First, in order to establish the behavior of the method in the absence of any irregularities, results are given for several different finite difference discretizations of the Laplace operator on the unit square with Dirichlet boundary conditions. In all cases, a uniform square grid with mesh size $h = 1/64$ is used. The stencils used are as follows:

$$(1)\quad \frac{1}{h^2}\begin{bmatrix} & -1 & \\ -1 & 4 & -1 \\ & -1 & \end{bmatrix}, \qquad (2)\quad \frac{1}{2h^2}\begin{bmatrix} -1 & & -1 \\ & 4 & \\ -1 & & -1 \end{bmatrix},$$

$$(3)\quad \frac{1}{8h^2}\begin{bmatrix} -1 & -1 & -1 \\ -1 & 8 & -1 \\ -1 & -1 & -1 \end{bmatrix}, \qquad (4)\quad \frac{1}{20h^2}\begin{bmatrix} -1 & -4 & -1 \\ -4 & 20 & -4 \\ -1 & -4 & -1 \end{bmatrix}.$$

Table 4.1 contains the results for these problems, which are typical for isotropic operators. There are some important points to note here.

First, although the 9-point stencils require more time for relaxation on the finest grid than the 5-point stencils, the overall times per cycle are somewhat less. The reason is that the complexity values and average numbers of interpolation points are lower for the 9-point stencil. Quantitatively, this can be seen when the complexities and average values shown in Table 4.1 are inserted in (6.6) to approximate the number of floating point operations. Both Problem 1, with a 5-point stencil, and Problem 3, with a 9-point stencil, require approximately 79 operations per cycle and per point on the finest grid. It is generally true that larger stencils result in faster coarsening, lowering the complexity, since (C1) can be satisfied with a smaller percentage of the strong connections of each F-point in C. For example, for the 5-point stencil, *all* the neighbors of each F-point must be in C, while for the 9-point stencil, as few as 2 can suffice.

Remark 6.6. For comparison, a standard geometric multigrid cycle for isotropic problems, using full weighting for the residual transfer, would require approximately 48 operations per fine grid point for a (general) 5-point stencil and 80 for a 9-point stencil. Thus, in these examples, the

TABLE 4.1

		Times		Complexity		Averages	
Problem	ρ	t_V	t_{prep}	σ^A	σ^Ω	κ^A	κ^I
1	0.054	0.29	1.63	2.21	1.69	6.46	3.29
2	0.067	0.27	1.52	2.12	1.64	6.28	3.12
3	0.078	0.26	1.83	1.30	1.31	8.73	2.51
4	0.109	0.26	1.83	1.30	1.31	8.73	2.51

work per AMG cycle is comparable for 5-point stencils and for 9-point stencils it is nearly optimal.

Second, AMG works well when applied to the skewed Laplacian, while geometric multigrid methods do not. This is because the grid actually consists of two sets of points with no connections between these sets. We cannot define any relaxation process that results in smoothing of the error between the two sets, so usual geometric coarsening and linear interpolation cannot give meaningful coarse grid corrections. AMG, on the other hand, automatically treats the problem as two decoupled problems, since the coarsening process is based on interpolation along strong connections.

Remark 6.7. The average ratio of setup time to V-cycle time for the above problems is about 6.3, while that of the corresponding numbers of floating point operations (computed from (6.5), (6.6)) is around 2. This clearly shows that real arithmetic actually takes up a relatively small amount of the total setup time.

Anisotropic operators and discontinuous diffusion coefficients. Although somewhat artificial, the example of the skewed Laplace operator demonstrates the ability of AMG to tailor coarsening to the problem at hand. It is easy to see from the algorithm described in § 4.6.2 that coarsening must occur in the direction of strong connections, so that, for example, 2-D problems that are strongly anisotropic in a direction aligned with the grid are actually treated as a number of separate 1-D problems.

Here, we report on experiments with three examples of operators involving "problem" coefficients. As before, the domain for each is the unit square, Dirichlet boundary conditions are imposed, and each is discretized on a uniform grid with $h = 1/64$, using 5-point stencils. The first problem is

$$-\varepsilon u_{xx} - u_{yy} = f,$$

and several choices for ε are used. In geometric multigrid, a line relaxation must be used in order to ensure smoothing when standard coarsening is used, with the direction of the relaxation depending on the size of ε. An alternative is to coarsen only in one direction (and to use point relaxation).

The second problem, more complex than the previous one, is

$$-(100^{x+y-1} u_x)_x - u_{yy} = f.$$

Here, the direction and strength of the anisotropy varies over the domain. In order to solve this problem by usual multigrid methods, alternating direction line relaxation is necessary. In this case coarsening in one direction is not useful, since the direction would have to change over the domain. This is not a problem in AMG, since the type of coarsening in each part of the domain is automatically adapted to the direction of strong connections there. Finally, we present a problem that we cannot solve with geometric multigrid methods simply by choosing the proper relaxation scheme. This is a diffusion problem with a discontinuous diffusion coefficient, and is as follows:

$$-\nabla(d(x,y)\nabla u) = f \text{ with } d(x,y) = \begin{cases} 1 & 0.0 \le x \le 0.5 \quad 0.0 \le y \le 0.5, \\ 10 & 0.0 \le x \le 0.5 \quad 0.5 < y \le 1.0, \\ 100 & 0.5 < x \le 1.0 \quad 0.5 < y \le 1.0, \\ 1000 & 0.5 < x \le 1.0 \quad 0.0 \le y \le 0.5. \end{cases}$$

In the application of usual multigrid to such problems, error is not smoothed across the discontinuities and interpolation must be adapted at interfaces (cf. [5]).

The results for the application of AMG to these problems are given in Table 4.2. Note that the convergence factors do not vary much, and are comparable to those obtained for the Laplace operator, even when the direction of strong connections changes or the coefficients are discontinuous.

For Problem 1, there are two points to note. First, the grid complexity tends to 2 as ε becomes very small or very large. This is because coarsening is basically in one direction only for such problems, so about half the grid points are chosen as C-points. On coarser grids, the strength of connections in all directions becomes more uniform, and faster coarsening, such as that obtained for the 9-point stencils, results, thereby decreasing the complexity. For problems with very small or large ε, however, this only occurs on very coarse grids, if at all, and the grid complexity stays around 2.

Second, the operator complexity, and thus the time per cycle, becomes quite high for $\varepsilon = .1$ or 10. This is mainly due to the coarsening obtained

TABLE 4.2

ε	Differential operator	ρ	Times		Complexity		Averages	
			t_V	t_{prep}	σ^A	σ^Ω	κ^A	κ^I
0.001	$-\varepsilon u_{xx} - u_{yy}$	0.082	0.31	1.05	2.54	1.92	6.55	1.84
0.01		0.094	0.32	1.20	2.72	1.93	6.96	2.03
0.1		0.063	0.37	1.82	3.33	1.87	8.80	2.64
0.5		0.071	0.28	1.62	2.19	1.68	6.44	3.27
1		0.054	0.29	1.63	2.21	1.69	6.46	3.29
2		0.059	0.28	1.61	2.19	1.68	6.44	3.28
10		0.079	0.36	1.84	3.37	1.87	8.90	2.62
100		0.095	0.31	1.18	2.68	1.92	6.90	2.00
1000		0.083	0.30	1.07	2.54	1.92	6.55	1.84
	$-(100^{x+y-1}u_x)_x - u_{yy}$	0.089	0.30	1.68	2.35	1.72	6.76	3.31
	$-\nabla(d(x,y)\nabla u)$	0.082	0.30	1.45	2.45	1.79	6.75	2.72

from grid 2 to grid 3, and is a result of the choice of θ in the definition of strong connections. For this problem, the complexity can be reduced to a more reasonable level by a different choice of the parameter θ.

The solution of the above problems requires an average of 89 operations per fine grid point per V-cycle. For comparison, a robust geometric multigrid algorithm capable of solving these problems without changes (with the exception of Problem 3), using a V-cycle, standard coarsening, full weighting, and two alternating line relaxation sweeps per level (assuming that the decompositions have been previously computed and stored), requires 70 operations per fine grid point per cycle.

Problem grids and domains. Irregular domains and grids can cause problems for usual geometric multigrid methods in several ways. The domain may contain features that cannot be resolved easily on coarse grids, so that the operators defined cannot provide good coarse grid corrections. Irregular grids are used for a variety of reasons, often in combination with finite elements. In such cases, it may not be possible to find a natural sequence of coarser grids, and it may be inconvenient (or impossible) to define a finest grid that does admit a natural coarsening. Here we give results for finite element discretizations of the Laplace operator on several such domains. These are shown in Figs. 4.1–4.4. (Some of the figures show grids coarser than the ones actually used so that the grid details are not obscured.) Each of these problems presents special difficulties for usual multigrid methods.

Problem 1. Figure 4.1 shows a regular discretization of a complex domain. Small holes in the region with a size equal to the mesh size, and

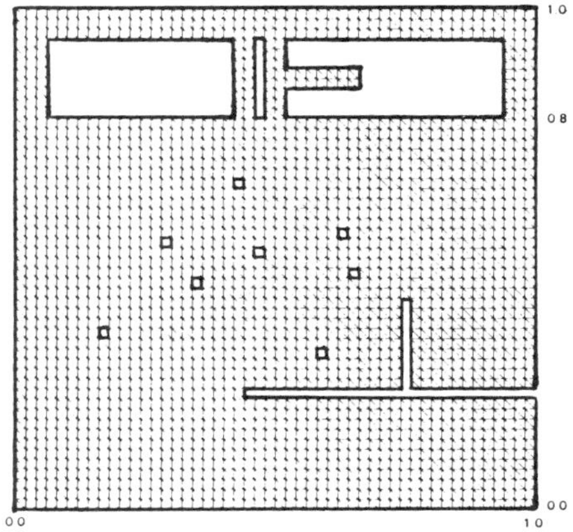

FIG. 4.1

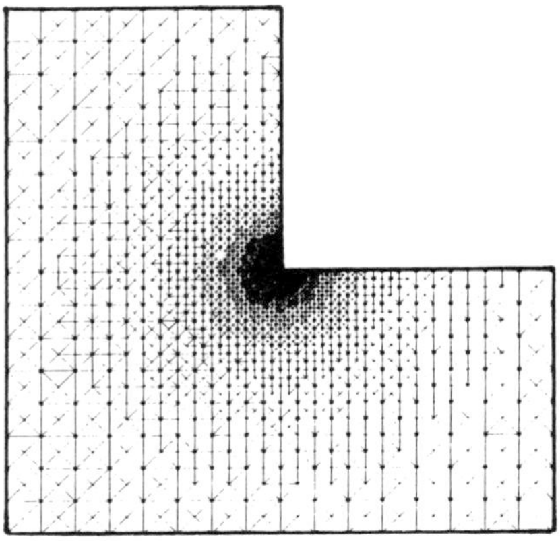

FIG. 4.2

detailed structures, make usual coarsening impossible. The re-entrant corners also present singularities.

Problem 2. The domain and grid in Fig. 4.2 cause several problems. First, there is a singularity due to the re-entrant corner. Also the grid near the corner is locally refined, resulting in irregular stencils at a number of points. The difference in mesh sizes over the region can cause problems for other iterative methods. In usual multigrid methods, the approach to local refinement is to define a sequence of successively finer uniform grids, each covering a smaller area. With the grid given, though, choosing the coarser grids will be difficult.

Problem 3. The triangle in Fig. 4.3 is discretized on a "random" grid, which would cause many problems in the geometric approach. This example actually does have "natural" coarser grids, but determining these grids can be impractical. AMG, however, uses no knowledge of the grid structure, so that finding a natural coarsening, if one even exists, is not necessary.

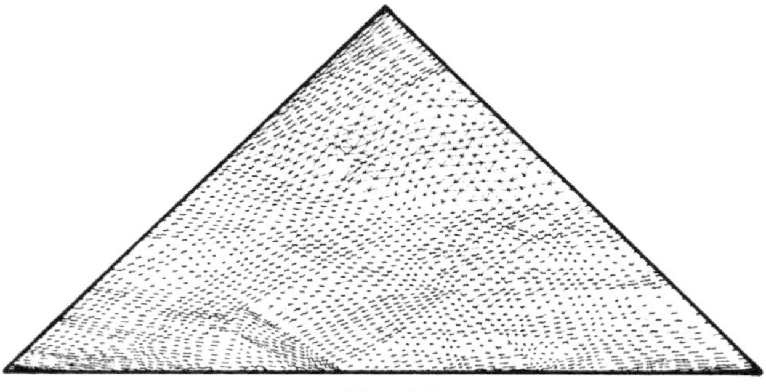

FIG. 4.3

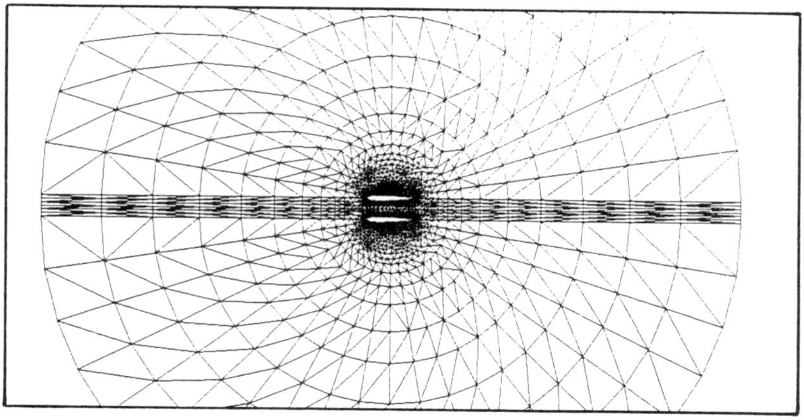

FIG. 4.4(a)

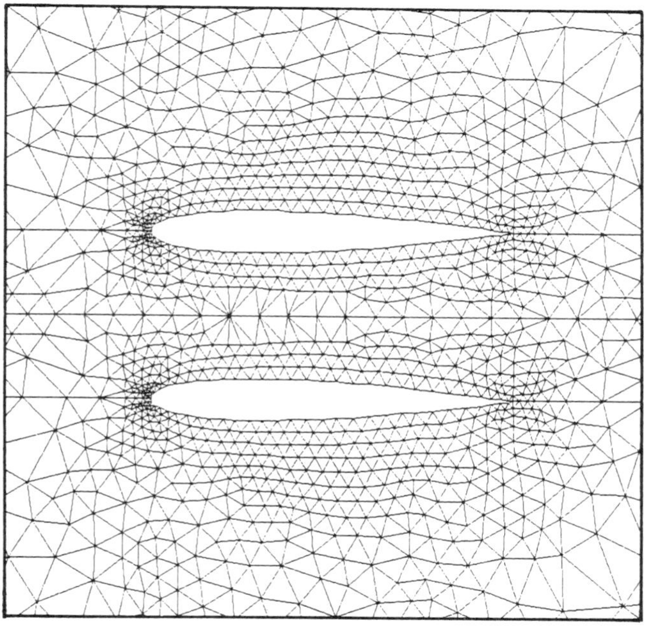

FIG. 4.4(b)

Problems 4 and 5. Finally, Figs. 4.4(a) and (b) show a grid generated at INRIA to model flow around two NACA012 airfoils (representing a cross section of an air intake). The outer boundary is a circle, and the figures show the details of the grid around the airfoils. This example incorporates many of the difficulties of the previous problems. The grid is refined close to the airfoils, the trailing edges cause singularities, and there is no natural coarsening. For Problem 5, another similar domain not shown here, is used to model flow around a single NACA012 airfoil. Neumann boundary conditions are imposed on the surface of the airfoils, and Dirichlet conditions are used at the outer boundary (to specify the flow at "infinity").

The results of the application of AMG to each of these problems is given in Table 4.3. The number of points on the finest grid is given by n. The increase of the convergence factors over those for the finite difference discretizations is small, in spite of the difficulties in the problems. It should be noted that the matrices obtained for Problems 3–5 are not M-matrices (off-diagonals of both signs result), so the assumptions made to motivate the coarsening process do not have to hold strictly.

Nonsymmetric operators. Although AMG was developed primarily for symmetric problems, some nonsymmetric problems can be handled as well with no modifications to the algorithm. We show this here using the convection-diffusion operator

$$-\varepsilon \Delta u + a(x, y) u_x + b(x, y) u_y = f(x, y).$$

Again, the domain is the unit square and Dirichlet boundary conditions are used. The problem is discretized on a uniform grid with $h = 1/64$. The discrete operator is the 5-point finite difference approximation of the form

$$\frac{1}{h^2}\begin{bmatrix} & -\varepsilon + bh\mu_y & \\ -\varepsilon + ah(\mu_x - 1) & -\Sigma & -\varepsilon + ah\mu_x \\ & -\varepsilon + bh(\mu_y - 1) & \end{bmatrix},$$

where

$$\mu_x = \begin{cases} \varepsilon/2ah & \text{if } ah > \varepsilon, \\ 1 + \varepsilon/2ah & \text{if } ah < -\varepsilon, \\ \frac{1}{2} & \text{if } |ah| \le \varepsilon, \end{cases} \quad \text{and} \quad \mu_y = \begin{cases} \varepsilon/2bh & \text{if } bh > \varepsilon, \\ 1 + \varepsilon/2bh & \text{if } bh < -\varepsilon, \\ \frac{1}{2} & \text{if } |bh| \le \varepsilon. \end{cases}$$

TABLE 4.3

Problem			Times		Complexity	
	n	ρ	t_V	t_{prep}	σ^A	σ^Ω
1	4064	0.089	0.29	2.22	1.93	1.71
2	2232	0.124	0.17	1.10	2.48	1.68
3	1953	0.153	0.20	1.26	2.84	1.87
4	1650	0.229	0.16	0.83	2.54	1.73
5	960	0.153	0.09	0.42	2.50	1.82

TABLE 4.4

Problem				Times		Complexity	
		ε	ρ	t_V	t_{prep}	σ^A	σ^Ω
1	$l=0$	10^{-5}	0.00005	0.29	0.78	2.30	1.97
	$l=1$	10^{-5}	0.005	0.45	1.67	4.45	2.19
	$l=2$	10^{-5}	0.0004	0.46	2.01	4.63	2.07
	$l=3$	10^{-5}	0.002	0.41	1.57	3.81	2.05
	$l=4$	10^{-5}	0.00005	0.29	0.78	2.30	1.97
2	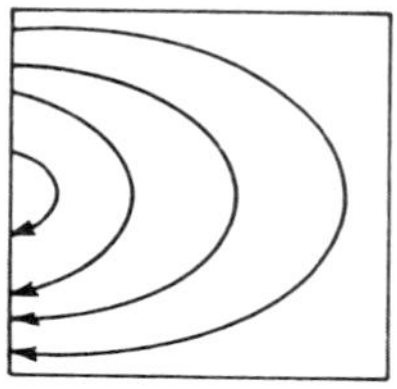	0.1	0.060	0.29	1.62	2.21	1.69
		0.001	0.055	0.40	1.49	3.68	2.10
		0.00001	0.030	0.39	1.41	3.57	2.10
3	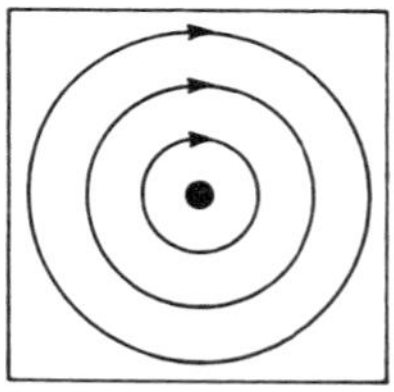	0.1	0.056	0.29	1.63	2.21	1.68
		0.001	0.160	0.40	1.51	3.76	2.10
		0.00001	0.173	0.40	1.46	3.72	2.13

Σ denotes the sum of the surrounding coefficients. Simple central differencing cannot be used when ah or bh become large compared to ε, since the operator is no longer stable.

Table 4.4 contains the convergence factors for AMG applied to several different choices of a, b, ε. The functions a and b used are as follows:

$$(1) \qquad a(x, y) \equiv \sin l\frac{\pi}{8}, \quad b(x, y) \equiv \cos l\frac{\pi}{8}, \quad (l = 0, 1, 2, 3, 4),$$

$$(2) \qquad a(x, y) = (2y-1)(1-x^2), \qquad b(x, y) = 2xy(y-1),$$

$$(3) \qquad a(x, y) = 4x(x-1)(1-2y), \qquad b(x, y) = -4y(y-1)(1-2x).$$

In Problem 1, $\varepsilon = 10^{-5}$ was used. The characteristic directions given in Problems 2 and 3 are shown in Table 4.4, and the values of ε used are .1, .001, and .00001.

For Problem 1, where the characteristic directions are straight lines, convergence is extremely fast. This is especially true when the characteristics are aligned with the grid lines. In this case, if ε were 0, AMG would act as a direct solver, giving the exact solution in one cycle. For small ε, the convergence factors actually decrease to zero with cycling, in contrast to the behavior for symmetric problems, where the factors generally increase to some asymptotic value. For this reason, the convergence factors given are geometric averages over a number of cycles.

For Problems 2 and 3, convergence factors are similar to those obtained for the symmetric problems studied when ε is large. As ε is decreased, Problem 2 behaves in a manner similar to Problem 1, with decreasing convergence factors. However, the factors increase in Problem 3. This may be due to the fact that as $\varepsilon \to 0$ the problem becomes more and more singular. The continuous problem for $\varepsilon = 0$ is not well defined, since the null space includes functions with arbitrary constant values on any characteristic circle that does not intersect the boundary.

Remark 6.8. The behavior of AMG as a direct solver for some problems with $\varepsilon = 0$ is a consequence of the coarsening. Since strong connections are in the upstream direction, for any point i, there are generally few, if any, strong connections between points in S_i. This results in choosing $C_i = S_i$ for all $i \in F$. The C/F-relaxation then produces error that lies exactly in the range of interpolation, and the coarse grid correction is exact. Thus, the reason for convergence is purely algebraic, and not a consequence of smoothing in the usual sense. When the characteristics change direction over the region, as in Problems 2 and 3, the directionality of the strong connections on coarser grids is lost, and AMG can no longer function as a direct solver.

Remark 6.9. In all cases, except in Problem 1 for $l = 0$ and $l = 4$, when ε is small, complexity is higher than for symmetric operators. This is because the algorithm produces coarse grid lines that are roughly perpendicular to the characteristic directions, and the number of grid points is reduced by around a factor of 2. Unless the flow is actually aligned with the grid lines, however, the strong connections of each point tend to "fan out" so that each point is strongly connected to an increasing number of points on the upstream grid line.

4.6.5. A posteriori error estimates. In this section we briefly recall a possibility of obtaining realistic a posteriori error bounds if A in (2.1) is an M-matrix. The idea behind the procedure described below is well known and relies on monotonicity principles (cf. [Sc1]). We first state the following lemma.

LEMMA 6.10. *For every M-matrix A,*

(i) $Au \geq 0 \Rightarrow u \geq 0$, *and*

(ii) $Au > 0 \Rightarrow u > 0$.

Let u now be any approximate solution of $A\mathbf{u} = b$ with corresponding error $e = \mathbf{u} - u$ and residual $r = b - Au = Ae$. Furthermore, suppose any z with $Az > 0$ to be given. Then an immediate consequence of the above lemma is that

$$\alpha Az \leq r \leq \beta Az \Rightarrow \alpha z \leq e \leq \beta z.$$

Thus, to obtain lower and upper bounds for the error e, we only have to find a vector z with $Az > 0$ and compute α and β. This can be done as sketched in the following procedure.

(1) Perform one AMG cycle to get an approximate solution z of the linear system $A\mathbf{u} = 1$ starting with the first approximation $z^{(0)} = 1$, say.

(2) Check whether $Az > 0$. If this is the case, go to (3). Otherwise improve z by performing another cycle and repeat the check. Usually one or two cycles will be enough to obtain $Az > 0$.

(3) The error estimate $\alpha z \leq e \leq \beta z$ holds with $\alpha = \min_i \{r_i/(Az)_i\}$ and $\beta = \max_i \{r_i/(Az)_i\}$.

The above procedure costs only little extra work and usually will give quite realistic error estimates.

4.6.6. Variations of the coarsening strategy. The coarsening procedure presented in § 4.6.2 gives satisfactory results for a wide variety of problems. However, for some cases we have not yet considered here, coarsening may not be satisfactory, either because the coarse grids are too fine, resulting in high complexity and increased work in cycling; or because the error after relaxation may not be approximated well enough by the range of interpolation, which results in slow convergence per cycle. For example, the application of AMG to finite difference equations in 3-D often results in high operator complexity. Also, some of the problems of § 4.7 require a more careful determination of the interpolation weights in order to maintain efficiency.

In the following, we discuss several modifications to the coarsening process given in § 4.6.2 which can be used to improve efficiency in such situations. These can be split into two groups: modifications in the coarse grid choice, and changes in the definition of the interpolation operator. Often, these are used in conjunction in an effort to get both faster coarsening and improved interpolation accuracy.

Although the trend in AMG has been to set up the necessary multigrid components with a minimal amount of user-supplied information, many problems do have an underlying geometric structure (although it may be quite irregular), and this information can often be used to improve performance. In particular, the coordinates of the points at which the variables are defined are often readily available, since they are frequently used in the generation of the problem itself. The use of this geometric information, along with that contained in the matrix, provides the basis for some of the approaches given below.

4.6.6.1. Modifications to the C-point choice.

Long-range interpolation. Long-range interpolation is simply the use, for interpolation to an F-point i, of points not necessarily in S_i, the set of points to which i is *directly* strongly connected. The primary reason for considering this generalization is to reduce the operator complexity. It is convenient from a programming standpoint to require that $C_i \subset S_i$, but this can be too restrictive in terms of coarsening. This is particularly true in the case of finite difference equations in 3-D. For example, the use of 7-point stencils allows only coarsening by a factor of 2, but the stencil size grows to about 19 points on the second grid. Coarsening becomes more efficient afterwards, but the overall operator complexity is quite high. In such problems it is possible to coarsen faster while not seriously hurting the quality of the correction.

Clearly, in enlarging the set of possible interpolation points, we cannot disregard the concept of strong connections without affecting the ability of interpolation to approximate algebraically smooth functions. For this reason, we introduce the idea of *long-range strong connections*. We say that i is strongly connected to j along a path of length l if there exists a sequence of points $i_0, i_1, i_2, \cdots, i_l$ with $i = i_0$ and $j = i_l$ such that $i_k \in S_{i_{k-1}}$ for $k = 1, 2, \cdots, l$. Let S_i^l denote the set of all points to which i is strongly connected along a path of length less than or equal to l.

The algorithm can be generalized to allow interpolation from points in S_i^l for some specified l. Generally, not all points in S_i^l are suitable for interpolation. We can see this easily by considering simple geometric cases: in geometric multigrid methods, if the mesh size is increased by more than a factor of 2 from one grid to the next, the work required for relaxation increases, and overall efficiency decreases even though less coarse grid work is required. For this reason, we may take only the "strongest" connections in S_i^l, where the strength of the connection from i to some point $j \in S_i^l$ is measured, for example, in terms of the number of paths of length less than or equal to l from i to j.

Once this restricted set of points, denoted by $\tilde{S}_i^l$, is determined, the algorithm can proceed as in § 4.6.2, with some minor changes, if we choose C and F so that, for each $i \in F$, each point $j \in S_i$ is either in C or is *directly* strongly connected to some point in $C \cap \tilde{S}_i^l$, which we take as C_i. In this case, interpolation can be defined as before. If this is not the case, one of the methods discussed in § 4.6.6.2 may be used.

Convex interpolation. As stated in § 4.6.1, in problems with an underlying geometric structure, the coarsening process tends to produce grids in which each F-point is "surrounded" by its interpolation points. This situation, where each F-point lies (at least approximately) in the convex hull of its interpolation points, is referred to as *convex interpolation*. Such an arrangement allows for $O(h^2)$ interpolation and is one of the main reasons for the efficiency of AMG. The coarsening algorithm of § 4.6.2, however, does not guarantee convex interpolation where strong connections are naturally one-sided, such as along boundaries. This does not essentially affect the behavior of AMG for the problems we have considered so far, but it must be corrected for some of the applications given in § 4.7. In those cases, it is sometimes useful to choose C and F so that we have convex interpolation even at the boundaries (by explicitly using the coordinates of the points at which the variables are defined). Of course, interpolation must still be from strong connections (either direct or long-range); otherwise the correction is meaningless, and convergence will deteriorate.

4.6.6.2. Modifications to the interpolation formula. The accuracy of interpolation can be only indirectly controlled in AMG. When interpolation is defined as in § 4.6.1, convex interpolation usually ensures good results. However, for some of the problems considered in § 4.7, for example, even convex interpolation is not sufficient for good convergence. This generally occurs when the algebraically smoothest components (i.e., those least affected by relaxation) are not well enough approximated by the range of interpolation. These components can be determined in several ways, either through use of the matrix itself or through more extensive use of geometrical information. Once these components are known, or approximated, it is possible to improve interpolation:

Suppose we have chosen C, F, and C_i for each $i \in F$ given, and we have q functions $z^1, z^2, \cdots, z^q$ which we want to interpolate exactly, if possible. (Note that here, superscripts indicate the function number, *not* the grid.) We restrict our discussion to a particular point $i \in F$. Let $C_i = \{k_1, k_2, \cdots, k_p\}$, and denote the value of the function z^μ at point k_ν by z^μ_ν, and the value of z^μ at the point i by z^μ_0. Assume that we have previously computed some interpolation weights $w_{ik_1}, w_{ik_2}, \cdots, w_{ik_p}$, which we would like to modify so that $z^1, z^2, \cdots, z^q$ can be approximated as closely as possible by the range of interpolation. For this, corrections $v_1, v_2, \cdots, v_p$ are computed that minimize the functional

$$\sum_{\mu=1}^{q} \left(z^\mu_0 - \sum_{\nu=1}^{p} (w_{ik_\nu} + v_\nu) z^\mu_\nu \right)^2,$$

with the added constraint that $\sum_\nu v_\nu^2$ is minimal in the event that there is more than one solution. The weights are then updated by $w_{ik_\nu} \leftarrow w_{ik_\nu} + v_\nu$ for $\nu = 1, 2, \cdots, p$. This ensures that, if the original interpolation weights already interpolate the desired functions, they are not changed. Otherwise, the new weights will differ from the old ones in a minimal way. Below, we discuss two ways to obtain functions of interest that may be used in this approach.

"Linear" interpolation. For many physically based problems, functions that are algebraically smooth are also smooth in the usual geometric sense. This is not always the case. For example, discontinuities in the system or structure being analyzed can result in a problem where algebraically smooth functions vary greatly from one point to another, even if the points are physically close to each other. However, connections between such points, as defined by the matrix entries, will be weak. Thus, if the coarse grids are chosen as described before, we can expect the smoothest functions in the neighborhood of a point i to be linear on C_i. So, for a 2-D problem, for example, we can improve the interpolation using the above method by taking $z^1 = 1$, $z^2 = x$, and $z^3 = y$. When we define the interpolation weights in this way, it is natural to enforce convex interpolation.

Note that this use of linear interpolation does not introduce the same kind of disadvantages as linear interpolation in geometric multigrid methods. There, the assumption that smooth functions are locally linear, even in the direction of weak connections, requires a careful choice of the relaxation method in order to smooth, since the interpolation points are fixed independently of the operator.

Interpolation of eigenvectors. The algebraically smoothest functions of a problem are generally the eigenvectors corresponding to the smallest eigenvalues. The smaller the eigenvalue, the better the corresponding eigenvector must be approximated by the range of interpolation. The interpolation definition given in § 4.6.1 is usually sufficient for this purpose. However, in some problems (cf. § 4.5.4.3), this must be specifically enforced in order to maintain satisfactory convergence. This approach requires the approximate calculation of the eigenvectors corresponding to the smallest eigenvalues (generally there are very few small enough to cause problems in convergence). Such accuracy is rarely necessary, since the linear interpolation described above usually ensures accurate enough interpolation of the needed eigenvectors when standard interpolation does not. Therefore we do not report on our experiences here, and give only a rough outline of the method for completeness.

One (inefficient) method for finding the eigenvectors is to first construct all the coarse grid components in the usual way, then use a modified cycle to obtain an approximation to the desired eigenvector. The Gauss–Seidel relaxation used in AMG is actually a functional minimization of the energy norm of the error. Instead, a relaxation process can be formulated to minimize the Rayleigh quotient of the grid function (this requires carrying some additional corrections to all levels). An AMG cycle using this relaxation produces an approximation to the eigenvector that corresponds to the smallest eigenvalue. This eigenvector approximation can be used to modify interpolation and the coarse grid matrices on all levels. The cycling and interpolation improvement can be repeated until the desired accuracy is reached.

The computation of the eigenvector and updating of the interpolation can also be integrated so that, before the coarse grid correction on each level during the Rayleigh quotient minimization cycle, the interpolation and the coarser grid operator are updated with the current eigenvector approximation *before* the operation proceeds to the next coarser grid. This can be done in a quite efficient way and, if necessary, several eigenvector approximations can be computed simultaneously in a similar way if they are kept (approximately) orthogonal on all levels. The eigenvector approximations are used only as a tool for the improvement of interpolation. Once they have been calculated to the desired accuracy (thereby producing the desired interpolation), the AMG components can be used in standard cycling to solve the given problem.

4.7. AMG for systems problems. The algorithm described in the previous section was developed under the assumption that the fine grid matrix was a symmetric M-matrix. This assumption is often strongly violated when a number of different quantities are being approximated. For this reason, that method described in § 4.6 can fail when applied directly to discretized systems of partial differential equations and other similar matrix problems.

This class of problems is simply too large to allow any one algebraic multigrid approach to work efficiently for all cases, but methods can be developed that apply to limited classes. For these approaches, the user must be required to supply some information concerning the nature of the problem. Here, for ease of motivation, we concentrate on discretized systems of PDEs, and assume that some of the details of the discretization can be easily supplied. However, as in AMG for scalar problems, uniform discretizations and regular domains are not required, so such approaches should still prove quite useful, even though much of the "black box" nature is lost.

Consider a system of partial differential equations in p unknown functions:

$$\begin{aligned} \mathcal{L}_{[11]}u_{[1]} + \mathcal{L}_{[12]}u_{[2]} + \cdots + \mathcal{L}_{[1p]}u_{[p]} &= f_{[1]}, \\ \mathcal{L}_{[21]}u_{[1]} + \mathcal{L}_{[22]}u_{[2]} + \cdots + \mathcal{L}_{[2p]}u_{[p]} &= f_{[2]} \\ &\vdots \\ \mathcal{L}_{[p1]}u_{[1]} + \mathcal{L}_{[p2]}u_{[2]} + \cdots + \mathcal{L}_{[pp]}u_{[p]} &= f_{[p]} \end{aligned} \quad \text{on } \Omega, \tag{7.1}$$

where $\mathcal{L}_{[ij]}$ is a linear differential operator and $u_{[i]}$ is a scalar function defined on Ω. We consider two approaches, obtained from (7.1), to discrete problems; both are closely related to AMG for scalar problems. The first, which is similar to the simplest usual multigrid approach, is to define interpolation separately for each of the unknown functions, and is called the "unknown" approach. The second, which can be applied when all unknown functions of the discrete problem are defined on the same set of points (i.e., on *a nonstaggered* grid), treats the problem in a point-oriented block fashion and is therefore called the "point" approach.

4.7.1. The "unknown" approach. A simple extension of the scalar algorithm allows the method to be used for a number of problems in which more than one unknown is involved. We consider the system (7.1). To discretize this problem, it is necessary to define p grids $\Omega_{[1]}, \Omega_{[2]}, \cdots, \Omega_{[p]}$, each a discretization of Ω. Let $u_{[i]}$, defined on $\Omega_{[i]}$, be the discrete approximation to the ith unknown function. Then, let $A_{[ij]}$ be a finite difference or finite element approximation to $\mathscr{L}_{[ij]}$. The discrete problem can be written as follows:

$$\begin{bmatrix} A_{[11]} & A_{[12]} & \cdots & A_{[1p]} \\ A_{[21]} & A_{[22]} & \cdots & A_{[2p]} \\ \vdots & \vdots & & \vdots \\ A_{[p1]} & A_{[p2]} & \cdots & A_{[pp]} \end{bmatrix} \begin{bmatrix} \mathbf{u}_{[1]} \\ \mathbf{u}_{[2]} \\ \vdots \\ \mathbf{u}_{[p]} \end{bmatrix} = \begin{bmatrix} b_{[1]} \\ b_{[2]} \\ \vdots \\ b_{[p]} \end{bmatrix}. \tag{7.2}$$

In order to avoid confusion in parts of our discussion, it is important to distinguish between what we call *points, unknowns,* and *variables.* An *unknown* is one of the functions being approximated (e.g., pressure, temperature, or a component of displacement), and each unknown is identified by a number. A *point* is a physical point of one of the grids $\Omega_{[1]}, \Omega_{[2]}, \cdots \Omega_{[p]}$. The value representing the unknown $u_{[i]}$ at a particular point of $\Omega_{[i]}$ is called a *variable.* Thus each variable is associated with both a point and an unknown. Here, the user must explicitly provide the correspondence between the variables and the unknowns.

The method we call the "unknown" approach can be simply described in terms of the scalar algorithm of § 4.6. In choosing the coarse grids and defining interpolation, the method of § 4.6.2 is used, but all off-diagonal blocks are ignored. This results in a block-diagonal interpolation operator, with each F-variable interpolating only from other variables corresponding to the same unknown. The coarse grid operator is constructed as before, using the *full* fine grid operator, so that connections between the different functions are not lost on coarser grids. In the cycling process, simple Gauss–Seidel is used, and the variable/unknown correspondence is no longer needed.

For this approach to be formally applicable, there are several restrictions on system (7.1) and the discretization chosen. The ith equation in (7.1) must be naturally associated with the ith unknown function, $u_{[i]}$. (This is not the case, for example, with the Cauchy–Riemann equations.) The problem should be discretized so that there is a natural one-to-one correspondence between variables and equations, since this is required for Gauss–Seidel relaxation. Finally, the resulting discrete problem must have diagonal blocks $A_{[ii]}$ to which scalar AMG can be applied.

This method works well in problems for which variable-wise relaxation produces algebraically smooth error separately in each unknown (i.e., if $e_{[i]}$ is the error in the ith unknown, $A_{[ii]}e_{[i]}$, properly scaled, is small compared to $e_{[i]}$). The character of the off-diagonal blocks determines whether or not there is such smoothing: it is sufficient that the off-diagonal blocks in each row are "small" compared to the corresponding diagonal block. However, the entries in the off-diagonal blocks do not necessarily have to be weak connections in the usual sense. For example, consider the case of the plane-stress problem of elasticity given in (7.3) below. When Poisson's ratio ν is not too large (say $\nu \leq .5$), the present approach works well even though the connections between unknowns in the discrete operator are relatively large.

When this method can be used, it is quite effective. Below, we give some results of the application of this approach to elasticity problems and a discrete problem in VLSI design.

Elasticity problems. The plane-stress elasticity problem can be written in terms of the displacements $\mathcal{u}$ and $\mathcal{v}$ as

$$\begin{aligned} &\mathcal{u}_{xx} + \frac{1-\nu}{2}\mathcal{u}_{yy} + \frac{1+\nu}{2}\mathcal{v}_{xy} = \mathcal{f}, \\ &\frac{1+\nu}{2}\mathcal{u}_{xy} + \frac{1-\nu}{2}\mathcal{v}_{xx} + \mathcal{v}_{yy} = \mathcal{g}, \end{aligned} \tag{7.3}$$

where ν is Poisson's ratio, chosen to be $\frac{1}{3}$ (which is a realistic value for many applications). The domain used is the unit square. A finite element discretization of the problem on a uniform grid with bilinear test functions is used, with $h = 1/16$, $1/32$, and $1/48$.

We first applied the "unknown" approach without any of the modifications described in § 4.6.6. Each side of the domain was specified as either fixed (Dirichlet boundary condition) or free. Table 4.5 shows the V-cycle convergence factors and CPU times per cycle obtained as a function of h

TABLE 4.5

	$l=0$		$l=1$		$l=2$		$l=3$	
h	ρ	t_V	ρ	t_V	ρ	t_V	ρ	t_V
1/16	0.13	0.07	0.20	0.08	0.36	0.09	0.46	0.10
1/32	0.18	0.34	0.30	0.36	0.52	0.38	0.65	0.39
1/48	0.18	0.82	0.31	0.84	0.49	0.87	0.75	0.88

and l, the number of free sides, where we take $l=0$, 1, 2, and 3 (for $l=2$ the free sides are adjacent).

For the full Dirichlet problem, convergence is good for all h tested. However, the behavior becomes worse as free boundaries are introduced, with the convergence clearly dependent on problem size for the case of 3 free boundaries. This indicates that the quality of interpolation is not good along the free boundaries. We then used two methods to improve the interpolation. These are as follows:

Method 1. Convex interpolation, as described in § 4.6.6.1, was forced, so that each F-point was in the convex hull of its set of interpolation points.

Method 2. In addition to enforcing convex interpolation, the interpolation weights were modified to interpolate linear functions exactly according to § 4.6.6.2.

The results for these approaches on the problem with 3 free boundaries are given in Table 4.6. Although there may still be a slight dependence on problem size when Method 1 is used, the convergence factors obtained are much better than those for the original method. With Method 2, the factors approach those for the original method on the full Dirichlet problem, and do not depend on h.

Remark 7.1. When convex interpolation is enforced, the times per cycle increase. This is because the use of convex interpolation generally increases the number of C-points somewhat. Although no additional C-points are forced in order to define the linear interpolation, the coarser grid operators produced are different from those when only convex interpolation is used, which explains the difference in cycle times between Method 1 and Method 2.

We also performed some preliminary experiments using a 3-D elasticity problem. The domain was the unit cube. We took $h=1/8$. Although the problem sounds small, there are 2187 variables and around 150,000 matrix entries stored. The 4 bottom corners are fixed, and the body is free elsewhere. In this case, Method 2 from above is used. Furthermore, long-range interpolation is used with path length 2 in order to lower the storage requirements. The results were quite promising. The asymptotic

TABLE 4.6

	Method 1		Method 2	
h	ρ	t_V	ρ	t_V
1/16	0.25	0.15	0.21	0.15
1/32	0.30	0.64	0.22	0.69
1/48	0.33	1.46	0.22	1.82

convergence factor obtained was .276, with an operator complexity of 1.50 for a (1, 1) V-cycle.

A VLSI design problem. The following problem occurs as part of a CAD-system for VLSI-design that is under development at the University of Saarbrücken (see [Be1]). As a first step in developing the physical layout of a chip, the chip components are regarded as nodes of a planar graph, some connected not only to certain neighbors but also to some geometrically fixed points at the boundary of the chip (*graph with fixed boundary*). Starting from this graph, which describes the logical connections of the chip components, we compute a first guess for a layout by computing geometrical locations of the nodes, minimizing the functional

$$\sum_l w_l f(\|s_l\|),$$

where s_l denote the edges of the graph (including those to the fixed boundary points), w_l are certain nonnegative weights, and $f(\xi)$ is some "cost" function.

In the simplest case, one considers $w_l = 1$ and $f(\xi) = \xi^p$ with some $p \geq 2$, the most interesting cases being $p = 2$ (which corresponds to requiring that the overall signal transmission time be minimum) and $p \to \infty$ (which corresponds to requiring the maximum signal transmission time to be minimum). The above minimization corresponds to solving the following system of equations:

$$\sum_{j \in N_i} \|\mathbf{P}_i - \mathbf{P}_j\|^{p-2}(\mathbf{x}_i - \mathbf{x}_j) = 0 \qquad (i = 1, \cdots, n),$$

$$\sum_{j \in N_i} \|\mathbf{P}_i - \mathbf{P}_j\|^{p-2}(\mathbf{y}_i - \mathbf{y}_j) = 0 \qquad (i = 1, \cdots, n),$$

where the $\mathbf{P}_i = (\mathbf{x}_i, \mathbf{y}_i)$ $(i = 1, \cdots, n)$ denote the nodes of the graph (including the fixed boundary nodes). N_i denotes the neighborhood of i, that is, the set of indices j such that $\mathbf{P}_i$ and $\mathbf{P}_j$ are connected by an edge of the graph. The unknowns are the coordinates of the *inner* nodes.

If $p = 2$, this system is linear and similar to two decoupled discrete Laplace equations, and we can apply the algorithm described in § 4.6 without any modification. For $p > 2$, the system is nonlinear, and the Jacobian matrix is symmetric and has the form

$$\begin{pmatrix} M_x & B \\ B & M_y \end{pmatrix},$$

with M_x and M_y being weakly diagonally dominant M-matrices. If p is not too large ($p \leq 10$, say), the "unknown" approach allows for a rapid solution if combined with Newton's method in a straightforward way. We found that, in order to keep the linearization work small, it was reasonable to

TABLE 4.7

Number of nodes	Convergence factor			
	$p=2$	$p=4$	$p=6$	$p=8$
250	0.16	0.15	0.17	0.20
500	0.17	0.15	0.18	0.22
1000	0.19	0.17	0.20	0.29
2000	0.21	0.20	0.23	0.32
4000	0.23	0.19	0.23	0.32
8000	0.24	0.19	0.24	0.33

make a simple "continuation" in p. Furthermore, instead of recomputing all coarse grid components in each Newton step, we can reduce the setup work of AMG by up to 50% (without affecting convergence essentially) by recomputing only the coarse grid operators, keeping the grids and the interpolation operators fixed.

In Table 4.7 we give numerically observed convergence factors per cycle, for different p, that typically are around 0.2, with a tendency to increase slightly as p becomes large. We mention that $p > 10$ is usually not needed in practice; solutions then change very little as p is increased further. Figures 4.5 and 4.6 show two explicit examples for demonstration, where the figures on the left and right show the given and optimized graphs, respectively (where $p=4$ for Fig. 4.5 and $p=8$ for Fig. 4.6).

4.7.2. The "point" approach. In many cases, systems of the form (7.1) are discretized on *nonstaggered grids* (i.e., $\Omega_{[1]} = \Omega_{[2]} = \cdots = \Omega_{[p]}$). The scalar algorithm of § 4.6 can, in principle, be extended to such problems in a straightforward way. The main idea is to apply the algorithm in a "block" manner, with all variables that correspond to the same point relaxed, interpolated, and coarsened together. However, a practical application can be quite expensive and difficult. We have not developed an efficient implementation of the ideas described below, but a preliminary test, presented below, shows that the method can be effective.

Clearly, the correspondence between variables and points must be provided before we can use this approach. We can then rewrite (7.2) so that all variables and equations associated with the same point are grouped together:

$$\begin{bmatrix} \underset{\sim}{A}_{(11)} & \underset{\sim}{A}_{(12)} & \cdots & \underset{\sim}{A}_{(1n)} \\ \underset{\sim}{A}_{(21)} & \underset{\sim}{A}_{(22)} & \cdots & \underset{\sim}{A}_{(2n)} \\ \vdots & \vdots & & \vdots \\ \underset{\sim}{A}_{(n1)} & \underset{\sim}{A}_{(n2)} & \cdots & \underset{\sim}{A}_{(nn)} \end{bmatrix} \begin{bmatrix} \underset{\sim}{\mathbf{u}}_{(1)} \\ \underset{\sim}{\mathbf{u}}_{(2)} \\ \vdots \\ \underset{\sim}{\mathbf{u}}_{(n)} \end{bmatrix} = \begin{bmatrix} \underset{\sim}{b}_{(1)} \\ \underset{\sim}{b}_{(2)} \\ \vdots \\ \underset{\sim}{b}_{(n)} \end{bmatrix}, \tag{7.4}$$

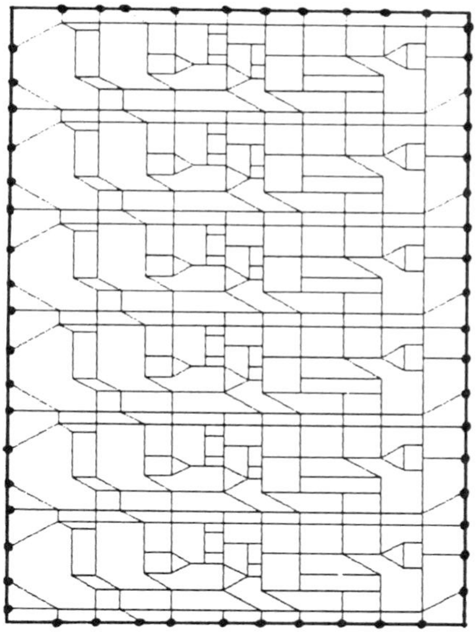

FIG. 4.5(a)

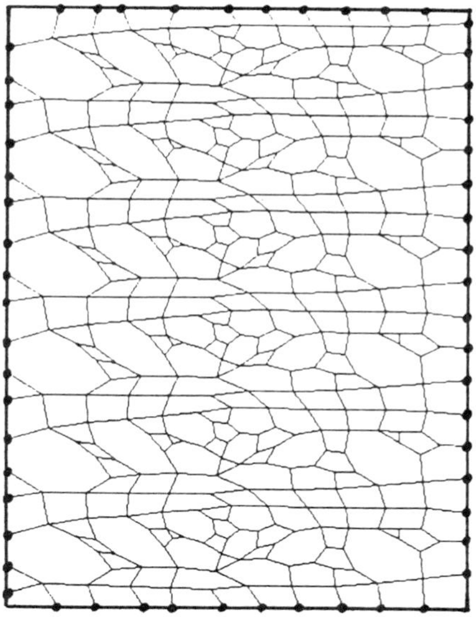

FIG. 4.5(b)

FIG. 4.6(a)

FIG. 4.6(b)

where $\mathbf{u}_{(i)}$ is the vector of variables defined at point i, $\underset{\sim}{b}_{(i)}$ is the appropriate vector of values for the right-hand side, and n is the number of points on the grid. Here, the obvious analogue to the scalar relaxation is a point-oriented block relaxation in which all the variables associated with a particular point are changed simultaneously to satisfy the equations defined at that point. The grid can then be coarsened pointwise, so that all the variables at each point are either C-variables or F-variables. We will call such points C-points and F-points, respectively. Interpolation is then defined so that all variables at a given F-point interpolate from all variables at the surrounding C-points.

As in the scalar case, the coarsening process should be defined in such a way that smooth error lies in the range of interpolation. For problems that exhibit the same type of algebraic smoothing behavior as M-matrices with usual Gauss–Seidel relaxation, the properly-scaled residual becomes small relative to the error. This property can be used to express the error at a point in terms of the error at surrounding C-points, in a way analogous to that in the scalar algorithm. Let i be an F-point. Then, letting $\underset{\sim}{e}_{(j)}$ denote the error vector at a point j, the error at point i after relaxation must satisfy

$$\underset{\sim}{A}_{(ii)}\underset{\sim}{e}_{(i)} \approx -\sum_{j\neq i} \underset{\sim}{A}_{(ij)}\underset{\sim}{e}_{(j)},$$

which is analogous to (6.2). We again wish to use this formula to construct the interpolation operator, and the coarse grid should be chosen in such a way that this is possible. We will say that a point i is strongly connected to j if $\underset{\sim}{A}_{(ij)}$ is large (in some chosen norm) compared to the other off-diagonal matrices in row i. As in the scalar case, we define S_i to be the set of strong connections of i, and given C and F, set $C_i = S_i \cap C$. We can now restate the goal in choosing the coarse grid (cf. Criterion (C1) in § 4.6.1): for each point $i \in F$, each point $j \in S_i$ must either be in C_i, or must strongly depend on C_i *in the proper way*.

This is where a difficulty arises. Simply requiring $\underset{\sim}{A}_{(jk)}$ to be large for some $k \in C_i$ is not sufficient. (For example, consider two unrelated scalar problems defined on the same grid. Such a criterion would allow the use of completely irrelevant information in choosing the coarse grid.) Instead, we note that $\underset{\sim}{e}_{(i)}$ is only affected by $\underset{\sim}{e}_{(j)}$ if $\underset{\sim}{A}_{(ij)}\underset{\sim}{e}_{(j)}$ is large. Then it is clear that we only need that part of $\underset{\sim}{e}_{(j)}$ corresponding to large eigenvalues of $\underset{\sim}{A}_{(ij)}$ to be well determined by C_i. Suppose that $C_i = \{k_1, k_2, \cdots, k_l\}$. Then we should require that these eigenvectors be well approximated by the range of the matrix $[\underset{\sim}{A}_{(jk_1)}\underset{\sim}{A}_{(jk_2)} \cdots \underset{\sim}{A}_{(jk_l)}]$. Once this is ensured, an approximation to $\underset{\sim}{e}_{(j)}$ can be derived in terms of the e at points in C_i, and interpolation can be defined in a manner similar to that in the scalar algorithm.

This method requires that, in the discretized system, the set of equations defined at each point be associated with the set of variables defined there. A strict variable/equation correspondence is not necessary, since this method performs relaxation inverting the diagonal blocks of (7.4).

The approach outlined above would clearly be more expensive than the "unknown" approach of § 4.7.1, although some simplifications are possible. However, this seems the most natural method for a number of problems with a true point-oriented character, such as those arising in the analysis of frame structures. An example of such a problem is the following.

A structural problem. One structural problem, a rigid frame, is illustrated in Fig. 4.7. The actual structure is twenty levels high, rather than the eight shown in the figure. The nodes are the locations where members (beams) are joined, and two nodes are connected whenever there is a member between them. (Although the members cross on the faces of the structure, there are no nodes there since the members are not actually joined.) The unknowns in this problem are the displacements of the nodes in the x, y, and z directions. The equations to be solved can be obtained by considering each member as a spring, with the end displacements related to the forces at the nodes by Hooke's Law; linearizing around the initial node locations; and setting the sum of all internal and external forces at each node to zero. In rigid frame problems, rotations of the joints are usually included as unknowns, but they have been eliminated here for simplicity. Here, the four points along the bottom are fixed, and the remainder of the structure is free. (For a more detailed discussion of such problems, see [Ch1].)

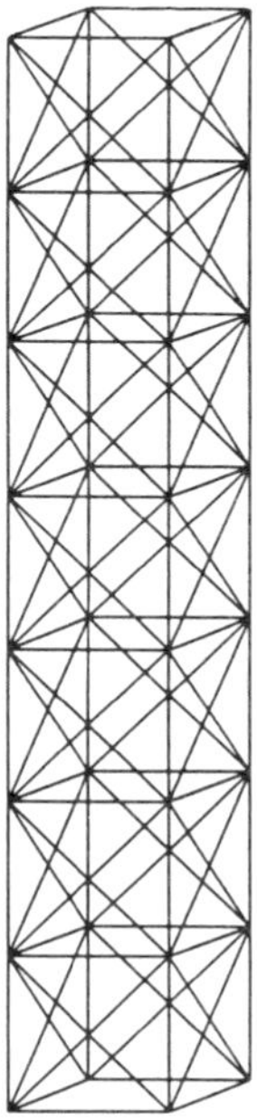

FIG. 4.7

There are difficulties in applying the "unknown" approach to this problem. Even with convex linear interpolation, asymptotic convergence factors of .8 and above were obtained. As a test, we then applied the point-oriented approach described above, with the coarse grids chosen to satisfy the given criterion. The V-cycle convergence factor obtained was .18. This clearly shows that the point approach may be essential for such problems.

4.8. Conclusions and further research. Algebraic multigrid has proven to be a robust and efficient black box solution method for several types of matrix equations. Although many of the numerical experiments reported here have been for problems defined on uniform discretizations of simple domains, we have tried to show that the method is certainly not restricted to these cases. For such "nice" problems, properly designed geometric multigrid algorithms can be more efficient, although the ease of use of AMG makes it an attractive tool even here. However, there are many problems that could benefit from multigrid ideas, but to which the application of usual multigrid methods is difficult or impossible. For example, many applications programs generate irregular grids tailored to particular domains or operators. Because AMG is insensitive to complexities in the domains and grids chosen, it is ideal for such problems. Another class of problems to which usual multigrid methods cannot be applied are those that are discrete in nature, since the necessary multigrid components cannot be defined. These are often similar to the matrix equations obtained from the discretization of PDEs on irregular grids, and AMG can be easily applied.

Current and planned work in AMG lies in two main directions. The primary goal is to extend the application range of the method and to develop robust, efficient algorithms for new classes of problems. The second and related goal is to develop methods for the efficient solution of chains of related problems, where there are smooth or local changes to the system, in either the right-hand sides or the fine grid operators, from one problem to the next.

New problem areas. There are a number of areas in which the application of AMG seems promising, both in terms of the algorithm's success and of its potential benefit. These include problems in fluid flows and structural mechanics. The ability of AMG to deal with complex domains makes it attractive in both of these areas. This is particularly true in structural problems. Here, there is also a need to solve large discrete problems, for which AMG is well suited.

One change in the algorithm that can make it applicable to a wider range of problems is the use of different relaxation methods. The use of block Gauss–Seidel in the method of § 4.7.2 is one example. Another would be the use of distributed relaxation methods, such as Kaczmarz iteration. This would allow the use of AMG for problems, such as the Cauchy–Riemann and Stokes equations, that lack the natural variable/equation correspondence necessary for Gauss–Seidel relaxation.

In the development of AMG algorithms for larger classes of problems, it is necessary to balance convenience and efficiency. Currently, it seems clear that different approaches are necessary for different types of problems. An algorithm capable of automatically determining the proper approach for a particular problem may spend too much time trying to find information that could be easily specified by the user. Another option is simply to develop, for certain types of problems, specialized algorithms that could be used over a wide range of changes in the domain or coefficients of the operator.

Chains of problems. The second area of research—developing efficient solutions to chains of related problems—is important in a number of applications, including nonlinear and time-dependent problems, as well as reliability analysis and design problems. A large amount of work in the

AMG algorithm is invested in the setup phase, and for the problems mentioned, the setup work for later problems can sometimes be reduced or avoided altogether, effectively reducing the time required for the solution of each problem. The extent of the changes in the matrix equation determines the amount of work necessary to resolve the problem. Local changes may require only a small amount of work in both the setup and the solution phase.

The simplest change is in the right-hand side. This is often done in time-dependent problems or in structural problems in order to study the effect of different loads on a structure. Then the setup phase need not be repeated at all. If the new right-hand side is a perturbation of the previous one, few cycles may be necessary. In addition, if the perturbation is only local, then relaxation on the finest level is only necessary in the neighborhood of the change, since the error introduced will normally be smooth far away. The area of relaxation increases on coarser grids, with global relaxation occurring only on the coarsest. In this way, the smooth changes in the solution are obtained by interpolation.

When the matrix itself changes, as in the solution of nonlinear problems by Newton's method, the entire setup process may not need to be repeated. If the change is small enough, all coarse grid components may be retained. It is more likely that some changes will be necessary. If the character of the connections in the matrix does not change drastically, it is possible to keep the coarse grids and the interpolation operators, while simply recomputing the coarse grid operators. This costs much less than repeating the entire setup phase, since a large amount of work is used in choosing the coarse grids. Another possibility is to recompute interpolation while again keeping the coarse grid structure.

It is also possible that the changes introduced in the matrix are only local. This is the case when local changes are made to determine their effect on the behavior of the entire structure under a load. In this case, the changes in the coarsening may also be only local. If necessary, new coarse grids or interpolation may be needed in the neighborhood of the change. The coarse grid matrices need to be only locally recomputed, since most of the matrix entries will be unaffected by the changes. The effect will spread on coarser grids, but the overall work necessary is small compared to the initial setup. Once the setup is completed, the solution may be quickly computed, again starting from the previous solution and using local relaxation.

It should be possible to automatically determine the amount of relaxation required and the extent of the changes needed in the coarse grid matrices. For example, during local relaxation, the area of relaxation can be increased until changes in the solution approximation are seen to be small. Deciding the changes necessary in the coarse grids, the interpolation, and the coarse grid matrices is more difficult, but feedback is available in the form of the resulting convergence factors.

Acknowledgments. This work was supported by the Air Force Office of Scientific Research under grant AFOSR-86-0126, the Department of Energy under grant DE-AC03-84ER80155, the NASA Langley Research Center under grant NAG-1-453, and the Gesellschaft für Mathematik und Datenverarbeitung.

REFERENCES

[Be1] B. Becker, G. Hotz, R. Kolla and P. Molitor, *Ein CAD-System zum Entwurf integrierter Schaltungen,* Report no. 16, Universität Saarbrücken, Fachbereich 10, 1984.

[Ch1] W. Chu-Kia and C. G. Salmon, *Introductory Structural Analysis,* Prentice–Hall, Englewood Cliffs, NJ, 1984.

[Ei1] S. C. Eisenstat, M. C. Gursky, M. H. Schultz and A. H. Sherman, *Yale sparse matrix package.* I: *The symmetric codes,* Internat. J. Numer. Methods Engrg., 18 (1982), pp. 1145–1151.

[Po1] A. Pope, *Geodetic computation and sparsity,* Proc. Sparse Matrix Symposium, Fairfield Glade, TN, October 1982.

[Sc1] J. Schröder, *Operator Inequalities,* Math. Sci. Engrg., 147, Academic Press, New York, 1980.

CHAPTER 5

Variational Multigrid Theory

J. MANDEL, S. McCORMICK AND R. BANK

5.1. Introduction. This chapter develops a convergence theory for multigrid methods applied to an important class of problems. We provide convergence bounds based on *abstract algebraic assumptions* about the hierarchy of discrete problems; these assumptions will be verified for some classes of elliptic boundary value problems discretized by finite elements. We think that strict division into parts concerned with the underlying differential equation and the algebraic process itself makes the theory more simple and applicable. Indeed, the theory can be applied, for example, to *finite difference discretizations,* to problems of *higher order,* or to elliptic *systems* (such as in linear elasticity) as soon as our algebraic assumptions are verified. Also, the estimates are quite sharp when good estimates of the constants in the algebraic assumptions are available. In particular, in the symmetric positive definite case, the assumptions reduce to a *single inequality,* which is closely related to a standard *error estimate* in the finite element method.

We consider multigrid methods in a variational setting only (cf. (2.2) and (2.3) below). This is natural to assume for finite element discretizations, and it substantially simplifies the arguments and makes stronger results possible. Using the *energy norm* as a suitable measure of error, we prove convergence for *any* positive number of smoothing steps. This applies even to the nonsymmetric and indefinite cases if the discretization is fine enough; an "energy norm" in this case is defined by a dominating symmetric, positive definite part of the discrete operator.

This chapter is organized as follows. In § 5.2, we introduce the abstract assumptions, which will be verified in § 5.3 for some finite element discretizations of elliptic boundary value problems. The symmetric, positive definite case with very general smoothers is treated in §§ 5.4 and 5.5. We extend the results to the general case in § 5.6 using perturbation arguments. Section 5.7 contains a short development of full multigrid and comments on complexity. Section 5.8 includes a Fourier analysis of certain ideal problems; this is used to compute values of some constants occurring in the theoretical estimates, thus yielding convergence estimates via our theory. In § 5.9, we make bibliographical comments on related work for each of the previous sections. Finally, § 5.10 contains several problems and exercises relevant to this chapter. Some of the exercises contain additional information about related results from the literature.

5.2. Notation and assumptions.

5.2.1. Problem hierarchy. Consider a hierarchy of real finite-dimensional spaces H^k, $k = 1, 2, \cdots$, related by given full-rank linear mappings I_{k-1}^k, called *prolongations*:

$$I_{k-1}^k \in [H^{k-1}, H^k], \qquad \ker(I_{k-1}^k) = \{0\}. \tag{2.1}$$

Each H^k is equipped with an inner product $\langle \cdot, \cdot \rangle_k$ and the associated norm $\|u\|_k = \langle u, u \rangle_k^{1/2}$. In most applications, H^k will be the Euclidean space $\mathbf{R}^{n_k}$ and

$$\langle u, v \rangle_k = \sum_{i=1}^{n_k} u_i v_i.$$

Adjoints relative to the inner product $\langle \cdot, \cdot \rangle_k$ will be denoted by ^{T} as transposes and linear mappings between the spaces H^k may be alternatively thought of as matrices.

Define the *restrictions* $I_k^{k-1} \in [H_k, H_{k-1}]$ by

$$I_k^{k-1} = (I_{k-1}^k)^T. \tag{2.2}$$

Let $m > 1$, A^m, $N^m \in [H^m]$, and A^m be symmetric, positive definite. Set

$$A^{k-1} = I_k^{k-1} A^k I_{k-1}^k, \qquad N^{k-1} = I_k^{k-1} N^k I_{k-1}^k \tag{2.3}$$

for $k = m, m-1, \cdots, 2$. Then all A^k are symmetric positive definite. We are interested in numerical solution of the problem

$$(A^k + N^k)\mathbf{u}^k = f^k \tag{2.4}$$

for $k = m$. We assume that $A^k + N^k$ is invertible, that is, that the problem (2.4) always has a unique solution.

For the analysis, however, it will be more convenient to consider a hierarchy of problems (2.4) for all positive k, with A^k and N^k satisfying (2.3). We will assume the operator N^k to be small in a suitable sense for large k. This will allow it to be considered as a perturbation.

5.2.2. Discrete norms. Because A^k is symmetric, positive definite, we may define another inner product on H^k by

$$\langle u^k, v^k \rangle_A = \langle A^k u^k, v^k \rangle_k, \qquad u^k, v^k \in H^k.$$

Adjoints relative to this inner product will be denoted by * and the associated norm by

$$|||u^k|||_{1,k} = \langle u^k, u^k \rangle_A^{1/2}, \qquad u^k \in H^k.$$

Note that because $A^{k-1} = (I_{k-1}^k)^T A^k I_{k-1}^k$ by (2.2) and (2.3), we have

$$|||u^{k-1}|||_{1,k-1} = |||I_{k-1}^k u^{k-1}|||_{1,k}, \qquad u^{k-1} \in H^{k-1}. \tag{2.5}$$

Thus, the $|||\cdot|||_{1,k}$ norm is a convenient common norm for all spaces H^k. The equality (2.5) states that the space $(H^{k-1}, |||\cdot|||_{1,k-1})$ is isometrically isomorphic to the subspace $R(I_{k-1}^k)$ of $(H^k, |||\cdot|||_{1,k})$. In the symmetric, positive definite case when $N^k = 0$ for all k, $|||\cdot|||_{1,k}$ is the *energy* norm associated with problem (2.4).

For all $k > 0$, let $B^k \in [H^k]$ be another symmetric, positive definite operator. This will be useful in the analysis of iterative methods, where B^k may be a scalar multiple of the identity, a suitable diagonal matrix, or in general some preconditioner (cf. § 5.4). Define another inner product and a norm on H^k by

$$\langle u^k, v^k \rangle_B = \langle B^k u^k, v^k \rangle_k, \qquad u^k, v^k \in H^k,$$
$$|||u^k|||_{0,k} = (\langle u^k, u^k \rangle_B)^{1/2}, \qquad u^k \in H^k.$$

The $|||\cdot|||_{0,k}$ norm is best thought of as an approximation of the $\mathscr{L}^2(\Omega)$ norm (see § 5.3).

The $|||\cdot|||_{0,k}$ and $|||\cdot|||_{1,k}$ norms are sufficient for much of our analysis. However, we will need a scale of norms defined as follows. (For a more detailed treatment of scales of spaces and norms, see the next section.)

Let

$$E^k = (B^k)^{-1} A^k. \tag{2.6}$$

The operator E^k is symmetric, positive definite relative to the inner product $\langle \cdot, \cdot \rangle_B$, since $\langle E^k u^k, v^k \rangle_B = \langle A^k u^k, v^k \rangle_k$. Thus the powers $(E^k)^s$, s a real number, are well defined.[1]

---

[1] The operator E^k has the *spectral decomposition*

$$E^k = \sum_{i=1}^{\dim H^k} \lambda_i P_i,$$

where $\lambda_i > 0$ are the eigenvalues of E^k and P_i are B^k-orthogonal projections satisfying

$$\sum_{i=1}^{\dim H^k} P_i = I, \quad P_i P_j = P_j P_i = 0 \quad \text{for } i \neq j.$$

Then

$$(E^k)^s = \sum_{i=1}^{\dim H^k} \lambda_i^s P_i.$$

The spectral projections P_i are given by $P_i u^k = \langle v_i, u^k \rangle_B v_i$, where v_i form a B^k-orthogonal basis of eigenvectors of E^k, $E^k v_i = \lambda_i v_i$, $\langle v_i, v_j \rangle_B = 0$ if $i \neq j$, $\langle v_i, v_i \rangle_B = 1$.

We can thus define a scale of norms on H^k by

$$\||u^k\||_{s,k} = \langle (E^k)^s u^k, u^k \rangle_B^{1/2}. \tag{2.7}$$

(The subscript k will be omitted whenever there is no danger of confusion. We shall thus write simply $\||u^k\||_s$.) Note that for $s=0$ and $s=1$ we recover the previous definitions of these norms.

5.2.3. The multigrid algorithm. Recall that one step of the multigrid algorithm $u^k \leftarrow MG_\mu^k(u^k, f^k)$ is defined as follows:

(a) If $k = k_1$, then MG_μ^k is defined by some direct or iterative method so that

$$\||\mathbf{u}^k - MG_\mu^k(u^k, f^k)\||_1^2 \le \varepsilon^k \, \||\mathbf{u}^k - u^k\||_1^2 \quad \text{with } \varepsilon^k < 1.$$

(b) If $k > k_1$ then:

Step 1: Perform $u^k \leftarrow G_1^k(u^k, f^k)$ ν_1 times.

Step 2: Set $u^{k-1} = 0$,

$$f^{k-1} = I_k^{k-1}(f^k - (A^k + N^k)u^k), \tag{2.8}$$

and perform $u^{k-1} \leftarrow MG_\mu^{k-1}(u^{k-1}, f^{k-1})$ μ times.

Step 3: Set $u^k \leftarrow u^k + I_{k-1}^k u^{k-1}$.

Step 4: Perform $u^k \leftarrow G_2^k(u^k, f^k)$ ν_2 times.

Here, $k_1 \ge 1$, $\nu_1, \nu_2 \ge 0$, $\nu_1 + \nu_2 > 0$, and $\mu > 0$ are integer parameters, and G_1^k and G_2^k are consistent iterative methods for the solution of (2.4), called *smoothers*.

Note that, because of (2.2) and (2.3), the problem

$$A^{k-1}u^{k-1} = f^{k-1},$$

with f^{k-1} given by (2.8), can be written as

$$(I_{k-1}^k)^T(A^k + N^k)I_{k-1}^k \mathbf{u}^{k-1} = (I_{k-1}^k)^T(f^k - (A^k + N^k)u^k).$$

Therefore, $I_{k-1}^k \mathbf{u}^{k-1}$ is the *Ritz–Galerkin* approximation in the range of I_{k-1}^k of the exact correction $\mathbf{v}^k$ such that

$$(A^k + N^k)\mathbf{v}^k = f^k - (A^k + N^k)u^k,$$

which would give $u^k + \mathbf{v}^k = \mathbf{u}^k$. This is usually referred to as the *variational* construction of the coarse grid problem, and therefore (2.2) and (2.3) are called variational conditions.

5.2.4. Assumptions. Define the subspaces $\hat{S}^k$, $\hat{T}^k$, and $\hat{U}^k$ of H^k by

$$\hat{S}^k = R(I_{k-1}^k),$$

$$\hat{T}^k = \{u^k \in H^k;\ \langle A^k u^k, v^k \rangle_k = 0 \text{ for all } v^k \in \hat{S}^k\},$$

$$\hat{U}^k = \{u^k \in H^k;\ \langle A^k u^k + N^k u^k, v^k \rangle_k = 0 \text{ for all } v^k \in \hat{S}^k\},$$

and define T^k, U^k as the projections along $\hat{S}^k$ onto $\hat{T}^k$, $\hat{U}^k$, respectively. That is, the operators T^k and U^k are uniquely determined by

$$R(T^k) = \hat{T}^k, \quad R(U^k) = \hat{U}^k, \quad \ker(T^k) = \ker(U^k) = \hat{S}^k.$$

Further, let $S^k = I - T^k$ be the projection onto $\hat{S}^k$ along $\hat{T}^k$.

Vectors from $\hat{S}^k$ can be thought of as representing smooth functions, which can be exactly interpolated on the coarse level $k-1$; vectors from $\hat{T}^k$ and $\hat{U}^k$ correspond to oscillatory functions.

Our convergence theory will be based on the following assumptions, which will be verified in the next section for some classes of problems arising from finite element discretizations of elliptic boundary value problems.

Basic Assumptions. We shall assume that there exist constants α, η, ψ, $0 < \alpha \le 1$, $0 < \eta \le 1$, $0 < \psi < 1$, and c_1, c_2, c_3, c_4 such that for all k,

$$\text{(2.9)} \qquad |||u^k|||_{1-\alpha}^2 \le c_1(\rho(E^k))^{-\alpha}\, |||u^k|||_1^2 \quad \text{for any } u^k \in \hat{U}^k,$$

$$\left.\begin{aligned} &\text{(2.10)} \qquad |\langle N^k u^k, v^k\rangle_k| \le c_2\, |||u^k|||_1\, |||v^k|||_{1-\eta} \\ &\text{(2.11)} \qquad |\langle N^k u^k, v^k\rangle_k| \le c_3\, |||u^k|||_{1-\eta}\, |||v^k|||_1 \end{aligned}\right\} \quad \text{for any } u^k, v^k \in H^k,$$

$$\text{(2.12)} \qquad \frac{1}{c_4}(\psi)^{-k} \le \rho(E^k) \le c_4(\psi)^{-k},$$

and

$$\text{(2.13)} \qquad E^k \ge I.$$

(The inequality $\ge$ between linear operators means that their difference is positive semi-definite.)

For $\alpha = 1$, assumption (2.9) can be interpreted as a requirement that E^k be well conditioned on the oscillatory subspace $\hat{U}^k$. Assumptions (2.10) and (2.12) mean that the bilinear form $\langle N^k u^k, v^k\rangle_k$ is of lower order; it will be shown later that (2.10) and (2.11) together with (2.12) and (2.13) imply that N^k is in a suitable sense small for large k. Actually, assumptions (2.10)–(2.13) will be needed only if $N^k \ne 0$.

Note that, from (2.13), we have for all $u^k \in H^k$ that

$$|||u^k|||_s \le |||u^k|||_t \quad \text{if } s \le t,$$

and, from (2.12), that

$$\max_{u^k \ne 0} \frac{|||u^k|||_t}{|||u^k|||_s} = (\rho(E^k))^{(t-s)/2} \approx \psi^{-k(t-s)/2}.$$

Thus, for $t > s$, the norm $|||\cdot|||_{t,k}$ is increasingly *stronger* than $|||\cdot|||_{s,k}$ for large k. It should be noted that, in usual discretizations of second order problems, $(\rho(E^k))^{-1/2}$ plays the role of the discretization parameter h (cf. (3.7)).

5.3. Motivation and verification of algebraic assumptions.

5.3.1. Hilbert spaces. We now summarize elements of interpolation theory of Hilbert spaces, which will be needed in the sequel. For more details and proofs, see [Kr1] and [Li1].

Let $\mathcal{H}^0$ be a Hilbert space with inner product $(\cdot,\cdot)_0$ and norm $\|\cdot\|_0$, and let $\mathcal{J}$ be an unbounded positive definite selfadjoint operator in $\mathcal{H}^0$ with a dense domain $\mathcal{D}(\mathcal{J})$, such that

$$\|u\|_0 \le \|\mathcal{J}u\|_0 \quad \text{for all } u \in D(\mathcal{J}).$$

According to the development in [Kr1, p. 236] and using the powers $\mathcal{J}^s$, $s>0$, the *Hilbert scale* of spaces $(\mathcal{H}^s, \|\cdot\|_s)$ is then given by

$$\mathcal{H}^s = \mathcal{D}(\mathcal{J}^s), \quad \|u\|_s = \|\mathcal{J}^s u\|_0, \quad s>0;$$

spaces $(\mathcal{H}^{-s}, \|\cdot\|_{-s})$, $s>0$, are defined as duals:

$$\mathcal{H}^{-s} = (\mathcal{H}^s)', \qquad s>0,$$
$$\|f\|_{-s} = \sup\{|f(u)| : u \in \mathcal{H}^s,\ \|u\|_s = 1\}.$$

Any element $f \in \mathcal{H}^0$ induces a bounded functional from $\mathcal{H}^{-s}$, $s \ge 0$, given by

$$u \to (u,f)_0, \qquad u \in \mathcal{H}^s,$$

with

$$|(u,f)_0| \le \|u\|_0\,\|f\|_0 \le \|u\|_s\,\|f\|_0.$$

We may thus identify $\mathcal{H}^0$ with a subspace of $\mathcal{H}^{-s}$, and the inner product in $\mathcal{H}^0$ with the dual pairing between $\mathcal{H}^s$ and $\mathcal{H}^{-s}$:

$$f(u) = (u,f)_0, \qquad u \in \mathcal{H}^s, \quad f \in \mathcal{H}^{-s}, \quad s \ge 0.$$

We will often be interested in Hilbert scales $\mathcal{H}^s$ with the values of s restricted to a certain interval, say, $0 \le s \le 1$.

Again following [Kr1], let $\mathcal{H}^0$ and $\mathcal{H}^1$ be *given* Hilbert spaces with inner products $(\cdot,\cdot)_0$, $(\cdot,\cdot)_1$ and norms $\|\cdot\|_0$, $\|\cdot\|_1$, respectively, $\mathcal{H}^1$ dense in $\mathcal{H}^0$, and $\mathcal{H}^1$ normally imbedded in $\mathcal{H}^0$; that is,

$$\mathcal{H}^1 \subset \mathcal{H}^0 \quad \text{and} \quad \|u\|_1 \ge \|u\|_0 \quad \text{for all } u \in \mathcal{H}^1.$$

Then there exists a unique Hilbert scale connecting these spaces. The operator $\mathcal{J}$ is given by $\mathcal{J} = \mathcal{E}^{1/2}$ with $\mathcal{E}$ determined by

$$(u, v)_1 = (\mathcal{E}u, v)_0 \quad \text{for all } u, v \in \mathcal{H}^1.$$

The spaces $\mathcal{H}^s$, $0<s<1$, are called *interpolation spaces* between $\mathcal{H}^0$ and $\mathcal{H}^1$ and they are denoted by $\mathcal{H}^s = [\mathcal{H}^0, \mathcal{H}^1]_s$. As an example of a Hilbert scale, we may take the Sobolev spaces $\mathcal{H}^s(\Omega)$, $\Omega \subset \mathbf{R}^N$ (see [Li1]).

Hilbert scales and interpolation spaces are defined in the same way for a finite-dimensional Hilbert space K^0 in place of $\mathcal{H}^0$. Of course, $\mathcal{J}$ is then a bounded operator and $K^s = K^0$ for all s, albeit with different norms. Note that the spaces $(H^k, \|\cdot\|_{s,k})$, $0 \le s \le 1$, as defined in the preceding section, are interpolation spaces in this sense for any k.

We have the following lemma, which is a particular case of the well-known strong interpolation property [Kr1].

LEMMA 3.1. *Let $\mathcal{H}^s$ and $\mathcal{K}^s$, $0 \le s \le 1$, be two Hilbert scales and let*

$$T \in [\mathcal{H}^0, \mathcal{K}^0] \cap [\mathcal{H}^1, \mathcal{K}^1].$$

Then $T \in [\mathcal{H}^s, \mathcal{K}^s]$ for all $0 < s < 1$, and

$$\|T\|_{\mathcal{H}^s \to \mathcal{K}^s} \le \|T\|_{\mathcal{H}^0 \to \mathcal{K}^0}^{1-s} \|T\|_{\mathcal{H}^1 \to \mathcal{K}^1}^{s}.$$

5.3.2. Discretization of elliptic variational problems. Let $(\mathcal{H}^s, \|\cdot\|_s)$, $-1 \le s \le 1$, be a given Hilbert scale. In addition, let $(\tilde{\mathcal{H}}^s, \|\cdot\|_s)$, $1 < s \le 2$, be additional Hilbert spaces; their relations to $\mathcal{H}^s$, $-1 \le s \le 1$, will be implied by the assumptions we make later (cf. (3.6) and (3.10)). In particular cases, we may have $\tilde{\mathcal{H}}^s = \mathcal{H}^s$, $1 < s \le 2$, where $\mathcal{H}^s$, $-1 \le s \le 2$, form a Hilbert scale.

Let $\mathcal{A}$, $\mathcal{N}$ be bilinear forms on $\mathcal{H}^1 \times \mathcal{H}^1$, $f \in \mathcal{H}^{-1}$, and consider the *abstract variational problem* of finding u so that

$$(3.1) \qquad u \in \mathcal{H}^1 \quad \text{and} \quad (\mathcal{A} + \mathcal{N})(u, v) = (f, v)_0 \quad \text{for all } v \in \mathcal{H}^1.$$

We shall give examples of problems of the form (3.1) later, along with the choices of the spaces $\mathcal{H}^s$, $\tilde{\mathcal{H}}^s$, and of discretizations.

We make the following assumptions:

The form $\mathcal{A}$ is *symmetric*, *$\mathcal{H}^1$-bounded*, and *$\mathcal{H}^1$-elliptic*:

$$(3.2) \qquad \left.\begin{array}{l} \mathcal{A}(u, v) = \mathcal{A}(v, u) \\ |\mathcal{A}(u, v)| \le c\, \|u\|_1\, \|v\|_1 \\ \mathcal{A}(u, u) \ge c\, \|u\|_1^2 \end{array}\right\} \quad \text{for all } u, v \in \mathcal{H}^1.$$

Remember that $c > 0$ is a generic constant so that it may represent different values in these and later formulas.

The form $\mathcal{N}$ is of *lower order*:

$$(3.3) \qquad \left.\begin{array}{l} |\mathcal{N}(u, v)| \le c\, \|u\|_1\, \|v\|_{1-\eta} \\ |\mathcal{N}(u, v)| \le c\, \|u\|_{1-\eta}\, \|v\|_1 \end{array}\right\} \quad \text{for all } u, v \in \mathcal{H}^1,$$

with some $0 < \eta \le 1$. If $\mathcal{N} = 0$, then, of course, this assumption is irrelevant.

Now let us consider discretizations of the problem (3.1). Suppose we are given a family of finite-dimensional subspaces $\mathcal{S}^k \subset \mathcal{S}^{k+1} \subset \mathcal{H}^1$, $k = 1, 2, \cdots$. Let a basis be given in each $\mathcal{S}^k$ and choose H^k to be the corresponding coefficient space equipped with the Euclidean inner product $\langle \cdot, \cdot \rangle_k$, so $H^k = \mathbf{R}^n$, $n = \dim \mathcal{S}^k$. Let $\mathcal{J}^k : H^k \to \mathcal{S}^k$ be the (algebraic) isomorphism between H^k and $\mathcal{S}^k$.

Define $A^k \in [H^k]$, $N^k \in [H^k]$, and $f^k \in H^k$ by

$$\left.\begin{array}{l} \langle A^k u^k, v^k \rangle_k = \mathcal{A}(\mathcal{J}^k u^k, \mathcal{J}^k v^k) \\ \langle N^k u^k, v^k \rangle_k = \mathcal{N}(\mathcal{J}^k u^k, \mathcal{J}^k v^k) \\ \langle f^k, v^k \rangle_k = (f, \mathcal{J}^k v^k)_0 \end{array}\right\} \quad \text{for all } u^k, v^k \in H^k.$$

The problem (3.1) now has the discretization

$$A^k \mathbf{u}^k + N^k \mathbf{u}^k = f^k, \qquad \mathbf{u}^k \in H^k. \tag{3.4}$$

The prolongation I_{k-1}^k is defined by the injection (identity) mapping $\mathcal{S}^{k-1}$ into $\mathcal{S}^k$, via the isomorphisms $\mathcal{J}^{k-1}$ and $\mathcal{J}^k$:

$$\mathcal{J}^{k-1} = \mathcal{J}^k \cdot I_{k-1}^k.$$

Thus, I_{k-1}^k is defined so that the following diagram commutes:

$$\begin{array}{ccc} H^k & \xrightarrow{\mathcal{J}^k} & \mathcal{S}^k \\ {\scriptstyle I_{k-1}^k}\uparrow & & \uparrow{\scriptstyle I} \\ H^{k-1} & \xrightarrow[\mathcal{J}^{k-1}]{} & \mathcal{S}^{k-1} \end{array}$$

The restriction I_k^{k-1} is then defined as the transpose:

$$\langle I_{k-1}^k u^{k-1}, v^k \rangle_k = \langle u^k, I_k^{k-1} v^{k-1} \rangle_{k-1} \quad \text{for all } u^{k-1} \in H^{k-1}, \quad v^k \in H^k.$$

Then the variational conditions (2.2) and (2.3) are satisfied.

For each k, let $B^k \in [H^k]$ be symmetric, positive definite and such that for all k,

$$c^{-1} \langle B^k u^k, u^k \rangle_k \le \|\mathcal{J}^k u^k\|_0^2 \le c \langle B^k u^k, u^k \rangle_k \quad \text{for all } u^k \in H^k. \tag{3.5}$$

We may then define $E^k = (B^k)^{-1} A^k$ and the scale of norms $|||\cdot|||_{s,k} = |||\cdot|||_s$ on H^k as in (2.6) and (2.7). Because of (3.2) and the continuous injection of $\mathcal{H}^1$ into $\mathcal{H}^0$, we may assume without loss of generality that $B^k \le A^k$. Suppose that the family $\{S^k\}$ satisfies the following *approximation property* for all $0 \le s \le \bar{\alpha}$ with some $0 < \bar{\alpha} \le 1$: *For all* $u \in \mathcal{H}^{1+s} \cap \mathcal{H}^1$ *there exists* $v^k \in \mathcal{S}^k$ *such that*

$$\|u - v^k\|_0 + (\rho(E^k))^{-1/2} \|u - v^k\|_1 \le c(\rho(E^k))^{-(1+s)/2} \|u\|_{1+s}. \tag{3.6}$$

In fact, the common estimate in the $\|\cdot\|_0$ and $\|\cdot\|_1$ norms will be used only with $s = 0$ in the proof of Lemma 3.2. The estimate in the $\|\cdot\|_1$ norm will be needed later for a fixed $s > 0$. But usual finite element estimates give (3.6) (see Ciarlet [Ci1] and Strang and Fix [St1]).

Then we have the following important lemma.

LEMMA 3.2. *There exists a constant c so that for all k, $u^k \in H^k$, and $0 \le s \le 1$, it holds that*

$$c^{-1} |||u^k|||_s \le \|\mathscr{J}^k u^k\|_s \le c\, |||u^k|||_s.$$

Proof. The proposition is true for $s=0$ by (3.5) directly, and for $s=1$ by (3.2), because $\mathscr{A}(\mathscr{J}^k u, \mathscr{J}^k u) = \langle A^k u^k, u^k\rangle_k = \langle E^k u^k, u^k\rangle_B = |||u^k|||_1^2$. The right-hand inequality then follows for $0 \le s \le 1$ if we interpolate the mappings $\mathscr{J}^k$, which are uniformly bounded (in k) both from $(H^k, |||\cdot|||_{0,k})$ to $\mathscr{H}^0$ and from $(H^k, |||\cdot|||_{1,k})$ to $\mathscr{H}^1$, using Lemma 3.1. For $u \in \mathscr{H}^0$, define $\mathscr{P}^k u \in H^k$ as the coefficients of the $\mathscr{H}^0$-orthogonal projection of u onto $\mathscr{S}^k$. From (3.5), $\mathscr{P}^k$ are uniformly bounded operators in $[\mathscr{H}^0, (H^k, |||\cdot|||_{0,k})]$. We show that the same is true in the 1-norm. To this end, let $u \in \mathscr{H}^1$ and, for each k, let $v^k \in H^k$ be such that u and $v^k = \mathscr{J}^k v^k \in \mathscr{S}^k$ satisfy (3.6) for $s=0$. Then $\mathscr{P}^k v^k = v^k$, and $|||\mathscr{P}^k(u - v^k)|||_{0,k} \le c\, \|u - v^k\|_0$ by the definition of $\mathscr{P}^k$ and by (3.5). Using (3.2) and (3.6) we then get

$$\begin{aligned}
|||\mathscr{P}^k u|||_{1,k} &\le |||\mathscr{P}^k(u - v^k)|||_{1,k} + |||v^k|||_{1,k}\\
&\le (\rho(E^k))^{1/2}\, |||\mathscr{P}^k(u - v^k)|||_{0,k} + c\,\|u\|_1 + c\,\|u - v^k\|_1\\
&\le c(\rho(E^k))^{1/2}(\rho(E^k))^{-1/2}\,\|u\|_1 + c\,\|u\|_1\\
&\le c\,\|u\|_1.
\end{aligned}$$

The proof is completed by interpolating the operators $\mathscr{P}^k$ using Lemma 3.1.

The quantity $(\rho(E^k))^{-1/2}$ used above satisfies the following:

$$c^{-1}(\rho(E^k))^{-1/2} \le \sup_{u^k \in \mathscr{S}^k} \frac{\|u^k\|_1}{\|u^k\|_0} \le c(\rho(E^k))^{-1/2}. \tag{3.7}$$

For problems of order 2 and quasi-uniform discretizations, it therefore plays the role of the discretization parameter h_k. For this reason, we may suppose that there is $0 < \psi < 1$ such that, for all k,

$$c^{-1}\psi^{-k} \le \rho(E^k) \le c\psi^{-k}. \tag{3.8}$$

Finally, we need to consider the *adjoint problem* of finding a v so that

$$v \in \mathscr{H}^1 \quad \text{and} \quad (\mathscr{A} + \mathscr{N})(u, v) = \langle u, f\rangle \quad \text{for all } u \in \mathscr{H}^1. \tag{3.9}$$

In particular, we assume that (3.9) is *regular* by supposing that it exhibits a unique solution $v \in \mathscr{H}^1$ for any $f \in \mathscr{H}^{-1}$ and that there is a constant $0 < \alpha \le 1$ such that

$$\text{if } f \in \mathscr{H}^{\alpha-1} \text{ then } v \in \bar{\mathscr{H}}^{1+\alpha} \text{ and } \|v\|_{\alpha+1} \le c\,\|f\|_{\alpha-1}. \tag{3.10}$$

Without loss of generality we suppose that $\alpha \le \bar{\alpha}$.

We are now in a position to verify our algebraic assumptions for the class of problems and discretizations considered in this section.

THEOREM 3.1. *Under the assumptions introduced here, the Basic Assumptions* (2.9)–(2.13) *are satisfied.*

Proof. It is nontrivial to verify only (2.9). But by Lemma 3.2, it suffices to show that $\|u^k\|_{1-\alpha} \leq c(\rho(E^k))^{-\alpha/2} \|u^k\|_1$ if $u^k \in \hat{U}^k$, that is, if

$$(3.11) \qquad u^k \in \mathcal{S}^k \quad \text{and} \quad (\mathcal{A}+\mathcal{N})(u^k, w^{k-1}) = 0 \quad \text{for all } w^{k-1} \in \mathcal{S}^{k-1}.$$

Let $f \in \mathcal{H}^{\alpha-1}$ and let v be the corresponding solution of the adjoint problem (3.9). Then with u^k satisfying (3.11) and for any $w^{k-1} \in \mathcal{S}^{k-1}$, we have

$$\begin{aligned}
\langle u^k, f \rangle &= (\mathcal{A}+\mathcal{N})(u^k, v) \\
&= (\mathcal{A}+\mathcal{N})(u, v - w^{k-1}) \\
&\leq c\, \|u\|_1 \inf_{w^{k-1} \in \mathcal{S}^{k-1}} \|v - w^{k-1}\|_1 \\
&\leq c(\rho(E^{k-1}))^{-\alpha/2} \|u^k\|_1 \|v\|_{1+\alpha} \\
&\leq c(\rho(E^k))^{-\alpha/2} \|u^k\|_1 \|f\|_{\alpha-1},
\end{aligned}$$

where we used (3.11), (3.2), (3.3), the continuous injection $\mathcal{H}^1 \subset \mathcal{H}^{1-\eta}$, the approximation assumption (3.6) for $s = \alpha$, the regularity assumption (3.10), and (3.18). We conclude the proof by noting that by duality

$$\|u^k\|_{1-\alpha} = \sup_{f \in (\mathcal{H}^{1-\alpha})'} \frac{|f(u^k)|}{\|f\|_{(\mathcal{H}^{1-\alpha})'}} = \sup_{f \in \mathcal{H}^{\alpha-1}} \frac{|(u^k, f)_0|}{\|f\|_{\alpha-1}}.$$

We now show examples of problems and discretizations in this setting.

Example 3.1. Let $\Omega \subset \mathbf{R}^d$ be a bounded domain with a Lipschitz boundary $\partial\Omega$, $\mathcal{H}^0 = \mathcal{L}^2(\Omega)$, $\mathcal{H}^1 = \mathcal{H}_0^1(\Omega)$, $\tilde{\mathcal{H}}^s = \mathcal{H}^s(\Omega)$ for $s > 1$, and

$$\mathcal{A}(u, v) = \int_\Omega \nabla v \cdot \underline{a} \nabla u,$$

where $\cdot$ is the $\mathbf{R}^d$ inner product and $\underline{a} = \underline{a}(\underline{x}) = (a_{ij})$, $i, j = 1, \cdots, d$, is a matrix of coefficients satisfying $a_{ij} \in \mathcal{C}^1(\bar{\Omega})$, $a_{ij} = a_{ji}$, $i, j = 1, \cdots, d$,

$$\underline{z} \cdot \underline{a}(\underline{x}) \underline{z} \geq c\, \|\underline{z}\|^2 \quad \text{for all } \underline{z} \in \mathbf{R}^d, \quad \underline{x} \in \bar{\Omega}.$$

Further, let

$$\mathcal{N}(u, v) = \int_\Omega v \underline{b} \cdot \nabla u + guv,$$

with $\underline{b} = (b_i)$, $b_i \in \mathcal{C}^1(\bar{\Omega})$, and $g \in \mathcal{C}(\bar{\Omega})$. Then (3.1) is a weak formulation of the *Dirichlet problem*

$$(3.12) \qquad \begin{aligned} -\nabla \cdot (\underline{a} \nabla u) + \underline{b} \cdot \nabla u + gu &= f \quad \text{in } \Omega, \\ u &= 0 \quad \text{on } \partial\Omega. \end{aligned}$$

The interpolated spaces $\mathscr{H}^s$ coincide with the Sobolev spaces $\mathscr{H}_0^s(\Omega)$ for $s \neq \frac{1}{2}$.

If zero is not an eigenvalue of the operator in (3.12), then the regularity assumption (3.10) holds with $0 < \alpha < \frac{1}{2}$ (at least) (cf. Nečas [Ne1]). (The smoothness assumptions about $\underset{\sim}{a}$, $\underset{\sim}{b}$, and g can be weakened.) By Kadlec [Ka1], (3.10) holds with $\alpha = 1$ if Ω is convex or $\partial\Omega$ is smooth. It is easy to see that (3.3) holds with $\eta = 1$.

For Ω polygonal, consider a family of usual piecewise linear finite element spaces $\mathscr{S}^k = \mathscr{V}^h \subset \mathscr{C}(\bar{\Omega}) \cap \mathscr{H}^1(\Omega)$ with the usual nodal bases (cf. Ciarlet [Ci1] and Strang and Fix [St1]). Assume that we accomplish the passage from $\mathscr{V}^{2h}$ to $\mathscr{V}^h$ by adding nodes at the midpoints of the sides of the triangles and decomposing each triangle into four triangles (if $d = 2$) or, generally, by some means that ensures that $\mathscr{V}^{2h} \subset \mathscr{V}^h$ and the usual *shape and size regularity* is maintained. Then for $B^k = c(h_k)^d I$ we have $c^{-1}(h_k)^{-2} \leq \rho(E^k) \leq c(h_k)^{-2}$, where $h_k = c2^{-k}$. Hence, $\psi = \frac{1}{4}$. The approximation property (3.7) holds for $\bar{\alpha} = 1$ (cf. [Ci1], [St1]).

Example 3.2. Let $\Omega \subset \mathbf{R}^3$ be bounded and polygonal, $\partial\Omega = \Gamma_1 \cup \Gamma_2$, $\Gamma_1 \cap \Gamma_2$ of measure zero, Γ_1 of positive measure, $\mathscr{V} = \{u \in \mathscr{H}^1(\Omega) : u = 0$ on $\Gamma_1\}$, $\mathscr{H}^0 = (\mathscr{L}^2(\Omega))^3$, and $\mathscr{H}^1 = (\mathscr{V})^3$ equipped with the norm

$$\|\underset{\sim}{u}\|_1 = \left(\sum_{i=1}^{3} \|u_{[i]}\|^2_{\mathscr{H}^1(\Omega)}\right)^{1/2}.$$

Further, let $\mathscr{N} = 0$ and

$$\mathscr{A}(\underset{\sim}{u}, \underset{\sim}{v}) = \int_\Omega \left(k - \frac{2}{3}\mu\right) \operatorname{div} \underset{\sim}{u} \operatorname{div} \underset{\sim}{v} + 2\mu \sum_{i,j=1}^{3} e_{ij}(\underset{\sim}{u}) e_{ij}(\underset{\sim}{v}),$$

where

$$e_{ij}(\underset{\sim}{u}) = \frac{1}{2}\left(\frac{\partial u_{[i]}}{\partial x_{[j]}} + \frac{\partial u_{[j]}}{\partial x_{[i]}}\right)$$

and $0 < \mu < (3/2)k$ are constants. (We use μ and k here in this example because of the prevailing usage.) The right-hand side $\underset{\sim}{f}$ is a functional defined by

$$\langle \underset{\sim}{v}, \underset{\sim}{f} \rangle = \int_\Omega \underset{\sim}{f}_V \cdot \underset{\sim}{v} + \int_{\Gamma_2} \underset{\sim}{f}_N \cdot \underset{\sim}{v},$$

with $\underset{\sim}{f}_V \in (\mathscr{H}^{-1}(\Omega))^3$, $\underset{\sim}{f}_N \in (\mathscr{H}^{-1/2}(\Gamma_2))^3$ given. (In the second integral, $\underset{\sim}{v}$ is understood to be the trace on Γ_2.)

This is a simple *elasticity problem* (cf. Nečas and Hlaváček [Ne2]). For Ω polygonal, we may put $\mathscr{S}^k = (\mathscr{V}^h)^3$, with $\mathscr{V}^h$ similar to that of the preceding example. Then (3.6) holds with $\bar{\alpha} = 1$, and (3.8) with $\psi = \frac{1}{4}$ as before.

When we set $\bar{\mathscr{H}}^s = (\mathscr{H}^s(\Omega))^3$ for $s > 1$, then the regularity assumption (3.10) holds with some $\alpha > 0$ for the *Dirichlet* problem, i.e., the case when

$\Gamma_1 = \partial\Omega$. The proof is a simple extension of that from Nečas [Ne1] to elliptic *systems* [Ne3]. Regularity in the general case is at present an open problem.

We conclude this section with some remarks.

(1) *All* constants in (2.10)–(2.12) can be estimated explicitly. Unfortunately, this is not true of the constant c_1 in (2.9) because the constant in the regularity assumption (3.10) is known only for model problems. The *regularity parameter* α, however, can be computed for regions with piecewise smooth boundaries, but it depends on the angles of corners (cf. Strang and Fix [St1]).

(2) Formula (2.9) suggests that we should choose B^k so that $\rho(E^k)$ is not large. In fact, this is just what we need in practice: since $E^k = (B^k)^{-1} A^k$ is the essential operator used in relaxation, E^k should be as well conditioned as possible (on the "oscillatory" space $\hat{U}^k$). For problems where the discrete equation coefficients are very homogeneous (e.g., Poisson's equation on a uniform grid), it is sufficient (though not necessarily best) to choose $B^k = cI$, a scalar multiple of the identity. But for less homogeneous problems (e.g., badly graded grids and widely varying coefficients), such a choice would give an unacceptably large value of $\rho(E^k)$. A better choice here is $B^k = cD^k$, where $A^k = D^k - L^k - U^k$ is the decomposition defined in Chapter 1. This will tend to make $\rho(E^k)$ insensitive to such heterogeneity. For a more detailed discussion, see McCormick [395].

(3) The convergence theory to be developed in the following sections depends crucially on the value of the regularity parameter α. Actually, for the V-cycle we shall need $\alpha = 1$. The regularity assumption (3.10), with $\mathscr{H}^s$ chosen to be the usual Sobolev spaces as above, does not hold with $\alpha = 1$ in the presence of singularities caused, e.g., by a re-entrant corner. But (2.9) with $\alpha = 1$ can be saved, at least in the two-dimensional case, by the use of suitable *weighted* Sobolev spaces and *nonuniform discretizations* that are gradually refined near such singular points. For such treatment of some second order selfadjoint problems in the multigrid framework, see Yserentant [600].

(4) We have omitted discretizations by finite differences, numerical integration, treatment of curved boundaries, etc., which do not fall within the framework of this section. *But note that the theory developed below can be applied if the variational conditions* (2.2) *and* (2.3) *hold, and if the discrete problems satisfy* (2.9)–(2.13), *by whichever means these assumptions may have been verified.*

(5) Elliptic variational problems of order $2m$ can be formulated as (3.1) with $\mathscr{H}^1$ as a suitable closed subspace of the Sobolev space $\mathscr{H}^m(\Omega)$, and $\mathscr{A} + \mathscr{N}$ a bilinear differential form of order m with the principal part $\mathscr{A}$. We then put $\mathscr{H}^0 = \mathscr{L}^2(\Omega)$ and $\tilde{\mathscr{H}}^s = \mathscr{H}^{ms}(\Omega)$ for $s > 1$. The integration by parts can verify that (3.3) holds with $\eta \geq 1/m$ if the coefficients are sufficiently

smooth. Regularity (3.10) with $0<\alpha<1/(2m)$ is again guaranteed by Nečas [Ne1] in the case of the Dirichlet problem and Lipschitz boundary $\partial\Omega$. It should be noted that (3.10) holds with $\alpha=1$ if the boundary $\partial\Omega$ and the coefficients are smooth (see Lions and Magenes [Li1]). Regularity in the general case is again an open problem.

(6) By passing to the factor space modulo the null space $\mathcal{K}$ of $\mathcal{A}$, we can treat also *semi-coercive* problems. For example, for the *Neumann problem* $-\Delta u=0$, $\partial u/\partial n=0$, $\mathcal{K}$ consists of constant functions. *But to do this, we must have* $\mathcal{K}\subset\mathcal{S}^k$ *for all* k, *and also* $\mathcal{N}(u,v)=0$ *whenever* $u\in\mathcal{K}$ *or* $v\in\mathcal{K}$. Compare Exercise 4.4 and § 5.8 for an example of such a process.

5.4. Convergence theory in the symmetric, positive definite case: approximation property and smoothing factors. In this section we consider the case $\mathcal{N}=0$, so that $N^k=0$ for all k. The purpose of this section is to derive the inequality (4.11) below, which will be used in the next section for obtaining convergence estimates for the multigrid process. The inequality (4.11) is in turn based on (2.9) and the so-called *smoothing factor*, defined by (4.6), which is determined from the properties of the smoother G_1^k or G_2^k alone.

From the Basic Assumptions, we need only (2.9), which we make more specific by introducing a constant δ and writing it as

$$\|\!|u^k|\!\|_{1-\alpha}^2\le\delta^\alpha(\rho(E^k))^{-\alpha}\,\|\!|u^k|\!\|_1^2\quad\text{for all } u\in\hat{T}^k. \tag{4.1}$$

This will be called the (*algebraic*) *approximation property*, which is justified by condition (4.2) in the following lemma.

LEMMA 4.1. *Condition* (4.1) *is equivalent to each of the following*:

(4.2) *For any* $u^k\in H^k$, *there exists* $v^k\in\hat{S}^k$ *so that*

$$\|\!|u^k-v^k|\!\|_1^2\le\delta^\alpha(\rho(E^k))^{-\alpha}\,\|\!|u^k|\!\|_{1+\alpha}^2.$$

(4.3) *For all* $u^k\in H^k$, $\quad\|\!|T^ku^k|\!\|_1^2\le\delta^\alpha(\rho(E^k))^{-\alpha}\,\|\!|u^k|\!\|_{1+\alpha}^2,$

(4.4) $\quad\delta^\alpha\ge\rho T^k(E^k)^{-\alpha}(\rho(E^k))^\alpha.$

Proof. The equivalence of (4.2) and (4.3) follows immediately from the definition of T^k. The equivalence of (4.3) and (4.4) follows from the observation that (we omit the superscripts k)

$$\begin{aligned}\sup_v\frac{\|\!|Tv|\!\|_1^2}{\|\!|v|\!\|_{1+\alpha}^2}&=\sup_v\frac{\langle Tv,v\rangle_A}{\langle E^\alpha v,v\rangle_A}\\&=\sup_u\frac{\langle TE^{-\alpha/2}u,E^{-\alpha/2}u\rangle_A}{\langle u,u\rangle_A}\\&=\rho(E^{-\alpha/2}TE^{-\alpha/2})\\&=\rho(TE^{-\alpha}).\end{aligned}$$

Finally, the equivalence of (4.1) and (4.4) follows from

$$\sup_{v \in \hat{T}} \frac{|||v|||_{1-\alpha}^2}{|||v|||_1^2} = \sup_{v \in H} \frac{\langle E^{-\alpha}Tv, Tv\rangle_A}{|||v|||_1^2}$$
$$= \rho(TE^{-\alpha}T) = \rho(TE^{-\alpha}),$$

where in the first equality we used the fact that $|||Tv|||_1 \le |||v|||_1$, with equality for $v \in \hat{T}$.

In the remainder of this section, we may omit the superscripts k without danger of confusion.

The smoother $G(u, f)$ is, in general, an iterative method for the discrete system $Au = f$. For convenience (cf. [377] for the general case), we restrict ourselves to *stationary linear* iterative methods of the following form:

$$u \leftarrow G(u, f) = u - Q^{-1}(Au - f), \qquad Q \in [H]. \tag{4.5}$$

Denoting by $F = I - Q^{-1}A$ the *linear part* of G, the error $e = \mathbf{u} - u$ is transformed by ν applications of G according to $e \leftarrow F^{\nu}e$. We define the *(algebraic) smoothing factor* $\sigma = \sigma(F)$ by

$$\sigma(F) = \rho(E) \inf_{e \in H} \frac{|||e|||_1^2 - |||Fe|||_1^2}{|||Fe|||_2^2} \tag{4.6}$$

and consider only smoothers having the property that $\sigma \ge 0$, i.e., smoothers that *do not increase the energy norm of the error.* Note that $\sigma(F^0) = \sigma(I) = 0$.

The smoothing factor for more than one smoothing step is estimated in the following lemma.

LEMMA 4.2. *Let $\nu \ge 2$ be an integer and assume $\sigma(F) \ge 0$. Then*

$$\sigma(F^{\nu}) \ge \sigma(F^{\nu-1}). \tag{4.7}$$

In addition, if F is a normal operator in $\langle \cdot, \cdot \rangle_A$ $F^*F = FF^*$; in particular, if F is symmetric in $\langle \cdot, \cdot \rangle_A$, $F^* = F$), then

$$\sigma(F^{\nu}) \ge \nu\sigma(F). \tag{4.8}$$

Proof. Let $e_{(\lambda)} = F^{\lambda}e$, $\lambda = 0, 1, \cdots, \nu$, and note that $\sigma(F)$ is the maximal number s such that

$$\frac{|||e_{(0)}|||_1^2}{|||e_{(1)}|||_1^2} \ge 1 + \frac{s}{\rho(E)} \frac{|||e_{(1)}|||_2^2}{|||e_{(1)}|||_1^2} \tag{4.9}$$

for all $e_{(0)} \in H$, $e_{(1)} \ne 0$. Without loss of generality, let $e_{(\nu)} \ne 0$. Then from this equivalent definition of σ,

$$\frac{|||e_{(0)}|||_1^2}{|||e_{(\nu)}|||_1^2} \geq \frac{|||e_{(1)}|||_1^2}{|||e_{(\nu)}|||_1^2}$$
$$\geq 1 + \frac{\sigma(F^{\nu-1})}{\rho(E)} \frac{|||e_{(\nu)}|||_2^2}{|||e_{(\nu)}|||_1^2},$$

which gives (4.7).

Assume that $F^*F = FF^*$. Let $u \in H$ and $v = F^*Fu$. Then $|||F^2u|||_1^2 = \langle (F^*)^2F^2u, u\rangle_A = |||v|||_1^2$ and the Cauchy–Schwarz inequality gives $|||Fu|||_1^2 = \langle u, v\rangle_A \leq |||u|||_1\, |||v|||_1 = |||u|||_1\, |||F^2u|||_1$. Setting $u = e_{(\lambda)}$, we obtain

$$\text{(4.10)} \qquad \frac{|||e_{(\lambda)}|||_1}{|||e_{(\lambda+1)}|||_1} \geq \frac{|||e_{(\lambda+1)}|||_1}{|||e_{(\lambda+2)}|||_1}.$$

So each iteration reduces the error at most by the same factor as the last one, which yields

$$\frac{|||e_{(0)}|||_1^2}{|||e_{(\nu)}|||_1^2} \geq \left(\frac{|||e_{(\nu-1)}|||_1^2}{|||e_{(\nu)}|||_1^2}\right)^\nu$$
$$\geq \left(1 + \frac{\sigma(F)}{\rho(E)} \frac{|||e_{(\nu)}|||_2^2}{|||e_{(\nu)}|||_1^2}\right)^\nu.$$

Hence, (4.8) now follows from the inequality $(1+x)^\nu \geq 1 + \nu x$, which holds for all $x \geq 0$.

Remark 4.1. Note that $F = F^*$ if $Q = Q^T$ in (4.5).

The convergence theory in the next section will be based on the condition verified by the following lemma.

LEMMA 4.3. *The approximation property* (4.1) *implies, for any smoother with a linear part F, that*

$$\text{(4.11)} \qquad \frac{|||e|||_1^2}{|||Fe|||_1^2} \geq 1 + \frac{\sigma(F)}{\delta}\left(\frac{|||TFe|||_1^2}{|||Fe|||_1^2}\right)^{1/\alpha} \quad \textit{for all } e \in H \textit{ such that } Fe \neq 0.$$

Proof. Let $e_{(0)} = e$, $e_{(1)} = Fe \neq 0$. By the Hölder inequality,[2]

$$|||e_{(1)}|||_{1+\alpha} \leq |||e_{(1)}|||_1^{1-\alpha}\, |||e_{(1)}|||_2^\alpha,$$

[2] Let $\{v_i\}$ be a B-orthonormal basis consisting of eigenvectors of E, $Ev_i = \lambda_i v_i$. From (2.7), we have for any $e \in H$, $|||e|||_s^2 = \sum \lambda_i^s e_i^2$, where $e = \sum e_i v_i$. The Hölder inequality gives

$$\sum (\lambda_i e_i^2)^{1-\alpha}(\lambda_i^2 e_i^2)^\alpha \leq \left(\sum \lambda_i e_i^2\right)^{1-\alpha}\left(\sum \lambda_i^2 e_i^2\right)^\alpha.$$

so

$$\left(\frac{|||e_{(1)}|||_{1+\alpha}^2}{|||e_{(1)}|||_1^2}\right)^{1/\alpha} \leq \frac{|||e_{(1)}|||_2^2}{|||e_{(1)}|||_1^2}.$$

Then by (4.9) and (4.3),

$$\begin{aligned}\frac{|||e_{(0)}|||_1^2}{|||e_{(1)}|||_1^2} &\geq 1 + \frac{\sigma(F)}{\rho(E)}\left(\frac{|||e_{(1)}|||_{1+\alpha}^2}{|||e_{(1)}|||_1^2}\right)^{1/\alpha} \\ &\geq 1 + \frac{\sigma(F)}{\delta}\left(\frac{|||Te_{(1)}|||_1^2}{|||e_{(1)}|||_1^2}\right)^{1/\alpha}.\end{aligned}$$

This is just (4.11), so the lemma is proved.

Now we estimate the smoothing factor $\sigma(F)$ for some smoothers. Perhaps the simplest is the *preconditioned Richardson's iteration*

$$R(u, f) = u - \frac{\omega}{\rho(B^{-1}A)} B^{-1}(Au - f), \tag{4.12}$$

with fixed real parameter $0 < \omega < 2$. Note that its linear part is $F = I - \omega(\rho(E))^{-1}E$.

THEOREM 4.1. *Let $0 < \omega < 2$. Then*

$$\sigma\left(I - \frac{\omega}{\rho(E)} E\right) \geq \min\{2\omega, (\omega - 1)^{-2} - 1\}.$$

Proof. Let $e \in H$, $F = I - \omega(\rho(E))^{-1}E$, $Fe \neq 0$. Then

$$\frac{|||e|||_1^2 - |||Fe|||_1^2}{|||Fe|||_2^2} = \frac{\langle I - (I - \omega(\rho(E))^{-1}E)^2 e, e\rangle_A}{\langle E(I - \omega(\rho(E))^{-1}E)^2 e, e\rangle_A},$$

so

$$\sigma \geq \inf_{0 \leq t \leq 1} \frac{1 - (1 - \omega t)^2}{t(1 - \omega t)^2}.$$

The minimum is attained at either $t = 0$ or $t = 1$.

Remark 4.2. The optimal value in this estimate is $\omega = 3/2$, which gives $\sigma = 3$. Lemma 4.2 applies here, so we have for ν steps of Richardson with $\omega \leq 3/2$, that $\sigma \geq 2\nu\omega$. But compare Exercise 4.2.

The following result applies when $Q \in [H]$ in (4.5) is invertible but otherwise arbitrary.

THEOREM 4.2. *Let $Q \in [H]$ be invertible. Then $\sigma(I - Q^{-1}A)$ is the maximal number s such that*

$$\rho(B^{-1}A)(Q + Q^T - A) \geq s(A - Q^T)B^{-1}(A - Q). \tag{4.13}$$

Proof. For any $u \in H$,

$$|||u|||_2^2 = \langle B(B^{-1}A)^2 u, u\rangle = \langle AB^{-1}Au, u\rangle.$$

Note that $s \le \sigma(I - Q^{-1}A)$ is equivalent to

$$|||e|||_1^2 - |||(I - Q^{-1}A)e|||_1^2 \ge s\ |||(I - Q^{-1}A)e|||_2^2/\rho(E)$$

for all $e \in H$, which is equivalent to

$$(I - Q^{-1}A)^T[A + s(\rho(E))^{-1}AB^{-1}A](I - Q^{-1}A) \le A.$$

Expanding the parentheses, subtracting A, and dividing out $A(Q^{-1})^T$ from the left and $Q^{-1}A$ from the right, we find that this is in turn equivalent to

$$s(\rho(E))^{-1}(Q^TB^{-1}Q - AB^{-1}Q - Q^TB^{-1}A + AB^{-1}A) \le Q + Q^T - A,$$

which is just (4.13).

Remark 4.3. Theorem 4.2 can also be used to estimate the smoothing factor of the adjoint $(I - Q^{-1}A)^* = I - (Q^T)^{-1}A$.

We can now use Theorem 4.2 to estimate the factor σ of smoothing by *Gauss–Seidel* (*GS*) *iterations*, written as $GS(u, f) = (D - L)^{-1}(L^Tu + f)$, where $A = D - L - L^T$, D is the diagonal of A, and $-L$ the strictly lower triangular part of A. Actually, for our next theorem we need not assume any such thing about D and L. Rather, we need only assume that

(4.14) $\quad A = D - L - L^T, \qquad D$ is symmetric, positive definite.

THEOREM 4.3. *Suppose* (4.14) *is true. Then*

$$\sigma((D - L)^{-1}L^T) = \frac{\rho(B^{-1}A)}{\rho(D^{-1}LB^{-1}L^T)}.$$

In particular, if $B = tD$ for some real t, then

$$\sigma((D - L)^{-1}L^T) = \frac{\|D^{-1/2}AD^{-1/2}\|}{\|D^{-1/2}LD^{-1/2}\|^2}.$$

Proof. If $Q = D - L$, then $Q + Q^T = D + A$ is symmetric, positive definite, so $D - L$ is invertible. Condition (4.13) now becomes $\rho(B^{-1}A)D \ge sLB^{-1}L^T$. This proves the theorem.

The estimate (4.8) from Lemma 4.2 does not generally apply here, so we cannot establish that more iterations improve the smoothing factor. However, (4.8) does apply to Gauss–Seidel, provided every other relaxation sweep is performed in a reverse order. This is the *symmetric Gauss–Seidel method* (*SGS*), which we write abstractly as

(4.15) $\quad (D - L)v = L^Tu + f,$

(4.16) $\quad (D - L^T)w = Lv + f, \qquad SGS(u, f) = w.$

THEOREM 4.4. *Suppose* (4.14) *holds. Then the linear part of SGS is*

(4.17) $\quad F = F^* = I - Q^{-1}A \quad$ *with* $Q = (D - L)D^{-1}(D - L^T).$

Its smoothing factor for ν *iterations satisfies*

$$\sigma(F^\nu) \ge 2\nu \frac{\rho(B^{-1}A)}{\rho(LD^{-1}L^TB^{-1})}.$$

In particular, if $B = tD$ *for some real* t, *then*

$$\sigma(F^\nu) \ge 2\nu \frac{\|D^{-1/2}AD^{-1/2}\|}{\|D^{-1/2}LD^{-1/2}\|^2}.$$

Proof. From (4.15),

$$(D - L)(v - u) = f - Au, \tag{4.18}$$

and similarly from (4.16),

$$(D - L^T)(w - v) = f - Av. \tag{4.19}$$

But $f - Av = f - Au - A(v - u) = (D - L)(v - u) - A(v - u)$, so by (4.19), $(D - L^T)(w - v) = L^T(v - u)$. Now,

$$\begin{aligned}(D - L^T)(w - u) &= (D - L^T)(w - v) + (D - L^T)(v - u)\\ &= L^T(v - u) + (D - L^T)(v - u)\\ &= D(v - u)\\ &= D(D - L)^{-1}(f - Au)\end{aligned}$$

from (4.18); hence, $w - u = (D - L^T)^{-1}D(D - L)^{-1}(f - Au)$, which proves (4.17).

Note that $Q = A + LD^{-1}L^T$ so that $Q = Q^T$ and hence, $F = F^*$. Moreover, condition (4.13) becomes

$$\rho(B^{-1}A)(A + 2LD^{-1}L^T) \ge s(LD^{-1}L^T)B^{-1}(LD^{-1}L^T).$$

This, together with Lemma 4.2 and the fact that $A \ge 0$, concludes the proof.

Remark 4.4. Note that $Q = A + R$, $R = LD^{-1}L^T$, so Q is just an incomplete factorization of A. General *ILU smoothers* are of the form (4.17), where D is not necessarily the diagonal of A but still satisfies (4.14). This includes SSOR and most of the commonly proposed symmetric ILU smoothers.

For our last theorem we need the following simple lemma.

LEMMA 4.4. *If* $|||F|||_1 < 1$, *then*

$$\sigma(F) = \frac{\rho(E)}{\rho(E[(I - FF^*)^{-1} - I])}.$$

Proof. Let $e \in H$, $Fe \ne 0$. Then

$$\frac{|||e|||_1^2 - |||Fe|||_1^2}{|||Fe|||_2^2} = \frac{\langle (I - F^*F)e, e\rangle_A}{\langle F^*EFe, e\rangle_A},$$

so, from (4.6), we have

$$\frac{\rho(E)}{\sigma(F)} = \rho(F^*EF(I-F^*F)^{-1}) = \rho(EF(I-F^*F)^{-1}F^*),$$

where

$$F(I-F^*F)^{-1}F^* = F\sum_{i=0}^{\infty}(F^*F)^iF^* = (I-FF^*)^{-1}-I.$$

This proves the lemma.

Often it is possible to write the matrix A in the block form

$$A = \begin{pmatrix} A_{11} & A_{12} \\ A_{12}^T & A_{22} \end{pmatrix}, \quad A_{11} = A_{11}^T, \quad A_{22} = A_{22}^T, \tag{4.20}$$

where the diagonal blocks A_{11} and A_{22} are easily inverted. This is the case, for example, with the 5-point discretization using a "*red-black*" ordering of nodes. For this case, A_{11} and A_{22} are diagonal. Let $u=(u_1, u_2)$ and $v=(v_1, v_2)$, corresponding to the block form, and consider the *two-block Gauss–Seidel method* ("red-black" relaxation) defined by

$$A_{11}v_1 = f - A_{12}u_2, \quad A_{22}v_2 = f - A_{12}^Tv_1, \quad RB(u,f) = v. \tag{4.21}$$

THEOREM 4.5. *Suppose that B has the block structure*

$$B = \begin{pmatrix} B_{11} & 0 \\ 0 & B_{22} \end{pmatrix},$$

and B_{ii} commutes with A_{ii}, $i=1,2$. (This is true, in particular, if B_{ii} is a multiple of A_{ii} or the identity.) Then ν steps of RB have the smoothing factors

$$\sigma(F^\nu) \ge (2\nu-1)\frac{\rho(B^{-1}A)}{\rho(B_{11}^{-1}A_{11})}, \qquad \sigma((F^*)^\nu) \ge (2\nu-1)\frac{\rho(B^{-1}A)}{\rho(B_{22}^{-1}A_{22})}.$$

Proof. The iteration matrix of RB is

$$F = \begin{pmatrix} 0 & -A_{11}^{-1}A_{12} \\ 0 & A_{22}^{-1}A_{12}^TA_{11}^{-1}A_{12} \end{pmatrix}, \qquad |||F|||_1 < 1.$$

Denote $W = A_{11}^{-1}A_{12}A_{22}^{-1}A_{12}^T$. Then

$$F^\nu = \begin{pmatrix} 0 & -W^{\nu-1}A_{11}^{-1}A_{12} \\ 0 & A_{22}^{-1}A_{12}^TW^{\nu-1}A_{11}^{-1}A_{12} \end{pmatrix},$$

and, because F^* corresponds to reverse order of block relaxation in (4.21),

$$(F^*)^\nu = \begin{pmatrix} W^\nu & 0 \\ -A_{22}^{-1}A_{12}^TW^{\nu-1} & 0 \end{pmatrix}.$$

Thus,

$$F^{\nu}(F^*)^{\nu}=\begin{pmatrix} W^{2\nu-1} & 0\\ -A_{22}^{-1}A_{12}^{T}W^{2\nu-1} & 0\end{pmatrix}.$$

It follows that $\rho(W)<1$ because $|||F|||_1<1$, and we can compute

$$A[(I-F^{\nu}(F^*)^{\nu})^{-1}-I]=\begin{pmatrix} A_{11}(I-W^{2\nu-1})^{-1}(I-W)W^{2\nu-1} & 0\\ 0 & 0\end{pmatrix}.$$

Now, because $A_{11}^{1/2}WA_{11}^{-1/2}$ is symmetric and positive definite, and A_{11} and B_{11} commute, we have

$$\begin{aligned}\rho(E[(I-F^{\nu}(F^*)^{\nu})^{-1}-I])&=\rho(B_{11}^{-1}A_{11}(I-W^{2\nu-1})^{-1}(I-W)W^{2\nu-1})\\ &\le \rho(B_{11}^{-1}A_{11})\sup_{0\le t\le 1}\frac{(1-t)t^{2\nu-1}}{1-t^{2\nu-1}}\\ &=\frac{\rho(B_{11}^{-1}A_{11})}{2\nu-1}.\end{aligned}$$

It remains to apply Lemma 4.4. The estimate for F^* is obtained by reversing the order of block relaxation.

Remark 4.5. Note that we always have $\rho(B_{ii}^{-1}A_{ii})\le\rho(B^{-1}A)$. However, for usual discretizations, we may often expect that $\rho(B^{-1}A)\approx 2\rho(B_{ii}^{-1}A_{ii})$.

5.5. Estimate of convergence factors. In this section, starting from the inequality (4.11), we analyze the convergence properties of the μ-cycle multigrid algorithm $u^k\leftarrow MG_\mu^k(u^k,f^k)$ defined in § 5.2.3. We shall estimate the *squared energy norm convergence factors* defined by

$$\varepsilon^k=\inf\{\varepsilon\colon |||\mathbf{u}^k-MG_\mu^k(u^k,f^k)|||^2\le\varepsilon\ |||\mathbf{u}^k-u^k|||^2,\text{ for all }u^k\}.$$

Here and below (we need no other norm here), $|||\cdot|||=|||\cdot|||_1$.

Let the smoothers G_1^k and G_2^k have linear parts F_1^k and $(F_2^k)^*$, respectively. The estimates will be based on condition (4.11), verified in Lemma 4.3, which we write as

$$\frac{|||e^k|||^2}{|||(F_i^k)^{\nu_i}e^k|||^2}\ge 1+\beta_i\left(\frac{|||T^k(F_i^k)^{\nu_i}e^k|||^2}{|||(F_i^k)^{\nu_i}e^k|||^2}\right)^{1/\alpha}\tag{5.1}$$

for all $e^k\in H^k$, $(F_i^k)^{\nu_i}e^k\ne 0$, $i=1,2$, and all $k=2,3,\cdots$. Here, $\beta_i\ge\sigma((F_i^k)^{\nu_i})/\delta$, and $\beta_i\ge\nu_i\sigma(F_i^k)/\delta$ if $(F_i^k)^*F_i^k=F_i^k(F_i^k)^*$ from Lemma 4.2.

The following lemma contains the essence of our theory, namely a recursive bound on ε^k. The estimate of ε^k is split into two components determined by the effects of pre-smoothing and post-smoothing. The recursion holds for each component separately.

LEMMA 5.1. *Let $k \geq 2$ and suppose we have already bounded ε^{k-1} in terms of level $k-1$ pre-smoothing and post-smoothing quantities ε_i^{k-1} according to*

$$\varepsilon^{k-1} \leq \varepsilon_1^{k-1}\varepsilon_2^{k-1}, \quad 0 \leq \varepsilon_i^{k-1} \leq 1, \quad i=1,2. \tag{5.2}$$

Then $\varepsilon^k \leq \varepsilon_1^k \varepsilon_2^k$, where

$$\begin{aligned} \varepsilon_i^k &= p(\alpha, \beta_i, (\varepsilon_i^{k-1})^\mu), \qquad i=1,2, \\ p(\alpha, \beta, \varepsilon) &= \sup_{0 \leq t \leq 1} \frac{t+\varepsilon(1-t)}{1+\beta t^{1/\alpha}}. \end{aligned} \tag{5.3}$$

Proof. Let $u_{(\lambda)}^k$, $\lambda = 0, 1, 2, 3$, be the current solution before Step 1 and after Steps 1, 3, and 4, respectively. Further, let $e_{(\lambda)} = \mathbf{u}^k - u_{(\lambda)}^k$. Then

$$e_{(1)}^k = (F_1^k)^{\nu_1} e_{(0)}. \tag{5.4}$$

By the variational condition (2.2), $\mathbf{u}^{k-1} = (A^{k-1})^{-1} f^{k-1}$ satisfies

$$\langle A^k(u_{(1)}^k + I_{k-1}^k \mathbf{u}^{k-1}) - f^k, v \rangle = 0 \quad \text{for all } v \in \hat{S}^k = R(I_{k-1}^k).$$

Consequently, $\langle e_{(1)}^k - I_{k-1}^k \mathbf{u}^{k-1}, v\rangle_A = 0$, for all $v \in \hat{S}^k$, so $e_{(1)}^k - I_{k-1}^k \mathbf{u}^{k-1} = T^k e_{(1)}^k$. It follows that

$$e_{(2)}^k = T^k e_{(1)}^k + e, \qquad e \in \hat{S}^k, \tag{5.5}$$

where, by (5.2),

$$|||e|||^2 \leq (\varepsilon_1^{k-1}\varepsilon_2^{k-1})^\mu \, |||S^k e_{(1)}^k|||^2 \tag{5.6}$$

because the norm of the initial error for solving $A^{k-1}\mathbf{u}^{k-1} = f^{k-1}$ is

$$|||\mathbf{u}^{k-1}|||_{1,k-1} = |||I_{k-1}^k \mathbf{u}^{k-1}|||_{1,k} = |||S^k e_{(1)}^k|||_{1,k}.$$

Note that we have used (2.5) here. Finally,

$$e_{(3)}^k = (F_2^{k*})^{\nu_2} e_{(2)}^k. \tag{5.7}$$

We may now omit the level superscript with the exception of the convergence factors. Suppose $e_{(3)} \neq 0$; let $v \in H$, $w = F_2^{\nu_2} v \neq 0$, and compute from (5.5) and (5.7) that

$$\langle e_{(3)}, v\rangle_A = \langle Te_{(1)} + e, w\rangle_A = \langle Te_{(1)}, Tw\rangle_A + \langle e, Sw\rangle_A$$

by orthogonality. Now, using (5.6) and the Cauchy–Schwarz inequality twice, we have

$$\begin{aligned} |\langle e_{(3)}, v\rangle_A|^2 &\leq (|||Te_{(1)}|||\;|||Tw||| + (\varepsilon_1^{k-1}\varepsilon_2^{k-1})^{\mu/2}\,|||Se_{(1)}|||\;|||Sw|||)^2 \\ &\leq [|||Te_{(1)}|||^2 + (\varepsilon_1^{k-1})^\mu\,|||Se_{(1)}|||^2][|||Tw|||^2 + (\varepsilon_2^{k-1})^\mu\,|||Sw|||^2]. \end{aligned}$$

Using this and (5.1) for $e=e_{(0)}$, $i=1$, and $e=v$, $i=2$, we get

$$\frac{|\langle e_{(3)}, v\rangle_A|^2}{|||e_{(0)}|||^2\,|||v|||^2} \le \frac{|||e_{(1)}|||^2}{|||e_{(0)}|||^2}\cdot\frac{|\langle e_{(3)}, v\rangle|^2}{|||e_{(1)}|||^2\,|||w|||^2}\cdot\frac{|||w|||^2}{|||v|||^2}$$

$$\le \frac{1}{1+\beta_1 t_1^{1/\alpha}}[t_1+(\varepsilon_1^{k-1})^\mu(1-t_1)][t_2+(\varepsilon_2^{k-1})^\mu(1-t_2)]\frac{1}{1+\beta_2 t_2^{1/\alpha}},$$

where we have set

$$t_1=\frac{|||Te_{(1)}|||^2}{|||e_{(1)}|||^2} \quad\text{and}\quad t_2=\frac{|||Tw|||^2}{|||w|||^2},$$

$0\le t_i\le 1$. This completes the proof.

The quantities t_i above can be interpreted as a "measure of coarseness" of the error; in fact, t_1 is small for smooth errors $e_{(1)}$, which can be well approximated in the subspace $\hat{S}$.

This lemma provides an estimate of two-level convergence with inexact correction. We are now ready for the main result.

THEOREM 5.1. *Define* $\bar\varepsilon=\bar\varepsilon_\mu(\alpha,\beta)$ *as the smallest* ε *satisfying* (*cf.* (5.3))

$$p(\alpha,\beta,(\varepsilon)^\mu)\le\varepsilon,\qquad 0\le\varepsilon\le 1.$$

Then $\varepsilon^1\le\bar\varepsilon_\mu(\alpha,\beta_1)\bar\varepsilon_\mu(\alpha,\beta_2)$ *implies that* $\varepsilon^k\le\bar\varepsilon_\mu(\alpha,\beta_1)\bar\varepsilon_\mu(\alpha,\beta_2)$ *for all* k.

Proof. Set $\varepsilon^1=\varepsilon_1^1\varepsilon_2^1$, $0\le\varepsilon_i^1\le\bar\varepsilon_\mu(\alpha,\beta_i)$, define ε_i^k, $k\ge 2$, by (5.3), and apply Lemma 5.1. The theorem then follows by our noting that $p(\alpha,\beta,\cdot)$ is nondecreasing.

It remains to determine the quantity $\bar\varepsilon=\bar\varepsilon_\mu(\alpha,\beta)$.

THEOREM 5.2. *If* $\mu=1$ *and* $\alpha=1$, *or* $\mu=2$ *and* $1\ge\alpha\ge(\beta+1)/(\beta+2)$, *then*

$$\bar\varepsilon=\frac{1}{1+\beta}. \tag{5.8}$$

For $\mu=2$ *and general* $0<\alpha\le 1$, *it holds that* $\bar\varepsilon\ge 1/(1+\beta)$, *and* $\bar\varepsilon$ *is uniquely determined by* $\Phi(\alpha,\bar\varepsilon)=\beta$, $0<\bar\varepsilon\le 1$, *where*

$$\Phi(\alpha,\varepsilon)=\begin{cases}\dfrac{1}{\varepsilon}-1 & \text{if } \alpha\ge\dfrac{1}{1+\varepsilon},\\[2ex] (1-\varepsilon)\left(1+\dfrac{1}{\varepsilon}\right)^{1/\alpha}\alpha(1-\alpha)^{(1/\alpha)-1} & \text{if } \alpha\le\dfrac{1}{1+\varepsilon}.\end{cases}$$

Proof. If $\alpha=1$, then $p(\alpha,\beta,(\varepsilon)^\mu)=\max\{(\varepsilon)^\mu, 1/(1+\beta)\}$. Let $\mu=2$, $0<\varepsilon\le 1$, and $0<\alpha<1$. Then $p(\alpha,\beta,(\varepsilon)^2)\le\varepsilon$ if and only if

$$t+(\varepsilon)^2(1-t)\le\varepsilon(1+\beta t^{1/\alpha})\quad\text{for all } 0\le t\le 1. \tag{5.9}$$

This is satisfied for $t=0$ because $0<\varepsilon\le 1$. Solving for β, we get that (5.9) is in turn equivalent to $\beta \ge \Psi(\alpha, \varepsilon, t)$ for all t, $0 < t \le 1$, where

$$\Psi(\alpha, \varepsilon, t) = \frac{(1-\varepsilon)[t(1+\varepsilon)-\varepsilon]}{\varepsilon t^{1/\alpha}}.$$

Noting that $d\Psi/dt = 0$ if and only if $t = \varepsilon/((1+\varepsilon)(1-\alpha))$, we can prove that $\Phi(\alpha, \varepsilon) = \sup_{0<t\le 1} \Psi(\alpha, \beta, t)$. Finally, $\Phi(\alpha, \cdot)$ is continuous and decreasing.

The first alternative in the definition of Φ yields (5.8) for $\mu = 2$ and $\alpha < 1$, which concludes the proof.

In Table 5.1, we give the values of $\beta = \Phi(\alpha, \varepsilon)$. This determines the value of β needed for attaining the estimate $\bar{\varepsilon}$.

In the last theorem, we provide asymptotic estimates of the W-cycle convergence factor for a large number of smoothing steps (cf. Lemma 4.2).

THEOREM 5.3. *Suppose* $0 < \alpha < 1$ *and* $\mu = 2$. *Then*

$$\bar{\varepsilon}_\mu(\alpha, \beta) < \frac{1}{\nu^\alpha \beta^\alpha \alpha^{-\alpha}(1-\alpha)^{\alpha-1} - 1} \le \frac{c}{\nu^\alpha}$$

for all sufficiently large ν, *where* $c = c(\alpha, \beta)$.

Proof. Let $\varepsilon \le (1/\alpha) - 1$. Then $1/(1+\varepsilon) \ge \alpha$ and, from the inequality $1 - \varepsilon < 1$, we obtain $\Phi(\alpha, \varepsilon) < (1 + 1/\varepsilon)^{1/\alpha}\alpha(1-\alpha)^{(1/\alpha)-1}$. Hence, $\varepsilon \ge (\nu^\alpha \beta^\alpha \alpha^{-\alpha}(1-\alpha)^{\alpha-1} - 1)^{-1}$ implies $\Phi(\alpha, \varepsilon) < \nu\beta$. It remains for us to note that $\Phi(\alpha, \cdot)$ is decreasing and $\Phi(\alpha, \varepsilon) \to \infty$ as $\varepsilon \to 0$, and to apply Theorem 5.2.

5.6. Convergence theory in the general case. We now study the mutigrid method for the problem $(A^k + N^k)\mathbf{u}^k = f^k$, where A^k is symmetric and positive definite, and N^k is a general perturbation. Recall that we are assuming $A^k + N^k$ to be invertible, $k = 1, 2, \cdots$. The following lemma summarizes the conditions we need in our analysis.

LEMMA 6.1. *Suppose the Basic Assumptions* (2.9)–(2.13) *are valid. Then the following holds for some* δ *and* ζ *and all* k *large enough:*

If $\alpha \le \eta$, *then*

(6.1) $$|||T^k u^k|||_1^2 \le \delta^\alpha (\rho(E^k))^{-\alpha} |||u^k|||_{1+\alpha}^2 \quad \textit{for all } u^k \in H^k.$$

In any case, we have

(6.2)
$$|\langle u^k, v^k \rangle_A| \le \zeta \psi^{k \min\{\alpha, \eta\}/2} |||u^k|||_1 \, |||v^k|||_1 \quad \textit{for all } u^k \in \hat{U}^k \textit{ and } \hat{v}^k \in \hat{S}^k$$

and

(6.3) $$\left|\left|\left| \frac{1}{\rho((B^k)^{-1}A^k)} (B^k)^{-1} N^k \right|\right|\right|_1 \le c\psi^{k\eta/2}.$$

TABLE 5.1
Values of $\beta = \Phi(\alpha, \varepsilon)$ from Theorem 5.2.

ε \ α	0.05	0.10	0.15	0.20	0.25	0.30	0.35	0.40	0.45	0.50	0.55	0.60	0.65	0.70	0.75	0.80	0.85	0.90	0.95	1.00
0.05						3168.15	895.60	356.91	178.59	104.74	68.93	49.46	37.96	30.73	26.01	22.85	20.76	19.50	19.00	19.00
0.10					1389.75	347.72	133.74	67.14	40.21	27.23	20.15	15.95	13.30	11.56	10.40	9.65	9.19	9.00	9.00	9.00
0.15				1844.35	309.72	98.58	45.03	25.72	17.03	12.49	9.87	8.25	7.21	6.52	6.07	5.80	5.68	5.67	5.67	5.67
0.20			7360.34	509.61	109.35	40.98	21.04	13.11	9.29	7.20	5.95	5.16	4.65	4.32	4.12	4.02	4.00	4.00	4.00	4.00
0.25			2046.42	192.00	49.44	20.92	11.71	7.79	5.81	4.69	4.00	3.57	3.29	3.12	3.03	3.00	3.00	3.00	3.00	3.00
0.30			735.72	87.62	26.03	12.12	7.26	5.09	3.95	3.29	2.88	2.63	2.47	2.38	2.34	2.33	2.33	2.33	2.33	2.33
0.35			314.40	45.46	15.17	7.64	4.84	3.53	2.83	2.42	2.17	2.01	1.92	1.87	1.86	1.86	1.86	1.86	1.86	1.86
0.40		6412.34	151.85	25.82	9.50	5.10	3.38	2.56	2.10	1.84	1.67	1.58	1.52	1.50	1.50	1.50	1.50	1.50	1.50	1.50
0.45		2571.00	80.20	15.65	6.25	3.55	2.45	1.91	1.61	1.43	1.32	1.26	1.23	1.22	1.22	1.22	1.22	1.22	1.22	1.22
0.50		1143.84	45.28	9.95	4.27	2.54	1.81	1.45	1.24	1.13	1.05	1.02	1.00	1.00	1.00	1.00	1.00	1.00	1.00	1.00
0.55		550.92	26.86	6.55	2.99	1.86	1.37	1.12	0.98	0.89	0.85	0.82	0.82	0.82	0.82	0.82	0.82	0.82	0.82	0.82
0.60		281.79	16.52	4.42	2.13	1.37	1.04	0.86	0.77	0.71	0.68	0.67	0.67	0.67	0.67	0.67	0.67	0.67	0.67	0.67
0.65		150.65	10.41	3.02	1.53	1.02	0.79	0.67	0.60	0.56	0.54	0.54	0.54	0.54	0.54	0.54	0.54	0.54	0.54	0.54
0.70		82.95	6.64	2.08	1.10	0.75	0.60	0.51	0.47	0.44	0.43	0.43	0.43	0.43	0.43	0.43	0.43	0.43	0.43	0.43
0.75		46.33	4.24	1.42	0.78	0.55	0.44	0.39	0.36	0.34	0.33	0.33	0.33	0.33	0.33	0.33	0.33	0.33	0.33	0.33
0.80		25.77	2.66	0.94	0.54	0.39	0.32	0.28	0.26	0.25	0.25	0.25	0.25	0.25	0.25	0.25	0.25	0.25	0.25	0.25
0.85		13.86	1.60	0.60	0.35	0.26	0.22	0.19	0.18	0.18	0.18	0.18	0.18	0.18	0.18	0.18	0.18	0.18	0.18	0.18
0.90	5833.66	6.81	0.87	0.34	0.21	0.16	0.13	0.12	0.11	0.11	0.11	0.11	0.11	0.11	0.11	0.11	0.11	0.11	0.11	0.11
0.95	1663.06	2.57	0.36	0.15	0.09	0.07	0.06	0.06	0.05	0.05	0.05	0.05	0.05	0.05	0.05	0.05	0.05	0.05	0.05	0.05

Proof. We omit the level index k here. Let $u \in H$ and define $v \in \hat{S}$ by requiring $u - v \in \hat{U}$, i.e., $v \in \hat{S}$ and $\langle (A+N)(u-v), w \rangle = 0$, for all $w \in \hat{S}$. Setting here $w = v$, we obtain that

$$|||u - v|||_1^2 = \langle A(u-v), u-v \rangle = \langle A(u-v), u \rangle + \langle N(u-v), v \rangle.$$

Now, from the Cauchy–Schwarz inequality and (2.11), using $\eta \geq \alpha$ and (2.9), and remembering that c is generic, we have that

$$\begin{aligned} |||u-v|||_1^2 &\leq |||u-v|||_{1-\alpha}\, |||u|||_{1+\alpha} + c\, |||u-v|||_{1-\eta}\, |||v|||_1 \\ &\leq |||u-v|||_{1-\alpha}\,(|||u|||_{1+\alpha} + c\, |||u|||_1 + c\, |||u-v|||_1) \\ &\leq c(\rho(E))^{-\alpha/2}\, |||u-v|||_1[(1+c)\, |||u|||_{1+\alpha} + c\, |||u-v|||_1]. \end{aligned}$$

Hence,

$$|||u-v|||_1 \leq \frac{c(\rho(E))^{-\alpha/2}}{1 - c(\rho(E))^{-\alpha/2}}\, |||u|||_{1+\alpha},$$

which, by (2.12), gives (6.1) for all sufficiently large k.

To prove (6.2) for $u \in \hat{U}$, $v \in \hat{S}$, we have by (2.11)

$$|\langle Au, v \rangle| = |-\langle Nu, v \rangle| \leq c\, |||u|||_{1-\eta}\, |||v|||_1.$$

If $\alpha \leq \eta$, we obtain from (2.13) and (2.9) that

$$|||u|||_{1-\eta} \leq |||u|||_{1-\alpha} \leq c(\rho(E))^{-\alpha/2}\, |||u|||_1.$$

If $\alpha > \eta$, then we get from the Hölder inequality and (2.9) that

$$|||u|||_{1-\eta} \leq |||u|||_{1-\alpha}^{\eta/\alpha}\, |||u|||_1^{1-\eta/\alpha} \leq c(\rho(E))^{-\eta/2}\, |||u|||_1.$$

It remains for us to use (2.12).

Finally, using (2.10), we have

$$\begin{aligned} |||B^{-1}N|||_1 &= \sup_{u,v} \frac{\langle E^{1/2}B^{-1}Nu, v \rangle_B}{|||u|||_1\, |||v|||_0} \\ &= \sup_{u,v} \frac{\langle B^{-1}Nu, E^{1/2}v \rangle_B}{|||u|||_1\, |||v|||_0} \\ &= \sup_{u,v} \frac{\langle Nu, E^{1/2}v \rangle}{|||u|||_1\, |||v|||_0} \\ &\leq \sup_{v} \frac{c\, |||E^{1/2}v|||_{1-\eta}}{|||v|||_0} \\ &= c(\rho(E))^{1-(\eta/2)}, \end{aligned}$$

thus proving (6.3), again with the use of (2.12).

Condition (6.1) is just the algebraic *approximation property* for the related problems $A^k\mathbf{u}^k = f^k$, which can be verified independently. Therefore we may admit $\eta < \alpha$, which can happen for higher order problems (cf. remark (5) at the end of § 5.3). We could have based the theory on (2.9) directly, but we want to have the quantity δ in the estimates.

Condition (6.2) means that the subspaces $\hat{U}^k$ and $\hat{S}^k$ are *nearly orthogonal in* $\langle\cdot,\cdot\rangle_A$ *for large* k. Condition (6.3) implies that the smoother

$$R^k(u^k, f^k) = u^k - \frac{\omega}{\rho((B^k)^{-1}A^k)}(B^k)^{-1}(A^k u^k + N^k u^k - f^k),$$

with the iteration matrix $F^k = W^k + Z^k$, where

$$W^k = \frac{1}{\rho((B^k)^{-1}A^k)}(B^k)^{-1}A^k, \qquad Z^k = \frac{1}{\rho((B^k)^{-1}A^k)}(B^k)^{-1}N^k,$$

can be considered as a perturbation of the case $N^k = 0$ for large k. This is done more generally in the following lemma. For the smoothing factor σ, compare (4.6). We may omit the level indices k here.

LEMMA 6.2. *Suppose* $F = W + Z \in [H]$, $|||W|||_1 < 1$, $\nu \geq 1$ *is an integer, and* $W^*W = WW^*$ *in the case* $\nu \geq 2$. *Then*

$$|||F^\nu e|||_1^2 + \frac{\nu\sigma(W)}{\rho(E)}|||F^\nu e|||_2^2 \leq |||e|||_1^2(1 + [(1 + |||Z|||_1)^{2\nu} - 1][1 + \nu\sigma(W)]) \tag{6.4}$$

for any $e \in H$.

Proof. Denote $\xi = |||Z|||_1$. It holds that

$$\begin{aligned} |||(W+Z)^\nu e|||_1 &\leq |||W^\nu e|||_1 + \sum_{i=1}^{\nu}\binom{\nu}{i}|||W|||_1^{\nu-i}|||Z|||_1^i|||e|||_1 \\ &\leq |||W^\nu e|||_1 + [(1+\xi)^\nu - 1]|||e|||_1 \end{aligned}$$

by the fact that $|||W|||_1 < 1$. Now

$$2[(1+\xi)^\nu - 1] + [(1+\xi)^\nu - 1]^2 = (1+\xi)^{2\nu} - 1. \tag{6.5}$$

From this we obtain

$$|||F^\nu e|||_1^2 \leq |||W^\nu e|||_1^2 + [(1+\xi)^{2\nu} - 1]|||e|||_1^2. \tag{6.6}$$

Similarly,

$$|||(W+Z)^\nu e|||_2 \leq |||W^\nu e|||_2 + (\rho(E))^{1/2}\sum_{i=1}^{\nu}\binom{\nu}{i}|||W|||_1^{\nu-i}|||Z|||_1^i|||e|||_1.$$

Now, using the inequality

$$|||W^\nu e|||_2 \leq (\rho(E))^{1/2}|||W^\nu e|||_1 \leq (\rho(E))^{1/2}|||e|||_1$$

and (6.5), we have that

$$\|\|F^{\nu}e\|\|_2^2 \le \|\|W^{\nu}e\|\|_2^2 + \rho(E)[(1+\xi)^{2\nu}-1]\,\|\|e\|\|_1^2. \tag{6.7}$$

Finally, from Lemma 4.2,

$$\|\|W^{\nu}e\|\|_1^2 + \frac{\nu\sigma(W)}{\rho(E)}\|\|W^{\nu}e\|\|_2^2 \le \|\|e\|\|_1^2,$$

which, together with (6.6) and (6.7), concludes the proof.

The following lemma relates the projections U^k and T^k (cf. § 5.2.4).

LEMMA 6.3. *Condition* (6.2) *implies, for all k large enough, that*

$$\|\|U^k e^k\|\|_1^2 \le \frac{\|\|T^k e^k\|\|_1^2}{1-\zeta^2\psi^{k\min\{\alpha,\eta\}}} \quad \textit{for all } e^k \in H^k.$$

Proof. Since $U^kT^k = U^k$, it suffices to consider the case $e^k \in \hat{T}^k$. Let ξ be the angle (in the inner product $\langle\cdot,\cdot\rangle_A$) between U^ke^k and $(I-U^k)e^k$. By (6.2), $|\cos\xi| \le \zeta\psi^{k\min\{\alpha,\eta\}/2}$. Because $e^k \in \hat{T}^k$, we have that $\langle e^k, (I-U^k)e^k\rangle_A = 0$, and thus $\|\|U^ke^k\|\|_1\,|\sin\xi| = \|\|e^k\|\|_1$, completing the proof.

We may now proceed to a result analogous to Lemma 5.1. Again, ε^k is the squared convergence factor in the $\|\|\cdot\|\|_1$ norm and G_1^k and G_2^k are pre-smoothing and post-smoothing with linear parts F_1^k and $(F_2^k)^*$, respectively.

LEMMA 6.4. *Let* $F_i^k = W_i^k + Z_i^k$, $i = 1, 2$, *satisfy the assumptions of Lemma* 6.2. *For all k sufficiently large, i.e.,* $k \ge k_0(\alpha,\zeta,\eta,\psi)$, *then* $\varepsilon^{k-1} \le \varepsilon_1^{k-1}\varepsilon_2^{k-1}$, $0 \le \varepsilon_i^{k-1} \le 1$, *implies* $\varepsilon^k \le \varepsilon_1^k\varepsilon_2^k$. *Here,*

$$\begin{aligned}\varepsilon_i^k = &\left[p\left(\alpha, \frac{\nu_i\sigma(W_i^k)}{\delta}, (\varepsilon_i^{k-1})^{\mu}\right) + c\psi^{k\min\{\alpha,\eta\}/2}\right]\\ &\cdot(1+[(1+\|\|Z_i^k\|\|_1)^{2\nu_i}-1][1+\nu_i\sigma(W_i^k)]),\end{aligned} \tag{6.8}$$

where $c = c(\zeta) > 0$ *and the function p was defined in* (5.3).

Proof. We proceed as in the proof of Lemma 5.1. With the errors denoted again by $e_{(\lambda)}^k$, we have

$$e_{(1)}^k = (F_1^k)^{\nu_1}e_{(0)}, \tag{6.9}$$

$$e_{(2)}^k = U^ke_{(1)}^k + e, \qquad e \in \hat{S}^k, \tag{6.10}$$

$$\|\|e\|\|_1^2 \le (\varepsilon_1^{k-1}\varepsilon_2^{k-1})^{\mu}\,\|\|(I-U^k)e_{(1)}^k\|\|_1^2, \tag{6.11}$$

$$e_{(3)}^k = (F_2^{k*})^{\nu_2}e_{(2)}^k. \tag{6.12}$$

We omit the level superscript from here on except for convergence factors ε and the powers of ψ. Suppose $e_{(3)} \ne 0$ and let $v \in H$, $w = F_2^{\nu_2}v \ne 0$. Denote $\tilde{\alpha} = \min\{\alpha,\eta\}$. Then from (6.10)–(6.12), (6.2), and using the Cauchy–

Schwarz inequality twice, we have

$$
\begin{aligned}
|\langle e_{(3)}, v\rangle_A| &= |\langle Ue_{(1)} + e, Uw + (I-U)w\rangle_A| \\
&\le |||Ue_{(1)}|||_1\, |||Uw|||_1 \\
&\quad + (\varepsilon_1^{k-1}\varepsilon_2^{k-1})^{\mu/2}\, |||(I-U)e_{(1)}|||_1\, |||(I-U)w|||_1 \\
&\quad + 2\zeta\psi^{k\bar{\alpha}/2}\, |||e_{(1)}|||_1\, |||w|||_1\, |||U|||_1\, |||I-U|||_1 \\
&\le [|||Ue_{(1)}|||_1^2 + (\varepsilon_1^{k-1})^{\mu}\, |||(I-U)e_{(1)}|||_1^2 + \tilde{c}\psi^{k\bar{\alpha}/2}\, |||e_{(1)}|||_1^2]^{1/2} \\
&\quad \cdot[|||Uw|||_1^2 + (\varepsilon_2^{k-1})^{\mu}\, |||(I-U)w|||_1^2 + \tilde{c}\psi^{k\bar{\alpha}/2}\, |||w|||_1^2]^{1/2},
\end{aligned} \tag{6.13}
$$

where $\tilde{c} = 2\zeta\, |||U|||_1\, |||I-U|||_1 \le c(\zeta)$ for k large enough, i.e., $k \ge c(\zeta, \alpha, \psi, \eta)$ (cf. Lemma 6.3). Now set

$$
t = \frac{|||Ue_{(1)}|||_1^2}{|||e_{(1)}|||_1^2}, \qquad s = \frac{|||(I-U)e_{(1)}|||_1^2}{|||e_{(1)}|||_1^2},
$$

with

$$
t + s \le 1 + 2\zeta\psi^{k\bar{\alpha}/2}(ts)^{1/2}, \qquad t \ge 0, \quad s \ge 0, \tag{6.14}
$$

from (6.2). But it follows from (6.14) that if $ts \ne 0$, then

$$
(ts)^{-1/2} \ge \left(\left(\frac{t}{s}\right)^{1/2} + \left(\frac{s}{t}\right)^{1/2} - 2\zeta\psi^{k\bar{\alpha}/2}\right) \ge 2(1 - \zeta\psi^{k\bar{\alpha}/2}),
$$

so, for k so large that $\zeta\psi^{k\bar{\alpha}/2} < 1$,

$$
\begin{aligned}
t + s &\le 1 + \zeta\psi^{k\bar{\alpha}/2}(1 - \zeta\psi^{k\bar{\alpha}/2})^{-1} \\
&= (1 - \zeta\psi^{k\bar{\alpha}/2})^{-1}.
\end{aligned} \tag{6.15}
$$

We bound the first term after the final inequality in (6.13) by

$$
|||e_{(1)}|||_1\, [t + (\varepsilon_1^{k-1})^{\mu} s + c\psi^{k\bar{\alpha}/2}]^{1/2}.
$$

By the Hölder inequality, (6.1), and Lemma 6.3,

$$
\begin{aligned}
\frac{|||e_{(1)}|||_2^2}{|||e_{(1)}|||_1^2} &\ge \left(\frac{|||e_{(1)}|||_{1+\alpha}^2}{|||e_{(1)}|||_1^2}\right)^{1/\alpha} \\
&\ge \frac{\rho(E)}{\delta}\left(\frac{|||Te_{(1)}|||_1^2}{|||e_{(1)}|||_1^2}\right)^{1/\alpha} \\
&\ge \frac{\rho(E)}{\delta}(1 - \zeta^2\psi^{k\bar{\alpha}})^{1/\alpha}\left(\frac{|||Ue_{(1)}|||_1^2}{|||e_{(1)}|||_1^2}\right)^{1/\alpha}.
\end{aligned}
$$

With the use of Lemma 6.2 and (6.9), the square of the first term in (6.13) can now be bounded by

$$|||e_{(0)}|||_1^2 \frac{t + (\varepsilon_1^{k-1})^\mu s + c\psi^{k\bar{\alpha}/2}}{1 + \dfrac{\nu_1 \sigma(W_1)}{\delta}(1 - \zeta^2 \psi^{k\bar{\alpha}})^{1/\alpha} t^{1/\alpha}}(1 + \gamma),$$

where $\gamma = [(1 + |||Z_1|||_1)^{2\nu_1} - 1][1 + \nu_1 \sigma(W_1)]$. Putting $t' = (1 - \zeta\psi^{k\bar{\alpha}/2})t$, $0 \leq t' \leq 1$ (cf. (6.15)) and requiring k so large that $\zeta^2 \psi^{k\bar{\alpha}} \leq \zeta\psi^{k\bar{\alpha}/2}$, we have that this expression is bounded by

$$|||e_0|||_1^2 \left(\frac{t' + (\varepsilon_1^{k-1})^\mu (1 - t')}{1 + (\nu_1 \sigma(W_1)/\delta)(t')^{1/\alpha}} + c\psi^{k\bar{\alpha}/2} \right)(1 + \gamma)$$

$$\leq |||e_{(0)}|||_1^2 \left(p\left(\alpha, \frac{\nu_1 \sigma(W_1)}{\delta}, (\varepsilon_1^{k-1})^\mu \right) + c\psi^{k\bar{\alpha}/2} \right)(1 + \gamma).$$

The use of the same argument for $w = (F_2^*)^{\nu_2} v$ in place of (6.9) and the dual argument

$$|||e_{(3)}|||_1^2 = \sup_v \frac{|\langle e_{(3)}, v \rangle_A|^2}{|||v|||_1^2}$$

completes the proof.

This lemma gives a rather explicit estimate. The following will suffice for our purpose.

COROLLARY 6.1. *Suppose in addition that* $|||Z_i|||_1 \leq \bar{\rho}\psi^{k\eta/2}$ *(cf. (6.3)) and* $\sigma(W_i) \leq \bar{\sigma}$ *for some constants* $\bar{\rho}$ *and* $\bar{\sigma}$. *Then*

$$\varepsilon_i^k \leq p\left(\alpha, \frac{\nu_i \sigma(W_i)}{\delta}, (\varepsilon_i^{k-1})^\mu \right) + c\psi^{k \min\{\alpha, \eta\}/2},$$

where $c = c(\alpha, \zeta, \nu_1, \nu_2, \bar{\rho}, \bar{\sigma}, \psi)$.

We may now proceed to the final results. As we shall see, the estimates approach those for the symmetric, positive definite case. We first consider V-cycles.

THEOREM 6.1. *Suppose* (6.1) *and* (6.2) *hold and let the smoothers satisfy the assumptions of Lemma* 6.2:

$$F_i^k = W_i^k + Z_i^k, \quad |||W_i^k|||_1 < 1, \quad \nu_i \geq 1 \text{ an integer},$$
$$(W_i^k)^* W_i^k = W_i^k (W_i^k)^* \quad \text{if } \nu_i \geq 2,$$

and of Corollary 6.1:

$$|||Z_i^k|||_1 \leq \bar{\rho}\psi^{k\eta/2}, \quad 0 < \eta \leq 1,$$
$$\sigma(W_i^k) \leq \bar{\sigma}$$

for all k.

Denote $\sigma(W_i) = \min_k \sigma(W_i^k)$. *If* $\alpha = 1$, $\mu = 1$, *and* $\bar{\varepsilon} = \bar{\varepsilon}_1\bar{\varepsilon}_2 < 1$, *where*

$$\bar{\varepsilon}_i = \frac{\delta}{\delta + \nu_i\sigma(W_i)}, \qquad i = 1, 2,$$

then $\varepsilon^{k_1} \le \bar{\varepsilon}$ *implies* $\varepsilon^k \le \bar{\varepsilon} + c\psi^{k_1\eta/2} < 1$, *for all* $k > k_1$, $c = c(\alpha, \zeta, \nu_1, \nu_2, \bar{\rho}, \bar{\sigma}, \psi)$, *if* k_1 *is sufficiently large,* i.e., $k_1 \ge c(\alpha, \zeta, \nu_1, \nu_2, \bar{\rho}, \bar{\sigma}, \psi, \eta, \bar{\varepsilon})$.

Proof. Let $\varepsilon^{k_1} = \varepsilon_1^{k_1}\varepsilon_2^{k_1}$, $0 \le \varepsilon_i^{k_1} \le \bar{\varepsilon}_i$. Note that because $\alpha = 1$, we have $\bar{\varepsilon}_i = \bar{\varepsilon}_\mu(\alpha, \beta_i)$, $\beta_i = \nu_i\sigma(W_i)/\sigma$ (cf. Theorem 5.2) and $p(\alpha, \beta, \varepsilon) = \max\{\varepsilon, 1/(1+\beta)\}$. So from Corollary 6.1 and by induction, we have for $k > k_1$ that

$$\varepsilon_i^k \le \bar{\varepsilon}_i + c(\psi^{(k_1+1)\eta/2} + \cdots + \psi^{k\eta/2}),$$

and

$$\psi^{(k_1+1)\eta/2} + \cdots + \psi^{k\eta/2} \le \psi^{k_1\eta/2}\frac{\psi^{\eta/2}}{1-\psi^{\eta/2}}.$$

This proves the theorem.

For the W-cycle, we have a sharper result.

THEOREM 6.2. *Let the assumptions of Theorem* 6.1 *hold, and suppose* $0 < \alpha \le 1$ *and* $\mu = 2$. *Define* $\bar{\varepsilon} = \bar{\varepsilon}_1\bar{\varepsilon}_2$, *where*

$$\bar{\varepsilon}_i = \bar{\varepsilon}_\mu\left(\alpha, \frac{\nu_i\sigma(W_i)}{\delta}\right),$$

with $\bar{\varepsilon}_\mu = \bar{\varepsilon}$ *given by Theorem* 5.2. *Finally let* $\varepsilon^{k_1} \le \bar{\varepsilon} < 1$. *Then*

$$\varepsilon^k \le \bar{\varepsilon} + c\psi^{k_1\min\{\alpha,\eta\}/2}\theta^{k-k_1} < 1$$

for all $k > k_1$, *with* $c = c(\alpha, \zeta, \nu_1, \nu_2, \bar{\rho}, \bar{\sigma}, \psi)$, $\theta = \theta(\alpha, \delta, \nu_1, \nu_2, \bar{\sigma}, \psi) < 1$, *if* k_1 *is sufficiently large, i.e.,* $k_1 \ge c(\alpha, \zeta, \delta, \nu_1, \nu_2, \bar{\rho}, \bar{\sigma}, \psi, \eta, \bar{\varepsilon})$.

Proof. Denote

$$\phi_i(\varepsilon) = p\left(\alpha, \frac{\nu_i\sigma(W_i)}{\delta}, (\varepsilon)^2\right).$$

This function is nondecreasing for $0 \le \varepsilon \le 1$, and it follows from the proof of Theorem 5.2 that $\phi_i(\varepsilon) < \varepsilon$ if $\bar{\varepsilon}_i < \varepsilon < 1$. In addition, there exist $\theta < 1$ and $\tau > 0$ such that

$$\phi_i(\bar{\varepsilon}_i + t) \le \bar{\varepsilon}_i + \theta t \quad \text{for all } 0 \le t \le \tau, \quad i = 1, 2. \tag{6.16}$$

Without loss of generality, let $\theta > \psi^{\min\{\alpha,\eta\}/2} \equiv \tilde{\psi}$. From Corollary 6.1,

$$\varepsilon_i^k \le \phi_i(\varepsilon_i^{k-1}) + c\tilde{\psi}^k, \qquad k > k_1 \tag{6.17}$$

with $\varepsilon^{k_1} = \varepsilon_1^{k_1}\varepsilon_2^{k_1}$, $0 \le \varepsilon_i^{k_1} \le \bar{\varepsilon}_i$. Define numbers $\varepsilon_{i,k_1} = 0$, $\varepsilon_{i,k} = \varepsilon_{i,k-1} + c\tilde{\psi}^k$,

$k > k_1$, with the same c as in (6.17). Then for all $k > k_1$,

$$\begin{aligned}\varepsilon_{i,k} &= c\tilde{\psi}^{k_1} \sum_{j=1}^{k-k_1} \theta^{k-k_1-j}\tilde{\psi}^j \\ &= c\tilde{\psi}^{k_1}\theta^{k-k_1} \sum_{j=1}^{k-k_1} \left(\frac{\tilde{\psi}}{\theta}\right)^j \\ &\leq \bar{c}\tilde{\psi}^{k_1}\theta^{k-k_1}.\end{aligned}$$

If k_1 is sufficiently large, then $\varepsilon_{i,k} \leq \tau$ for all $k > k_1$, and (6.16) and (6.17) imply that

$$\varepsilon_i^k \leq \tilde{\varepsilon}_i + \varepsilon_{i,k} \leq \tilde{\varepsilon}_i + c\psi^{k_1 \min\{\alpha,\eta\}/2}\theta^{k-k_1}.$$

COROLLARY 6.2. *Under the hypotheses of the theorem, if k_1 is large enough, then* $\limsup_{k\to\infty} \varepsilon^k \leq \tilde{\varepsilon}$. *Note that k_1 is fixed here; we thus take the limit for a large number of levels and a fixed coarsest level. Further,* $\lim_{k_1\to\infty} \sup\{\varepsilon^k : k > k_1\} \leq \tilde{\varepsilon}$, *that is, all convergence factors ε^k are asymptotically bounded by $\tilde{\varepsilon}$ for large k_1, regardless of the actual number of levels in the multigrid algorithm. The latter proposition holds also for the V-cycle (cf. Theorem* 6.1).

5.7. Full multigrid. For many applications, full multigrid (*FMG*) algorithms can produce results with errors comparable to the global error. More precisely, when implemented properly, FMG_μ can produce vectors that approximate the solution of the discrete equations at least as well as those discrete solutions can be expected to approximate the solution of the PDE. That is, if we use general *bounds* on the global error, we can construct *FMG* algorithms that give approximations with comparable algebraic error. (It is a subtle but very important point that we use global error bounds here, not the global error itself. In fact, the global error may be fortuitously small or even zero for special right-hand sides—we cannot expect this of the algebraic error.)

To make this rigorous, we have the following theorem, which for simplicity is restricted to the norm $|||\cdot||| = |||\cdot|||_1$.

We shall consider the FMG_μ^k algorithm from Chapter 1 with the initial approximation $u^k = 0$. With ν_0 the number of basic multigrid cycles, one full multigrid cycle is denoted by

$$u^k \leftarrow FMG_\mu^k(f^k)$$

and defined as follows:

Step 1: If $k = k_1$, go to Step 2. Otherwise, set $u^{k-1} \leftarrow FMG_\mu^{k-1}(f^{k-1})$ *and* $u^k \leftarrow I_{k-1}^k u^{k-1}$.

Step 2: Perform $u^k \leftarrow MG_\mu^k(u^k, f^k)$ ν_0 *times.*

THEOREM 7.1. *Let $(A^k + N^k)\mathbf{u}^k = f^k$ for all k, where the right-hand sides satisfy the variational conditions $f^{k-1} = I_k^{k-1} f^k$, $k > 1$. Suppose there exists a $\kappa < \infty$ so that for all k and some $\bar{\eta} > 0$*

$$\|\|I_{k-1}^k \mathbf{u}^{k-1} - \mathbf{u}^k\|\| \le \kappa(\rho(E^{k-1}))^{-\bar{\eta}/2}. \tag{7.1}$$

Let ε_μ^k be the squared convergence factor of MG_μ^k and suppose MG_μ^k is implemented so well that

$$(\varepsilon^k)^{\nu_0} \le \frac{1}{2}\left(\frac{\rho(E^{k-1})}{\rho(E^k)}\right)^{\bar{\eta}}. \tag{7.2}$$

(Note that by (2.12) *the right-hand side term is in some fixed interval* $[c^{-1}, c]$.) *Finally, assume that relaxation on the coarsest level is sufficiently good, namely, that the result of relaxation with initial guess $u^{k_1} = 0$ satisfies*

$$\|\|u^k - \mathbf{u}^k\|\| \le \kappa(\rho(E^k))^{-\bar{\eta}/2}, \tag{7.3}$$

for $k = k_1$. (This can be ensured by one or a combination of the following design choices: small k_1; large ν_1, ν_2; and effective "relaxation," e.g., Gaussian elimination.) Then the final approximation $u^k \leftarrow FMG_\mu(f^k)$ satisfies (7.3) *for all k.*

Proof. Let $u_{(0)}^k$ be the initial approximation on level k, and let $u_{(1)}^k$ be the final approximation. Thus, $u_{(0)}^k = I_{k-1}^k u_{(1)}^{k-1}$. We now show that (7.3) is true with $u^k = u_{(1)}^k$.

Condition (7.3) is true for $k = k_0$ by assumption. Suppose it is true for $k-1$ for some $k > k_0$. Using orthogonality and (2.5), we obtain

$$\begin{aligned}
\|\|u_{(1)}^k - \mathbf{u}^k\|\|^2 &\le (\varepsilon^k)^{\nu_0}\, \|\|u_{(0)}^k - \mathbf{u}^k\|\|^2 \\
&= (\varepsilon^k)^{\nu_0}(\|\|u_{(0)}^k - I_{k-1}^k \mathbf{u}^{k-1}\|\|^2 + \|\|I_{k-1}^k \mathbf{u}^{k-1} - \mathbf{u}^k\|\|^2) \\
&= (\varepsilon^k)^{\nu_0}(\|\|u_{(1)}^{k-1} - \mathbf{u}^{k-1}\|\|^2 + \|\|I_{k-1}^k \mathbf{u}^{k-1} - \mathbf{u}^k\|\|^2) \\
&\le 2(\varepsilon^k)^{\nu_0}\kappa^2(\rho(E^{k-1}))^{-\bar{\eta}} \\
&\le \kappa^2(\rho(E^k))^{-\bar{\eta}}.
\end{aligned}$$

This is just (7.3), so the theorem now follows by induction.

The actual computational complexity of FMG_μ now depends on various factors, including the values of μ, d, ν_0, ν_1, and ν_2 and the cost of other computations (residuals, intergrid transfers, etc.). Typically, the cost of these computations at level k is bounded by $c2^{dk}$. If $\mu < 2^d$, then it is easy to see that the cost of FMG_μ is bounded as well by $c2^{dk}$ (with a different c, of course), which is usually the cost of a few relaxation sweeps. For more details, see Chapter 1, § 5. Thus, in such circumstances and when (7.2) is satisfied, *FMG* can produce acceptable results at a cost of a few floating point operations per grid point.

5.8. Fourier analysis estimates. In this section, we give explicit convergence estimates for several model problems by way of estimating the constant δ from the preceding theory. For simplicity, we consider the symmetric, positive definite case only. We also comment briefly on the use of local mode analysis.

5.8.1. A model problem with periodic boundary conditions. Consider the Laplace equation

$$-\Delta u = f \quad \text{in } \Omega = (0, 1) \times (0, 1), \tag{8.1}$$

with periodic boundary conditions, i.e., the periodic extensions of u and f satisfy $-\Delta u = f$ in all of $\mathbf{R}^2$. This problem is singular, so we require $\int_\Omega f = 0$.

Let $n > 0$ be divisible by 4, put $h = 1/n$, and define the computational grid

$$\Omega^h = \{\underset{\sim}{x} = h\underset{\sim}{m};\, \underset{\sim}{m} \in \mathbf{Z}^2,\, 1 \le \underset{\sim}{m} \le n\}.$$

Here, $\mathbf{Z}$ is the set of all integers, and inequalities like $1 \le \underset{\sim}{m} \le n$ should be understood by components: $\underset{\sim}{m} = (m_{[1]}, m_{[2]})$, $1 \le m_{[i]} \le n$, $i = 1, 2$. The space $\tilde{H}^h$ of grid functions is defined as the space of all real-valued functions on Ω^h. Grid functions are considered periodically extended: $u(\underset{\sim}{x} + \underset{\sim}{m}) = u(\underset{\sim}{x})$, $\underset{\sim}{x} \in \Omega^h$, $\underset{\sim}{m} \in \mathbf{Z}^2$. The space $\tilde{H}^h$ is equipped with the inner product

$$\langle u, v \rangle_h = h^2 \sum_{\underset{\sim}{x} \in \Omega^h} u(\underset{\sim}{x}) v(\underset{\sim}{x}), \tag{8.2}$$

which approximates the $\mathcal{L}^2(\Omega)$ inner product.

In place of the level indices k and $k-1$, we shall use h and $2h$, respectively. We shall use linear operators on $\tilde{H}^h$ defined by the 9-point symmetric stencils

$$L^h \equiv \begin{bmatrix} s_{-1,1} & s_{0,1} & s_{1,1} \\ s_{-1,0} & s_{0,0} & s_{1,0} \\ s_{-1,-1} & s_{0,-1} & s_{1,-1} \end{bmatrix} \equiv [s_{\underset{\sim}{k}}], \qquad s_{\underset{\sim}{k}} = s_{-\underset{\sim}{k}}, \tag{8.3}$$

which means that, for $u^h \in \tilde{H}^h$,

$$L^h u^h(\underset{\sim}{x}) = \sum_{-1 \le \underset{\sim}{k} \le 1} s_{\underset{\sim}{k}} u^h(\underset{\sim}{x} + h\underset{\sim}{k}).$$

Ω^{2h} and $\tilde{H}^{2h}$ are defined analogously, with n replaced by $n/2$ and h by $2h$. Prolongations will be given by 9-point symmetric stencils $I_{2h}^h \equiv]s_{\underset{\sim}{k}}[$, $s_{\underset{\sim}{k}} = s_{-\underset{\sim}{k}}$. This means that, for $\underset{\sim}{z} \in \Omega^h$,

$$I_{2h}^h u^{2h}(\underset{\sim}{z}) = \sum s_{\underset{\sim}{k}} u^{2h}(\underset{\sim}{x}),$$

where the sum is taken over all $\underset{\sim}{x}$ such that

$$\underset{\sim}{z} = \underset{\sim}{x} + h\underset{\sim}{k}, \quad \underset{\sim}{x} = 2h\underset{\sim}{m}, \quad \underset{\sim}{k}, \underset{\sim}{m} \in \mathbf{Z}^2, \quad -1 \le \underset{\sim}{k} \le 1.$$

Now it can be verified that if we define the restriction by the stencil evaluated at $\underset{\sim}{x} \in \Omega^{2h}$, $I_h^{2h} \equiv \frac{1}{4}[s_{\underset{\sim}{k}}]$, with the same $s_{\underset{\sim}{k}}$ as in I_{2h}^h, then I_{2h}^h is the adjoint of I_h^{2h}:

$$\langle I_{2h}^h \underset{\sim}{u}^{2h}, \underset{\sim}{u}^h \rangle_h = \langle \underset{\sim}{u}^{2h}, I_h^{2h} \underset{\sim}{u}^h \rangle_{2h} \quad \text{for all } \underset{\sim}{u}^h \in \tilde{H}^h,\ \underset{\sim}{u}^{2h} \in \tilde{H}^{2h}.$$

Let $\underset{\sim}{\theta} = 2\pi h \underset{\sim}{m}$, $\underset{\sim}{m} \in \mathbf{Z}^2$, and consider the periodic complex functions

$$v_{\underset{\sim}{\theta}}^h(\underset{\sim}{x}) = e^{i\underset{\sim}{\theta}\underset{\sim}{x}/h}, \quad \underset{\sim}{x} = h\underset{\sim}{m}, \quad \underset{\sim}{m} \in \mathbf{Z}^2,$$

where $\underset{\sim}{\theta}\underset{\sim}{x} = \theta_{[1]}x_{[1]} + \theta_{[2]}x_{[2]}$. These functions are eigenfunctions of the operators of the form (8.3)

$$(8.4) \qquad \begin{aligned} L^h v_{\underset{\sim}{\theta}}^h(\underset{\sim}{x}) &= \sum_{-1 \le \underset{\sim}{k} \le 1} s_{\underset{\sim}{k}} e^{i\underset{\sim}{\theta}(\underset{\sim}{x}+\underset{\sim}{k}h)/h} \\ &= v_{\underset{\sim}{\theta}}^h(\underset{\sim}{x}) \sum_{-1 \le \underset{\sim}{k} \le 1} s_{\underset{\sim}{k}} \cos \underset{\sim}{\theta}\underset{\sim}{k}, \end{aligned}$$

where the symmetry $s_{\underset{\sim}{k}} = s_{-\underset{\sim}{k}}$ has been used. (Otherwise the imaginary parts would not cancel.) By periodicity

$$v_{\underset{\sim}{\theta}+2\pi\underset{\sim}{j}}^h(\underset{\sim}{x}) = v_{\underset{\sim}{\theta}}^h(\underset{\sim}{x}) \quad \text{for all } \underset{\sim}{x} = h\underset{\sim}{m}, \quad \underset{\sim}{j}, \underset{\sim}{m} \in \mathbf{Z}^2.$$

The n^2 functions $v_{\underset{\sim}{\theta}}^h$, $\underset{\sim}{\theta} = 2\pi h \underset{\sim}{m}$, $-\pi < \underset{\sim}{\theta} \le \pi$, form an orthonormal basis of the complex extension of $\tilde{H}^h$. Note that we could have considered the real eigenfunctions $\sin(\underset{\sim}{\theta}\underset{\sim}{x}/h)$ and $\cos(\underset{\sim}{\theta}\underset{\sim}{x}/h)$ instead, but computations in the complex extension are easier and more natural.

5.8.2. Discretizations. We consider two discretizations of problem (8.1). The *5-point* discretization is given by the stencils

$$(8.5) \qquad A^h \equiv \frac{1}{h^2}\begin{bmatrix} & -1 & \\ -1 & 4 & -1 \\ & -1 & \end{bmatrix}, \qquad I_{2h}^h = \frac{1}{2}\left]\begin{matrix} & 1 & 1 \\ 1 & 2 & 1 \\ 1 & 1 & \end{matrix}\right[,$$

and the *9-point* discretization by

$$(8.6) \qquad A^h \equiv \frac{1}{3h^2}\begin{bmatrix} -1 & -1 & -1 \\ -1 & 8 & -1 \\ -1 & -1 & -1 \end{bmatrix}, \qquad I_{2h}^h = \frac{1}{4}\left]\begin{matrix} 1 & 2 & 1 \\ 2 & 4 & 2 \\ 1 & 2 & 1 \end{matrix}\right[.$$

In both cases, we have $A^{2h} = I_h^{2h} A^h I_{2h}^h$, with the restriction I_h^{2h} defined as the adjoint of I_{2h}^h with respect to the inner products (8.2). This can be verified either immediately or from a variational principle (cf. Exercise 8.1), where A^h is given by the stiffness matrix divided by h^2 and I_{2h}^h by the natural embedding of the finite element spaces. The 5-point and 9-point discretizations correspond to linear and bilinear finite elements, respectively. The following discussion pertains to both (8.5) and (8.6).

The operator A^h is singular and its nullspace $\ker(A^h)$ consists of all constant functions. It holds that $\ker(A^h) \subset R(I_{2h}^h)$, and we can thus pass to the factorspace

$$H^h = \tilde{H}^h/\ker(A^h).$$

Elements of H^h are classes of functions

$$[f^h] = \{g^h \in \tilde{H}^h : f^h - g^h \in \ker(A^h)\}.$$

There is an isomorphism of the orthogonal complement of $\ker(A^h)$ and $\tilde{H}^h/\ker(A^h)$ given by $f^h \to [f^h]$, and H^h is equipped with the inner product (8.2) in the complement of $\ker(A^h)$. In the following, we simply write f^h for $[f^h]$ when there is no danger of confusion. Because the operators A^h, I_h^{2h}, and I_{2h}^h map constant functions into constant functions, they define in a natural way the operators between the factorspaces: $[A^h]:[u^h] \to [A^h u^h]$, etc. Again we omit the brackets.

5.8.3. Reduction of the estimate of δ to diagonal blocks. Now we shall find subspaces $\hat{E}_\theta^h$ of H^h which will be invariant under all operators concerned. This will make it possible to reduce the analysis to such subspaces.

We shall use the orthogonal bases

$$\{v_{\underset{\sim}{\theta}}^h;\ \underset{\sim}{\theta} = 2\pi h \underset{\sim}{m},\ -\pi < \underset{\sim}{\theta} \le \pi,\ \underset{\sim}{\theta} \ne (0,0),\ \underset{\sim}{m} \in \mathbf{Z}^2\} \tag{8.7}$$

and

$$\{v_{\underset{\sim}{\theta}}^{2h};\ \underset{\sim}{\theta} = 2\pi h \underset{\sim}{m};\ -\pi/2 < \underset{\sim}{\theta} \le \pi/2,\ \underset{\sim}{\theta} \ne (0,0),\ \underset{\sim}{m} \in \mathbf{Z}^2\} \tag{8.8}$$

of the complex extension of H^h and H^{2h}, respectively. Here, $v_{\underset{\sim}{\theta}}^{2h}$ are defined by restriction of $v_{\underset{\sim}{\theta}}^h$ onto Ω^{2h}:

$$v_{\underset{\sim}{\theta}}^{2h}(\underset{\sim}{x}) = v_{\underset{\sim}{\theta}}^h(\underset{\sim}{x}), \quad \underset{\sim}{x} = 2h\underset{\sim}{m}, \quad \underset{\sim}{m} \in \mathbf{Z}^2. \tag{8.9}$$

Because the $v_{\underset{\sim}{\theta}}^{2h}(\underset{\sim}{x})$ are defined for $\underset{\sim}{x} = 2h\underset{\sim}{m}$ only, it holds that

$$v_{\underset{\sim}{\theta}+\pi\underset{\sim}{m}}^{2h} = v_{\underset{\sim}{\theta}}^{2h}, \qquad \underset{\sim}{m} \in \mathbf{Z}^2. \tag{8.10}$$

Let $[a_k]$ and $[b_k]$ be the stencils of A^h and I_h^{2h}, respectively. Then by (8.4),

$$A^h v_{\underset{\sim}{\theta}}^h = \lambda(\underset{\sim}{\theta}) v_{\underset{\sim}{\theta}}^h, \qquad \lambda(\underset{\sim}{\theta}) = \sum_{-1 \le \underset{\sim}{k} \le 1} a_{\underset{\sim}{k}} \cos \underset{\sim}{\theta}\underset{\sim}{k}. \tag{8.11}$$

By (8.4), (8.9), and (8.10), we have, for $\underset{\sim}{m} \in \mathbf{Z}^2$,

$$I_h^{2h} v_{\underset{\sim}{\theta}+\pi\underset{\sim}{m}}^h = \mu(\underset{\sim}{\theta} + \pi\underset{\sim}{m}) v_{\underset{\sim}{\theta}}^{2h}, \qquad \mu(\underset{\sim}{\xi}) = \sum_{-1 \le \underset{\sim}{k} \le 1} b_{\underset{\sim}{k}} \cos \underset{\sim}{\xi}\underset{\sim}{k}. \tag{8.12}$$

We associate with each $\underset{\sim}{\theta} = 2\pi h\underset{\sim}{m}$, $-\pi/2 < \underset{\sim}{\theta} \le \pi/2$, the subspace $\hat{E}_{\underset{\sim}{\theta}}^h \subset H^h$ defined by

$$\hat{E}_{\underset{\sim}{\theta}}^h = \operatorname{span}\{v_{\underset{\sim}{\xi}}^h;\ \underset{\sim}{\xi} = \underset{\sim}{\theta} + \pi\underset{\sim}{n}_i,\ i = 1, 2, 3, 4\},$$

TABLE 5.2
Numerically computed values of δ for discretizations of the Laplace equation on the unit square with periodic boundary conditions.

	$h=1/8$	$h=1/16$	$h=1/32$	$h=1/64$	$h=1/128$	$h=1/256$
5-point	4.73	6.22	6.67	6.79	6.82	6.83
9-point	2.00	2.00	2.00	2.00	2.00	2.00

where

(8.13) $\qquad \underline{n}_1=(0,0), \quad \underline{n}_2=(1,0), \quad \underline{n}_3=(0,1), \quad \underline{n}_4=(1,1).$

Then $A^h\hat{E}^h_\theta \subset \hat{E}^h_\theta$, and, for $\underline{\theta} \neq (0,0)$, $\dim E^h_\theta = 4$ and $I^{2h}_h E^h_\theta = \operatorname{span}\{v^{2h}_\theta\}$; for $\underline{\theta}=(0,0)$, $\dim E^h_\theta = 3$, and $I^{2h}_h E^h_\theta = \{0\}$. The case $\underline{\theta}=(0,0)$ requires special attention because then $[v^h_\theta]=[0]$ and $[v^{2h}_\theta]=[0]$.

Let A_θ and R_θ be the matrix representations of the restrictions of A^h and I^{2h}_h to $\hat{E}^h_\theta$, respectively. Then by (8.11) and (8.12), $A_\theta = \operatorname{diag}(\lambda(\underline{\theta}+\pi\underline{n}_i))_{i=1,2,3,4}$ and $R_\theta = (\mu(\underline{\theta}+\pi\underline{n}_i))_{i=1,2,3,4}$ for $\underline{\theta}\neq(0,0)$, $\underline{\theta}=2\pi h\underline{m}$, $\underline{m}\in\mathbf{Z}^2$, $-\pi/2<\underline{\theta}\le\pi/2$; and $A_\theta=\operatorname{diag}(\lambda(\underline{\theta}+\pi\underline{n}_i))_{i=2,3,4}$ and $R_\theta=(0,0,0)$ for $\underline{\theta}=(0,0)$.

Now we choose $B=I$, $\alpha=1$, and compute for the discretizations (8.5) and (8.6) the quantity $\delta^h=\rho(A^h)\rho(T^h(A^h)^{-1})$. The operator $T^h(A^h)^{-1} = (I-I^h_{2h}(A^{2h})^{-1}I^{2h}_hA^h)(A^h)^{-1}$ reduces in the basis (8.7) to the diagonal blocks

$$K_\theta = \begin{cases} A_\theta^{-1} - R_\theta^T(R_\theta A_\theta R_\theta^T)^{-1}R_\theta & \text{for } \underline{\theta}\neq(0,0), \\ A_\theta^{-1} & \text{for } \underline{\theta}=(0,0). \end{cases}$$

Therefore

$$\begin{aligned}\delta^h &= \max\{\lambda(\underline{\theta}) : \underline{\theta}=2\pi h\underline{m},\ \underline{m}\in\mathbf{Z}^2,\ -\pi<\underline{\theta}\le\pi\} \\ &\quad\cdot \max\{\rho(K_\theta) : \underline{\theta}=2\pi h\underline{m},\ \underline{m}\in\mathbf{Z}^2,\ -\pi/2\le\underline{\theta}\le\pi/2\},\end{aligned}$$

which was computed numerically. The resulting values of δ are given in Table 5.2. The maximum of $\rho(K_\theta)$ is attained at $\underline{\theta}=\pm(2\pi h, 2\pi h)$ for the 5-point scheme, and at $\underline{\theta}=(0,\theta),(\theta,0)$, $\theta\neq0$, for the 9-point scheme. Moreover, $\rho(A^h)=8h^{-2}$ for the 5-point scheme, and $\rho(A^h)=4h^{-2}$ for the 9-point scheme.

TABLE 5.3
Estimates of V-cycle convergence factors for $h=1/64$ and $\nu_1=\nu_2=\nu$ smoothing steps by Richardson's iteration with optimal ω, where δ is obtained from Table 5.2.

	$\nu=1$	$\nu=2$	$\nu=3$	$\nu=4$
5-point	.69	.53	.43	.36
9-point	.40	.25	.18	.14

Smoothing by the Richardson iteration (4.12) has the optimal smoothing factor of at least $\sigma \geq 3$ for $\omega = 3/2$ by Theorem 4.1. By Lemma 4.2 and Theorems 5.1 and 5.2, the convergence factor in the energy norm of the V-cycle with $\nu_1 = \nu_2 = \nu$ smoothing steps is bounded by $1/(1+\beta)$, where $\beta = \nu\sigma/\delta$, assuming that the coarsest level is solved exactly. Table 5.3 gives some resulting estimates of convergence factors.

5.8.4. A model problem with Dirichlet boundary conditions. Consider the Dirichlet problem

$$(8.14) \qquad -\Delta u = f \quad \text{in } \Omega = (0,1) \times (0,1), \qquad u = 0 \quad \text{on } \partial\Omega.$$

Proceeding more briefly in analogy with the case of periodic boundary conditions, let $\Omega^h = \{\underline{x} = h\underline{m};\, \underline{m} \in \mathbf{Z}^2, 1 \leq \underline{m} < n\}$, $h = 1/n$, for $n > 0$, n divisible by 4, and define H^h as the space of all real functions on Ω^h. Grid functions $u \in H^h$ are considered extended by zero at the boundary nodes: $u(\underline{x}) = 0$, $\underline{x} \in \partial\Omega$, $\underline{x} = h\underline{m}$, $\underline{m} \in \mathbf{Z}^2$. The problem (8.14) is discretized by the 9-point stencils (8.6)

$$A^h \equiv [a_{\underline{k}}] = \frac{1}{3h^2}\begin{bmatrix} -1 & -1 & -1 \\ -1 & 8 & -1 \\ -1 & -1 & -1 \end{bmatrix}.$$

The space H^h is equipped with the inner product (8.2). Then the prolongation I_{2h}^h from (8.6) gives the restriction

$$I_h^{2h} \equiv [b_{\underline{k}}] = \frac{1}{16}\begin{bmatrix} 1 & 2 & 1 \\ 2 & 4 & 2 \\ 1 & 2 & 1 \end{bmatrix}$$

as its adjoint. We have $A^{2h} = I_h^{2h} A^h I_{2h}^h$.

Let $\underline{\theta} = \pi h \underline{m}$, $0 < \underline{\theta} < \pi$, $\underline{m} \in \mathbf{Z}^2$, and consider the $(n-1)^2$ orthogonal basis functions

$$\begin{aligned} w_{\underline{\theta}}^h(\underline{x}) &= 2 \sin \frac{\theta_{[1]} x_{[1]}}{h} \sin \frac{\theta_{[2]} x_{[2]}}{h} \\ &= \operatorname{Re}\,(e^{i\underline{\theta}^* \underline{x}/h} - e^{i\underline{\theta}\underline{x}/h}) \end{aligned}$$

where $\underline{\theta}^* = (\theta_{[1]}, -\theta_{[2]})$. Then for any stencil operator $L^h \equiv [s_{\underline{k}}]$ such that $s_{\underline{k}} = s_{-\underline{k}}$, $s_{\underline{k}^*} = s_{\underline{k}}$, where $\underline{k}^* = (k_{[1]}, -k_{[2]})$, we obtain as in (8.4) that

$$(8.15) \qquad L^h w_{\underline{\theta}}^h = w_{\underline{\theta}}^k \sum_{-1 \leq \underline{k} \leq 1} s_{\underline{k}} \cos \underline{\theta}\underline{k}.$$

Note that more symmetry, namely the property $s_{\underline{k}^*} = s_{\underline{k}}$, is needed here. This excludes the 5-point scheme (8.6) from consideration because the stencil of I_h^{2h} does not have this property.

In particular, we have (8.11) with $w_{\underline{\theta}}^h$ in place of $v_{\underline{\theta}}^h$. The basis functions $w_{\underline{\theta}}^{2h} \in H^{2h}$ are again defined as the restrictions of $w_{\underline{\theta}}^h$ to Ω^{2h}:

$$w_{\underline{\theta}}^{2h}(\underline{x}) = w_{\underline{\theta}}^h(\underline{x}), \quad \underline{x} \in \Omega^{2h}, \quad \underline{\theta} = \pi h \underline{m}, \quad \underline{m} \in \mathbf{Z}^2, \quad 0 < \underline{\theta} < \pi/2.$$

Let $\underline{n}_i$, $i = 1, 2, 3, 4$, be as in (8.13). Denote

$$\underline{\theta}(\underline{m}) = (\xi_{[1]}, \xi_{[2]}),$$

where for $j = 1, 2$,

$$\xi_{[j]} = \theta_{[j]} \quad \text{if } m_{[j]} = 0, \qquad \xi_{[j]} = \pi - \theta_{[j]} \quad \text{if } m_{[j]} = 1.$$

Then, by the properties of the sine function,

$$w_{\underline{\theta}(\underline{m})}^h(\underline{x}) = (-1)^{m_{[1]}+m_{[2]}} w_{\underline{\theta}}^h(x) \quad \text{for all } \underline{x} \in \Omega^{2h}.$$

Thus, using (8.15) we obtain for $0 < \underline{\theta} < \pi/2$ that

$$I_h^{2h} w_{\underline{\theta}(\underline{m})}^h = \mu(\underline{\theta}, \underline{m}) w_{\underline{\theta}}^{2h}, \quad \underline{m} = \underline{n}_i, \quad i = 1, 2, 3, 4,$$

where

$$\mu(\underline{\theta}, \underline{m}) = (-1)^{m_{[1]}+m_{[2]}} \sum_{-1 \le \underline{k} \le 1} b_{\underline{k}} \cos \underline{k}\underline{\theta}(\underline{m}).$$

By a direct computation, $I_h^{2h} w_{\underline{\theta}(\underline{m})}^h = 0$, $\underline{m} = \underline{n}_i$, $i = 1, 2, 3, 4$, if $\theta_{[1]} = \pi/2$ or $\theta_{[2]} = \pi/2$.

It follows that

$$\rho(T^h(A^h)^{-1}) = \max \{\rho(K_{\underline{\theta}}) : \underline{\theta} = \pi h \underline{m}, \ 0 < \underline{\theta} \le \pi/2\},$$

where

$$K_{\underline{\theta}} = \begin{cases} A_{\underline{\theta}}^{-1} - R_{\underline{\theta}}^T (R_{\underline{\theta}} A_{\underline{\theta}} R_{\underline{\theta}}^T)^{-1} R_{\underline{\theta}} & \text{if } \underline{\theta} < \pi/2, \\ A_{\underline{\theta}}^{-1} & \text{if } \theta_{[1]} = \pi/2 \quad \text{or} \quad \theta_{[2]} = \pi/2, \end{cases}$$

with

$$A_{\underline{\theta}} = \operatorname{diag}(\lambda(\underline{\theta}(\underline{n}_i)))_{i=1,2,3,4},$$
$$R_{\underline{\theta}} = (\mu(\underline{\theta}, \underline{n}_i))_{i=1,2,3,4}, \qquad \underline{\theta} < \pi/2.$$

We again choose $\alpha = 1$ and $B = I$. Numerically computed values of $\rho(A^h)$ and of $\delta^h = \rho(A^h)\rho(T^h(A^h)^{-1})$ are given in Table 5.4. The maximum of $\rho(K_{\underline{\theta}})$ was attained at $\underline{\theta} = (\pi h, \pi/2)$ and $\underline{\theta} = (\pi/2, \pi h)$.

TABLE 5.4
Numerically computed values of $\rho(A^h)h^2$ and δ for the Dirichlet problem in the unit square and the 9-point scheme.

	$h = 1/8$	$h = 1/16$	$h = 1/32$	$h = 1/64$	$h = 1/128$	$h = 1/256$
$\rho(A^h)h^2$	3.80	3.95	3.99	4.00	4.00	4.00
δ	1.86	1.96	1.99	2.00	2.00	2.00

TABLE 5.5
Estimates of V-cycle convergence factors for the Dirichlet problem in the unit square, the 9-point scheme, and $\nu_1 = \nu_2 = \nu$ smoothing steps.

	$h = 1/8$	$h = 1/16$	$h = 1/32$	$h = 1/64$	$h = 1/128$	$h = 1/256$
Richardson, $\omega = 3/2$, $\nu = 1$	0.38	0.40	0.40	0.40	0.40	0.40
GS, $\nu = 1$	0.25	0.25	0.25	0.25	0.25	0.25
SGS, $\nu = 1$	0.140	0.142	0.143	0.143	0.143	0.143
SGS, $\nu = 2$	0.075	0.077	0.077	0.077	0.077	0.077

Let us consider smoothing by Gauss–Seidel iteration (GS) with lexicographic ordering. Then

$$D^{-1}L \equiv \frac{1}{8}\begin{bmatrix} 0 & 0 & 0 \\ 0 & 0 & 1 \\ 1 & 1 & 1 \end{bmatrix},$$

where $D = (8/3)h^{-2}I$ is the diagonal and $-L$ the strictly lower triangular part of A^h. Now $D^{-1}L^TD^{-1}L$ is nonnegative and irreducible, and since $D^{-1}L^TD^{-1}Le \le \frac{1}{4}e$, $e(\underset{\sim}{x}) = 1$, $\underset{\sim}{x} \in \Omega^h$, it holds that

$$\|D^{-1/2}LD^{-1/2}\|^2 = \rho(D^{-1}L^TD^{-1}L) < \tfrac{1}{4}. \tag{8.16}$$

From Theorem 4.3, we obtain that the smoothing factor of GS satisfies

$$\sigma > \frac{\rho(A^h)}{(8/3)h^{-2}\cdot\frac{1}{4}} = \frac{3}{2}\rho(A^h)h^2.$$

Consequently, we have from Theorem 4.4, for ν steps of symmetric Gauss–Seidel (SGS), that $\sigma > 3\nu\rho(A^h)h^2$. Corresponding convergence factor estimates $1/(1 + \sigma/\delta)$ for the V-cycle are given in Table 5.5.

5.8.5. A remark on local mode analysis. Fourier analysis estimates can be used *heuristically* for more general problems when we disregard the effects of nonconstant coefficients and of the boundary. In essence, we can suppose that the discrete problem, in small rectangular regions inside the domain, is reasonably well approximated by a constant coefficient one with periodic boundary conditions if the coefficients of the operator vary "slowly," the region is "far from the boundary," and h is "sufficiently small." Considering a continuous range of frequencies $-\pi < \underset{\sim}{\theta} \le \pi$, instead of the discrete values of $\underset{\sim}{\theta}$, somehow simplifies the analysis; this corresponds to a formal analysis of a problem in $\Omega = \mathbf{R}^2$.

In practice, we often approximate H^{2h} by the span of smooth modes, i.e., such that $-\pi/2 < \underset{\sim}{\theta} < \pi/2$ (cf. Chapter 1, § 5). It is interesting to note that in this approximation the quantity $1/\delta$ becomes the *h-ellipticity measure* introduced by Brandt [97]. For more details and formalizations, see Brandt [88], Hemker [255], and Stüben and Trottenberg [538].

Results of local mode analysis generally have no mathematical validity (note that we may alway choose $\alpha = 1$, which can be *proved* not to be the case, e.g., for a problem with a re-entrant corner), but they provide invaluable heuristic estimates and guidelines for the design of multigrid methods. It should be noted that local mode analysis may give completely erroneous results if the effects of boundaries and/or nonconstant coefficients prevail.

5.9. Bibliographical remarks.

Section 5.1. For theory in a *nonvariational* framework, compare Hackbusch [220], [229], [233], [244]. Note that such theory requires that there be *sufficiently many* smoothing steps to guarantee convergence, and the required number of smoothing steps is not known except in model problems. This is true also for earlier theories in the variational setting (cf. Nicolaides [430], [431], Bank and Dupont [34], Bank [26], Bank and Douglas [32]; see also Exercise 5.5). One step was found to be sufficient in the V-cycle proofs for $\alpha = 1$ and for symmetric, positive definite problems (cf. Braess [80], [81], Braess and Hackbusch [83], Hackbusch [233], Maitre and Musy [362], McCormick [395], Verfürth [572], and Yserentant [601]). This observation was extended to $\alpha < 1$ and W-cycles by Mandel [375] and Mandel, McCormick and Ruge [377] for symmetric, definite problems, and by Mandel [373] for the general case.

Section 5.3. The arguments used here combine the approaches from Bank [26], Bank and Douglas [32], Hackbusch [229], and Mandel [373]; Lemma 4.1 is adapted from Bank and Dupont [34] (cf. also Hackbusch [244]).

Sections 5.4 *and* 5.5. This is taken, with minor changes, mostly from Mandel, McCormick, and Ruge [377], who also developed a *dual theory* based on the analogous assumption that (cf. McCormick [395] for $\alpha = 1$)

$$\frac{|||e_{(3)}|||_1^2}{|||e_{(2)}|||_1^2} \le 1 - \gamma\left(\frac{|||Te_{(2)}|||_1^2}{|||e_{(2)}|||_1^2}\right)^{1/\alpha}$$

for the post-smoothing. The theory in [377] covers, e.g., smoothing by conjugate gradients (cf. Exercise 5.2). Note that *nonlinear smoothers* can be treated in the same way.

The algebraic approximation property (4.1) is present in some equivalent or closely related form in nearly all multigrid convergence proofs. For a more detailed discussion, see [377]; for related assumptions, see Hackbusch [244].

Theorem 4.4 is new.

Lemma 4.4 relates the present theory to that of Maitre and Musy [365], who considered only the case $\alpha = 1$. Theorem 4.5 and its proof are adapted from theirs.

Theorem 5.3 shows that the present estimates exhibit the usual asymptotic behavior (cf. the works referenced above and Exercise 5.5).

Section 5.6. This section extends the result of Mandel [373], where only the case of one pre-smoothing step was considered. This was in turn based on a combination of the approaches from Bank [26] and Mandel [375].

Section 5.7. Full multigrid was perhaps first developed as an algorithm by Brandt [86]. By virtue of its practical importance and very simplicity, it is a component of many theoretical works too numerous to mention.

Section 5.8. The technique used is adapted from Stüben and Trottenberg [538]. The results, however, are new.

5.10. Exercises and problems. The following exercises contain relatively easy and/or known extensions of the theory. The problems are rather much more difficult and suggest directions of further research. (We would much appreciate receiving results of any work that these problems inspire.)

Section 5.3.

Exercise 3.1. Verify (2.9)–(2.13) for the *Neumann* problem $-\nabla \cdot (a\nabla u) + b \cdot \nabla u + gu = f$ in Ω, $\partial u/\partial n = 0$ on $\partial\Omega$, under suitable assumptions (e.g., Ω is polygonal) and discretized by $\mathscr{C}^0$ piecewise linear triangular finite elements. For the symmetric, positive definite, and semidefinite cases, compare Bank and Dupont [34]; for the nonsymmetric and indefinite cases, compare Bank [26].

Exercise 3.2. Extend the result of Exercise 3.1 to a domain with a piecewise smooth boundary. Boundary triangles may have one curved edge; basis functions remain linear. (This is the so-called natural extension.) Compare Bank and Dupont [34]. Note that we consider the variational formulation of a Neumann problem, which is set in $\mathscr{H}^1 = \mathscr{H}^1(\Omega)$. Thus such basis functions are admissible.

Problem 3.1. Verify (2.9)–(2.13) for the *Dirichlet* problem of order 2 in a domain $\Omega \subset \mathbf{R}^2$ with a piecewise smooth boundary, discretized by the $\mathscr{C}^0$ piecewise linear triangular finite elements, with *nonlinear* basis functions near the boundary, (cf. Zlámal [Zl1]). Design suitable prolongations. Compare also Problem 3.3 below.

Problem 3.2. Verify (2.9)–(2.13) for *isoparametric* finite elements (cf. Mansfield [379]).

Problem 3.3. Verify (2.9)–(2.13) for the case when the stiffness matrix on the highest level is obtained by *numerical integration* (cf. Zlámal [Zl1] and Hackbusch [229]).

Problem 3.4. Verify (2.9)–(2.13) for discrete problems arising from the *nonconforming* finite element method (cf. Ciarlet [Ci1]). Find suitable prolongations.

Problem 3.5. Is it possible to verify (2.9)–(2.13) when the problem on the highest level is singular, and the range of prolongation does not contain the nullspace? What can be said about the multigrid method in such a case?

Exercise 3.3. If (2.9)–(2.13) are true for discretizations of a homogeneous boundary value problem, then they remain valid in the inhomogeneous case. In fact, the inhomogeneous boundary condition affects only the right-hand side. Show this.

Problem 3.6. Investigate verification of (2.9)–(2.13) for problems of order 2 with the *mixed* boundary condition. Generalize to elliptic problems of order $2m$ with general boundary operators.

Problem 3.7. Verify (2.9)–(2.13) for *finite difference* approximations of elliptic problems of order $2m$. Use the results of Hackbusch [230], [237].

Section 5.4. Exercises 4.1 to 4.3 contain results from Mandel, McCormick and Ruge [377].

Exercise 4.1. Prove that if $q(\cdot)$ is a polynomial such that $q(0)=1$, then

$$\sigma\left(q\left(\frac{E}{\rho(E)}\right)\right) \geq \inf_{0\leq t\leq 1} \frac{1-q^2(t)}{tq^2(t)} \equiv \tilde{\sigma}(q).$$

Use Lemma 4.4 or the definition of σ.

Exercise 4.2. Using the result of Exercise 4.1, prove that

$$\tilde{\sigma}\left(\left(I-\frac{\omega E}{\rho(E)}\right)^{\nu}\right) = \max\ \{2\nu\omega, (\omega-1)^{-2\nu}-1\}.$$

This gives $\omega_{\text{opt}} \to 2$ as $\nu \to \infty$ (cf. also Maitre and Musy [365]).

Exercise 4.3. Find the polynomial q_ν of degree $\nu \geq 1$ that gives the optimal $\tilde{\sigma}$. Hint: $2(\tilde{\sigma}t+1)q_\nu^2(t)-1$ is the Chebyshev polynomial of degree $2\nu+1$ on the interval $[-1/\tilde{\sigma}, 1]$. This can be implemented as a sequence of Richardson's iterations with varying parameters ω. Show that $\tilde{\sigma}(q_\nu) \geq c\nu^2$.

Problem 4.1. Investigate the smoothing factors for block Jacobi and block Gauss–Seidel for systems such as the elasticity problem (Example 3.2) or discretizations by Hermite finite elements. In such cases, it is natural to relax together all variables belonging to one node. Let B be the corresponding block diagonal matrix and verify (2.9)–(2.13).

Problem 4.2. Compare the estimate of $\beta = \sigma(F)/\delta$ for SSOR or ILU, with $\sigma(F)$ from Theorem 4.4 and $B = tD$, against the estimate from Theorem 4.1 with $B = t(D-L)D^{-1}(D-L^T)$.

Exercise 4.4. The definition of the smoothing factor σ as well as the statements of Theorems 4.2–4.5 make sense even if A is only positive semi-definite. Change the proofs so that they remain valid for singular A. Hint: Use $I-(Q^T)^{-1}A$ instead of $(I-Q^{-1}A)^*$ and consider pseudo-inverses instead of inverses where necessary.

Section 5.5. Exercises 5.1, 5.2, and 5.4 contain results from Mandel, McCormick, and Ruge [377]. Exercise 5.3 is from Mandel [375].

Exercise 5.1. It is possible to prove an analogue of (5.1) for nonlinear smoothers, with σ defined by (4.9). But note that the theory works only for such smoothers when they are used for pre-smoothing. Prove that, for the *steepest descent* method applied to the preconditioned system $B^{-1}(Au - f) = 0$, we have (4.9) with $\sigma = 1$. For more results in the case $\alpha = 1$, see McCormick [395].

Problem 5.1. Investigate ν steps of steepest descent or *preconditioned conjugate gradients* as a pre-smoother by estimating σ in (4.9). Note that the case $\nu = 1$ reduces to Exercise 5.1.

Exercise 5.2. If G_2' is *any* smoother, even nonlinear, such that $|||\mathbf{u} - G_2'(u, f)|||_1 \le |||\mathbf{u} - G_2(u, f)|||_1$ for all $u \in H$, then replacing G_2 by G_2' in the post-smoothing step does not increase ε^k. Using this observation and Exercise 4.3, estimate ε^k for ν steps of *preconditioned conjugate gradients* as a post-smoother. In particular, show that for smoothing by steepest descent, $\alpha = 1$, $\mu = 1$, $\nu_1 = 0$, $\nu_2 = 1$, we have $\varepsilon^k \le \delta/(3 + \delta)$. (Compare Exercise 5.1.)

Exercise 5.3. Let $\mu = 2$, $0 < \alpha \le 1$, $\beta > 0$, $0 \le \bar{\varepsilon}^1 < 1$, and define $\bar{\varepsilon}^k = p(\alpha, \beta, (\bar{\varepsilon}^{k-1})^\mu)$. Prove that $\bar{\varepsilon}^k \to \bar{\varepsilon}_\mu(\alpha, \beta)$ monotonically as $k \to \infty$. Hint: $p(\alpha, \beta, (\varepsilon)^2)$ is nondecreasing in ε and the fixed point $p(\alpha, \beta, (\bar{\varepsilon})^2) = \bar{\varepsilon}$ is unique in $[0, 1)$.

Exercise 5.4. Let $\mu = 1$ and $0 < \alpha < 1$ and define $\bar{\varepsilon}^k$ as in Exercise 5.3. Prove that $\bar{\varepsilon}^k \to 1$ as $k \to \infty$. So the estimates for the V-cycle are not uniformly bounded away from one if $\alpha < 1$. This is true for all known V-cycle proofs.

Exercise 5.5. Assume $|||F_i|||_1 \le 1$ and develop a "classical" theory for the W-cycle based on the estimate $\varepsilon^k \le \varepsilon_1^k \varepsilon_2^k$ if $\varepsilon^{k-1} \le \varepsilon_1^{k-1} \varepsilon_2^{k-1}$, where

$$\varepsilon_i^k = |||TF_i^{\nu_i}|||_1^2 + (\varepsilon_i^{k-1})^\mu.$$

To prove this recursion, let $|||e_{(0)}|||_1 = 1$, $v \in H$, $|||v|||_1 = 1$. Then with the same notation as in the proof of Lemma 5.1,

$$\begin{aligned}
\langle e_{(3)}, v\rangle_A &= \langle e_{(2)}, F_2^{\nu_2} v\rangle_A \\
&= \langle TF_1^{\nu_1} e_{(0)} + e, F_2^{\nu_2} v\rangle_A \\
&\le |||TF_1^{\nu_1}|||_1\, |||TF_2^{\nu_2}|||_1 + (\varepsilon_1^{k-1}\varepsilon_2^{k-1})^{\mu/2} \\
&\le (|||TF_1^{\nu_1}|||_1^2 + (\varepsilon_1^{k-1})^\mu)^{1/2} (|||TF_2^{\nu_2}|||_1^2 + (\varepsilon_2^{k-1})^\mu)^{1/2}.
\end{aligned}$$

This estimate depends on the quantity $|||TF^\nu|||$, called the *two-level convergence factor*. Prove from (4.3), (4.6), and the Hölder inequality that

$$|||TF^\nu|||_1^2 \le \left(\frac{\delta}{\sigma(F^\nu)}\right)^\alpha.$$

It is possible to prove sharper estimates for $|||TF^{\nu}|||$ directly when F is given by Richardson iteration, $F = I - (\omega/\rho(E))E$ (cf. Hackbusch [233] and Bank and Douglas [32]). Their W-cycle estimates are better than ours for large ν_i, but they do not guarantee convergence if it is not large enough, depending on δ, and this is usually not known. Compare Maitre and Musy [363] and Musy [426] for a study of recursions of this type.

Another estimate can be obtained from Lemma 5.1: $|||TF^{\nu}|||_1^2 \leq p(\alpha, \sigma(F^{\nu})/\delta, 0)$, with p from (5.3) (see Musy [426]).

Problem 5.2. Estimate $|||TF^{\nu}|||_1^2$ directly for smoothers *GS, SGS, RB*. Compare Maitre and Musy [363] when $\alpha = 1$.

Problem 5.3. Estimate $|||Tq(E/\rho(E))|||_1^2$ for a polynomial q of degree ν such that $q(0) = 1$. Find an optimal polynomial, and use the same argument as in Exercise 5.2 to obtain an estimate for conjugate gradients as a post-smoother. Consider also the conjugate residual method as a pre-smoother. Compare Bank and Douglas [32].

Problem 5.4. Compare the estimates from Theorem 5.1, Exercise 5.5, and Problems 5.2 and 5.3.

Section 5.6.

Exercise 6.1. Estimate explicitly the constants in Theorems 6.1 and 6.2. Try to make the bounds as sharp as possible without complicating the argument too much.

Problem 6.1. Develop classical W-cycle estimates as in Exercise 5.5 (cf. Bank [26]).

Problem 6.2. Develop estimates for the W-cycle with smoothing by iterating on the normal system $(A + N)^T((A + N)\mathbf{u} - f) = 0$. Consider preconditioned Richardson and *GS* iterations. Show that $\nu \geq 1$ can be arbitrary independently of k (if k_1 is large enough, of course). Hint: In the definition of the smoothing factor, use the $|||\cdot|||_4$ norm instead of the $|||\cdot|||_2$ norm. This results in replacing α by $\alpha/2$. For a "classical" theory which requires ν large enough depending on δ, see Bank [26].

Use a similar reasoning for smoothing by the *Kaczmarz relaxation*, which is equivalent to *GS* applied to the system $(A + N)((A + N)^T\mathbf{v} - f) = 0$, where $\mathbf{u} = (A + N)^T\mathbf{v}$. For smoothing properties of Kaczmarz relaxation, see Brandt [103].

Problem 6.3. Verify the assumptions of Theorem 6.1 for smoothers other than Richardson's iteration.

Section 5.7.

Exercise 7.1. Make (7.1), (7.2) and (7.3) for $k = 1$ more explicit by using the various convergence bounds developed in earlier sections, and whatever natural problem assumptions are necessary. The best approach may be to consider certain general problem categories. In any case, these bounds should clearly depict the role of the various parameters like d, the ratios h^k/h^{k-1}, the degree of the element basis functions, and problem dependent parameters.

Problem 7.1. Using the results of Exercise 7.1 and usual assumptions (cf. Chapter 1) on the cost of the various multigrid processes, develop explicit estimates for the computational complexity of FMG_μ. This should be expressed in terms of the most basic parameters possible, clearly reflecting the role of the constants in (7.1) (i.e., κ and $\bar{\eta}$ or their more explicit representations).

Problem 7.2. It is perhaps a fairly common belief that *FMG* is effective even for deteriorating approximation order (i.e., decreasing $\bar{\eta}$). Study the bounds developed in Problem 7.1 as they relate to the general validity of this assertion. It has also been suggested (cf. Greenbaum [201]) that *FMG* complexity deteriorates with increasing size of the coefficient in the approximation property (i.e., κ). This may in fact be due to the use of Richardson's iteration, which does not account for special variations in coefficients, mesh size, etc. Note that there is a relation between α and $\bar{\eta}$ and between δ and κ, because these quantities are derived from error estimates. Study this question first by way of a specific model problem (e.g., Example 3.1) parameterized by the grade of the finest mesh (i.e., the ratio of the largest to the smallest mesh size on the finest level). Develop sharp estimates of the convergence factors using Theorems 4.1 and 4.3 for this model problem and analyze the resulting effects on *FMG* complexity. Extend these results as far as possible to the general case.

Section 5.8.

Exercise 8.1 Define

$\mathcal{V}^h = \{u \in \mathscr{C}(\mathbf{R}^2)$: u is linear on all triangles $(\underline{x}, \underline{x}+(h,0), \underline{x}+(h,h))$ and $(\underline{x}, \underline{x}+(0,h), \underline{x}+(h,h))$, $\underline{x}=h\underline{m}$, $\underline{m}\in\mathbf{Z}^2$, and u is periodic with period $(1,1)$, i.e., $u(\underline{x}) = u(\underline{x}+\underline{m})$ for all $\underline{x}\in\mathbf{R}^2$, $\underline{m}\in\mathbf{Z}^2\}$.

The mapping $\mathscr{I}^h$, defined by $\mathscr{I}^h\colon \tilde{H}^h \to \mathcal{V}^h$, $\mathscr{I}^h u(\underline{x}) = u(\underline{x})$, $\underline{x}=h\underline{m}$, $\underline{m}\in\mathbf{Z}^2$, is an isomorphism. Let $\mathscr{A}(u, v) = \int_\Omega \nabla u \cdot \nabla v$. Prove for the 5-point scheme (8.5) that $\langle A^h u, v\rangle_h = \mathscr{A}(\mathscr{I}^h u, \mathscr{I}^h v)$ for all $u, v \in \tilde{H}^h$, and that $\mathscr{I}^{2h} = \mathscr{I}^h I^h_{2h}$. Hence, $A^{2h} = I^{2h}_h A^h I^h_{2h}$. Compare § 5.3.

For the 9-point scheme, the definition of $\mathcal{V}^h$ is changed so that u is linear in each variable on all squares $(\underline{x}, \underline{x}+(h,0), \underline{x}+(h,h), \underline{x}+(0,h))$, $\underline{x}=\underline{m}h$, $\underline{m}\in\mathbf{Z}^2$.

Exercise 8.2. Extend the results of § 5.8 to discretizations of a general selfadjoint problem of order 2 with constant coefficients in a rectangle with step sizes different in each direction. Hint: $\underline{h} = (h_{[1]}, h_{[2]})$, $h\underline{m}$ becomes $\underline{m}\cdot\underline{h} = (m_{[1]}h_{[1]}, m_{[2]}h_{[2]})$ and $\underline{x}/\underline{h}$ is defined similarly. Of course, $\underline{n} = (n_{[1]}, n_{[2]})$.

Exercise 8.3. Write a program to compute $\rho(A^h)h^2$ and δ^h for general symmetric stencils $A^h \equiv [a_k]$, $I_h^{2h} \equiv [b_k]$ and either periodic or Dirichlet boundary conditions. By maximizing over a continuous range of frequencies it is possible to compute $\sup_{h>0} \rho(A^h)h^2$ and $\delta = \sup_{h>0} \delta^h$.

Exercise 8.4. Compute δ for model problems with Neumann and mixed boundary conditions. For the required techniques, see Stüben and Trottenberg [538].

Problem 8.1. Estimate σ for *GS, SGS,* and *RB* smoothers for the model problems with periodic boundary conditions. The main obstacle here is that L does not map constants into constants and thus does not induce an operator on H^h. Heuristically, we may expect, e.g., that $\|D^{-1/2}LD^{-1/2}\|^2 \approx \frac{1}{4}$ for lexicographic ordering. Hint: Use Exercise 4.4.

Problem 8.2. Study the transformation of the modes $v_{\underline{\theta}}^h$ and $w_{\underline{\theta}}^h$ by *GS, SGS,* and *RB* (cf. Brandt [88] and Stüben and Trottenberg [538]). Use the result to compute σ.

Problem 8.3. Using the results of Problem 8.2, compute β from (5.1) with $\alpha = 1$ directly for *GS, SGS,* and *RB.*

Problem 8.4. Stüben and Trottenberg [538] show how the modes are transformed by ZEBRA relaxation. Use this together with the result of Exercise 8.2 to compute β from (5.1) with $\alpha = 1$ for an anisotropic problem.

Exercise 8.5. Use the present techniques to compute $|||TF^\nu|||$, i.e., the two-grid convergence factor, for the model problems and the smoothers considered here (cf. Stüben and Trottenberg [538]).

Problem 8.5. Use Fourier analysis to compute δ, σ, and β for an elasticity problem in three dimensions discretized by finite elements. Analyze a suitable model problem.

Problem 8.6. Compute δ^h and $\rho(A^h)$ analytically for some of the model problems considered here. Hint: We know that $\underline{\theta}$ maximizes $\rho(K_{\underline{\theta}})$ numerically, which gives a lower bound for $\rho(T^h(A^h)^{-1})$. Prove that it is also an upper bound. Apply similar reasoning to $\rho(A^h)$.

Problem 8.7. The quantities δ^h computed numerically for the Dirichlet problem, and the 9-point scheme, approach $\delta^h = 2$ for the 9-point scheme with periodic boundary conditions. Musy [426] reports numerically computed values of δ^h, for the Dirichlet problem and the 5-point scheme, which are close to our computed δ^h for periodic boundary conditions. Is that only a coincidence? Try to formulate and prove something—this may be a step towards justification of the local mode analysis in some cases. Compare Hackbusch [240] for a one-dimensional problem.

Problem 8.8. Develop a Fourier analysis for the 5-point scheme (8.6) and Dirichlet boundary conditions.

Problem 8.9. Estimate the constants in the nonsymmetric theory by way of Fourier analysis of model problems. Compute convergence bounds using the theorems from § 5.6.

General.

Exercise. Change the theory so that it applies to the *complex* case. This requires the use of complex conjugates in some formulas, and a different definition of ellipticity.

Acknowledgment. This work was supported by the Air Force Office of Scientific Research under grant AFOSR-86-0126 and the Department of Energy under grant DE-AC03-84-ER80155.

REFERENCES

[Ci1] PH. CIARLET, *The Finite Element Method for Elliptic Problems.* North-Holland, Amsterdam, 1978.

[Ka1] J. KADLEC, *On the regularity of the solution of the Poisson equation on a domain with boundary locally similar to the boundary of a convex domain,* Czechoslovak Math. J., 14 (1964), pp. 386–393.

[Kr1] S. G. KREIN, JU. I. PETUNIN AND E. M. SEMENOV, *Interpolation of Linear Operators,* Translations of Mathematical Monographs 54, American Mathematical Society, Providence, RI, 1982.

[Li1] J. L. LIONS AND E. MAGENES, *Non-homogeneous Boundary Value Problems and Applications,* Vol. I, Springer, Berlin, 1972.

[Ne1] J. NEČAS, *Sur la coercivité des formes sesqui-linéaires elliptiques,* Rev. Roumaine Math. Pures Appl., 9 (1964), pp. 47–69.

[Ne2] J. NEČAS AND I. HLAVÁČEK, *Mathematical Theory of Elastic and Elasto-Plastic Bodies: An Introduction,* Elsevier, Amsterdam, 1981.

[Ne3] J. NEČAS, Private communication.

[St1] G. STRANG AND G. J. FIX, *An Analysis of the Finite Element Method,* Prentice–Hall, Englewood Cliffs, NJ, 1973.

[Zl1] M. ZLÁMAL, *Curved elements in the finite element method* I, II, SIAM J. Numer. Anal., 10 (1973), pp. 229–260; 11 (1974), pp. 347–362.

APPENDIX 1

FMV Program Listing

```
      PROGRAM FMV¹
C
C     THIS CODE USES A (2,1) FMV CYCLE WITH
C     RED-BLACK GAUSS-SEIDEL TO SOLVE POISSON'S
C     EQUATION ON AN (N+1)**2 GRID, N = 2**NLEV
C
C     LEVEL NLEV IS FINEST GRID . . H = 1/2**NLEV
C     LEVEL 1 IS COARSEST GRID . . H = 1/2
C
C     U HOLDS CURRENT APPROXIMATION TO FINEST
C     GRID SOLUTION AND COARSE GRID ERRORS
C     F HOLDS FINEST GRID RIGHT-HAND SIDE AND
C     COARSE GRID RESIDUALS
C     EX HOLDS SOLUTION TO PDE ON ALL GRIDS
C     KS(I) IS THE INITIAL INDEX FOR ITH GRID IN
C     ARRAYS U AND F
C
      DIMENSION U(8000),F(8000),KS(10),EX(8000)
         DO 500 I = 1,8000
         U(I) = 0.
         F(I) = 0.
  500    EX(I) = 0.
      H = .5
      KST = 1
      NLEV = 6
      N = 2**NLEV
      WRITE (6,5) N
    5 FORMAT (//'  ONE FMV CYCLE FOR N = ',I3,//)
C
C     INITIALIZE ARRAYS F AND KS
C
```

[1] No attempt was made either to fully document or to streamline this code. Its primary purpose is to illustrate the basic structure and simplicity of the multigrid algorithm.

```
      DO 10 K = 1,NLEV
        KS(K) = KST
        N = 2**K + 1
        CALL FSET (F(KST),EX(KST),N,H)
        KST = KST + N*N
        H = H/2.
   10 CONTINUE
      DO 300 N = 1,NLEV
      IF (N.GT.1) THEN
C
C     DESCEND FROM FINEST GRID (LEVEL N) TO COARSE
C     GRID (LEVEL 2)
C
        DO 100 K = N,2,-1
          KST = KS(K)
          KSTC = KS(K-1)
          NF = 2**K + 1
          NC = 2**(K-1) + 1
C
C     RELAX ON PRESENT GRID (NF**2), INJECT
C     RESIDUAL TO COARSER GRID (NC**2)
C
          CALL RELAX (U(KST),F(KST),NF)
          CALL RELAX (U(KST),F(KST),NF)
          CALL RESINJ (U(KST),F(KST),
          F(KSTC),NF,NC)
  100   CONTINUE
      ENDIF
C
C     ASCEND FROM COARSEST GRID (LEVEL 1) TO
C     FINEST GRID (LEVEL N)
C
      DO 200 K = 1,N
        KST = KS(K)
        KSTF = KS(K+1)
        NC = 2**K + 1
        NF = 2**(K+1) + 1
C
C     RELAX ON PRESENT GRID (NC**2), INTERPOLATE
C     TO FINER GRID (NF**2)
C
        CALL RELAX (U(KST),F(KST),NC)
        IF (K.EQ.N) CALL ERR (U(KST),EX(KST),
        F(KST),NC)
        IF (K.LT.NLEV) CALL INTADD (U(KST),
        U(KSTF),NC,NF)
```

```
  200   CONTINUE
  300   CONTINUE
        END
C
C
        SUBROUTINE ERR (U,EX,F,N)
        DIMENSION U(N,N),F(N,N),EX(N,N)
C
C       COMPUTE MAX NORM AND  DISCRETE L2 NORM OF
C       RESIDUAL AND GLOBAL ERROR
C
        H = 1./FLOAT(N-1)
        EMAX = 0.
        ESUM = 0.
        RMAX = 0.
        RSUM = 0.
        DO 10 I = 2,N-1
          DO 10 J = 2,N-1
            ER = ABS (U(I,J)-EX(I,J))
            IF (ER.GT.EMAX) EMAX = ER
            R = ABS (F(I,J)+U(I+1,J)+U(I-1,J)
            +U(I,J+1)+U(I,J-1)-4*U(I,J))
            IF (R.GT.RMAX) RMAX = R
            ESUM = ESUM + ER*ER
            RSUM = RSUM + R*R
   10   CONTINUE
        EH = SQRT (H*ESUM)
        RH = SQRT (H*RSUM)
        WRITE (6,20) N,RMAX,RH,EMAX,EH
   20   FORMAT ('  N = ',13,4E12.3)
        RETURN
        END
C
C
        SUBROUTINE FSET (G,EX,N,H)
        DIMENSION G(N,N),EX(N,N)
C
C       GENERATE RHS ARRAY AND EXACT SOLUTION ON
C       (N-1)**2 GRID
C
        H2 = H*H
        N1 = N - 1
        DO 10 I = 2,N1
          X = H*FLOAT (I-1)
          XX = X*X*(X*X-1.)
          X6 = 6.*X*X - 1.
```

```
            DO 10 J = 2,N1
              Y = H*FLOAT (J-1)
              YY = Y*Y*(Y*Y-1.)
              EX(I,J) = XX*YY
              G(I,J) = -2.*H2*(X6*YY + (6.*Y*Y-1.)*XX)
   10         CONTINUE
              RETURN
              END
C
C
        SUBROUTINE RELAX (V,G,N)
        DIMENSION V(N,N),G(N,N)
C
C       RED-BLACK GAUSS-SEIDEL ON (N+1)**2 GRID
C
        N1 = N-1
        DO 10 KRB = 1,-1,-2
          IRB = KRB
          DO 20 I = 2,N1
            IRB = -IRB
            JDEL = (IRB + 1)/2
            J0 = JDEL + 2
            DO 20 J = J0,N1,2
            V(I,J) = (G(I,J)+V(I+1,J)+V(I-1,J)
            +V(I,J+1)+V(I,J-1))/4.
   20     CONTINUE
   10   CONTINUE
        RETURN
        END
C
C
        SUBROUTINE RESINJ (V,F,G,N,NC)
        DIMENSION F(N,N),V(N,N),G(NC,NC)
C
C       COMPUTE RESIDUAL AND INJECT TO COARSER
C       (NC+1)**2 GRID
C
        N1 = NC - 1
        DO 10 I = 2,N1
          I2 = 2*I-1
          DO 10 J = 2,N1
            J2 = 2*J-1
            T = F(I2,J2)+V(I2+1,J2)+V(I2-1,J2)
            +V(I2,J2+1)+V(I2,J2-1)
            G(I,J) = 2.*T - 8.*V(I2,J2)
   10   CONTINUE
        RETURN
        END
```

```
C
C
      SUBROUTINE INTADD (V,W,N,NF)
      DIMENSION V(N,N),W(NF,NF)
C
C     INTERPOLATE CORRECTION AND ADD TO
C     APPROXIMATION AT FINE GRID RED POINTS
C
      DO 10 I = 2,N
        I2 = 2*I -1
        DO 10 J = 2,N
          J2 = 2*J - 1
          W(I2,J2-1) = W(I2,J2-1) + (V(I,J)
          + V(I,J-1))/2.
          W(I2-1,J2) = W(I2-1,J2) + (V(I,J)
          + V(I-1,J))/2.
   10 CONTINUE
C
C     ZERO OUT PRESENT GRID APPROXIMATION
C
      DO 20 I = 1,N
        DO 20 J = 1,N
          V(I,J) = 0.
   20 CONTINUE
      RETURN
      END
```

APPENDIX 2

Multigrid Bibliography

The intent of the Multigrid Bibliography of the GMD is to give as complete a survey as possible of the growing literature on multigrid methods. All papers and books that deal with multigrid techniques in the broadest sense are included: theoretical papers as well as those on practical applications in widely scattered fields (fluid and aerodynamics, pattern recognition, statistical physics, supercomputer applications, etc.). The references may have been published in scientific journals, conference proceedings or only as internal reports.

Entries in the GMD Multigrid Bibliography are arranged alphabetically according to author. This fourth edition contains more than 300 new entries compared with the third edition compiled in September 1983. These new entries are marked by an $*$. To facilitate a search by topic, we have included a KWIC reference guide to the bibliography in which papers are arranged according to key words in their titles.

The editors of this bibliography would be grateful for information about any missing references as well as any errors found by the reader. Furthermore, as we intend to continue the bibliography, information about new publications in the multigrid area is welcome.

Technical assistance in the completion of the fourth edition has been given by Cläre Cassirer. Her help is gratefully acknowledged.

K. Brand, M. Lemke, J. Linden
Gesellschaft für Mathematik und Datenverarbeitung
GMD-F1/T
Postfach 1240
D-5205 St. Augustin 1
Schloß Birlinghoven
West Germany

Multigrid Bibliography*,†

*[1] A. M. ABDALASS, J. F. MAITRE AND F. MUSY, *A multigrid solver for a stabilized finite element discretization of the Stokes problem,* Proc. 2nd European Conference on Multigrid Methods, Cologne, October 1–4, 1985, W. Hackbusch, U. Trottenberg, eds., Lecture Notes in Mathematics, Springer-Verlag, Berlin, to appear.

[2] R. K. AGARWAL, *Unigrid and multigrid algorithms for the solution of coupled, partial-differential equations using fourth-order-accurate compact differencing,* Report, Symposium on Numerical Boundary Condition Procedures and Multigrid Methods, NASA-Ames Research Center, Moffett Field, CA, 1981.

*[3] R. E. ALCOUFFE, *The multigrid method for solving the two-dimensional multigroup diffusion equation,* in Advances in Reactor Computations, Proc. Topical Meeting, Salt Lake City, UT, March 1983, pp. 340–351.

*[4] R. E. ALCOUFFE, F. W. BRINKLEY AND D. R. MARR, *User's guide for Twodant: A code package for two-dimensional, diffusion-accelerated, neutral-particle transport, Rev. 1,* Report 10018-M-Rev1, Los Alamos National Lab., Los Alamos, NM, 1984.

[5] R. E. ALCOUFFE, A. BRANDT, J. E. DENDY JR. AND J. W. PAINTER, *The multi-grid methods for the diffusion equation with strongly discontinuous coefficients,* SIAM J. Sci. Stat. Comput., 2 (1981), pp. 430–454.

[6] M. ALEF, *Mehrgittermethoden: Diskretisierungen höherer Ordnungen für Dirichlet-Standardaufgaben in Rechtecksgebieten,* Diplomarbeit, Institut für Angewandte Mathematik, Universität Bonn, 1982.

[7] S. AMINI, *Multi-grid methods for non-linear boundary-value problems,* Report CS-82-02, School of Mathematics, University of Bristol, 1982.

D. AMIT, cf. [116].

[8] B. ARLINGER, *Multigrid technique applied to lifting transonic flow using full potential equation,* Report L-0-1 B439, Saab-Scania, 1978.

[9] ———, *Axisymmetric transonic flow computations using a multigrid method,* Proc. Seventh International Conference on Numerical Methods in Fluid Dynamics, Stanford, 1980, W. C. Reynolds, R. W. MacCormack, eds., Lecture Notes in Physics 141, Springer-Verlag, Berlin, 1981, pp. 55–60.

*[10] M. T. ARTHUR, *A generalisation of Hall's scheme for solving the Euler equations for two-dimensional flows,* Proc. 2nd European Conference on Multigrid Methods, Cologne, October 1–4, 1985, W. Hackbusch and U. Trottenberg, eds., GMD-Studien Nr. 110, Gesellschaft für Mathematik und Datenverarbeitung, St. Augustin, to appear.

A. ASKAR, cf. [203], [204].

E. J. VAN ASSELT, cf. [607].

[11] E. J. VAN ASSELT, *A survey of multi-grid methods for nonlinear problems,* Report NN 22/80, Dept. of Numerical Mathematics, Mathematical Centre, Amsterdam, 1980.

[12] ———, *Application of the Osher–Engquist difference scheme and the full multi-grid method to a two dimensional nonlinear elliptic model equation,* Preprint NW 103/81, Dept. of Numerical Mathematics, Mathematical Centre, Amsterdam, 1981.

[13] ———, *The multi grid method and artificial viscosity,* in Multigrid Methods, Proc. conference held at Köln-Porz, November 23–27, 1981, W. Hackbusch and U. Trottenberg, eds., Lecture Notes in Mathematics 960, Springer-Verlag, Berlin, 1982, pp. 313–326.

* The fourth edition of the GMD Multigrid Bibliography compiled in May 1986 by K. Brand, M. Lemke, J. Linden, Gesellschaft für Mathematik und Datenverarbeitung, Birlinghoven, West Germany.

† Note to the reader: To avoid possible confusion, please note that the same article is cited twice in references [342] and [343] and references [428] and [474].

[14] E. J. VAN ASSELT, *Termination strategies for Newton iteration in full multigrid methods,* Report NW 159/83, Dept. of Numerical Mathematics, Mathematical Centre, Amsterdam, 1983.

[15] G. P. ASTRAKHANTSEV, *An iterative method of solving elliptic net problems,* U.S.S.R. Comput. Math. and Math. Phys., 11 (1971), pp. 171–182.

[16] G. P. ASTRAKHANTSEV AND L. A. RUKHOVETS, *A relaxation method in a sequence of grids for elliptic equations with natural boundary condition,* Z. Vycisl. Mat. i Mat. Fiz., 21 (1981), pp. 926–944.

*[17] ———, *Fedorenkos method for variational difference schemes with extrapolation,* in Variational Difference Methods in Mathematical Physics, Proceedings, October 1980, Novosibirsk, V. I. Lebedev, ed., AN SSSR, 1981, pp. 20–26.

*[18] W. AUZINGER, *DCMG01: A multigrid code with defect correction to solve Delta* $U - c(x, y)$, $U = f(x, y)$ *(on Omega),* $U = g(x, y)$ *(on D-Omega) on nonrectangular bounded domains Omega with high accuracy,* Arbeitspapiere der GMD, Nr. 127, Gesellschaft für Mathematik und Datenverarbeitung, St. Augustin, 1985.

[19] W. AUZINGER AND H. J. STETTER, *Defect correction and multigrid iterations,* in Multigrid Methods, Proc. conference held at Köln-Porz, November 23–27, 1981, W. Hackbusch and U. Trottenberg, eds., Lecture Notes in Mathematics 960, Springer-Verlag, Berlin, 1982, pp. 327–351.

[20] O. AXELSSON, *On multigrid methods of the two-level type,* in Multigrid Methods, Proc. conference held at Köln-Porz, November 23–27, 1981, W. Hackbusch and U. Trottenberg, eds., Lecture Notes in Mathematics 960, Springer-Verlag, Berlin, 1982, pp. 352–367.

*[21] ———, *A mixed variable finite element method for the efficient solution of nonlinear diffusion and potential flow equations,* in Advances in Multi-Grid Methods, Proc. conference held in Oberwolfach, December 8–13, 1984, D. Braess, W. Hackbusch and U. Trottenberg, eds., Notes on Numerical Fluid Mechanics, Vol. 11, Vieweg, Braunschweig, 1985, pp. 1–11.

*[22] ———, *Incomplete matrix factorizations as multigrid smoothers for vector and parallel computers,* in Appl. Math. Comp., Proc. 2nd Internat. Multigrid Conference, April 1985, Copper Mountain, CO, S. F. McCormick, ed., North-Holland, Amsterdam, to appear.

[23] O. AXELSSON AND I. GUSTAFSSON, *Preconditioning and two-level multigrid methods of arbitrary degree of approximation,* Math. Comp., 40 (1983), pp. 219–242.

*[24] D. BAI AND A. BRANDT, *Local mesh refinement multilevel techniques,* SIAM J. Sci. Stat. Comput., 8 (1987), pp. 109–134.

T. J. BAKER, cf. [303].

[25] N. S. BAKHVALOV, *On the convergence of a relaxation method with natural constraints on the elliptic operator,* U.S.S.R. Computational Math. and Math. Phys., 6 (1966), pp. 101–135.

[26] R. E. BANK, *A comparison of two multi-level iterative methods for nonsymmetric and indefinite elliptic finite element equations,* SIAM J. Numer. Anal., 18 (1981), pp. 724–743.

[27] ———, *A multi-level iterative method for nonlinear elliptic equations,* in Elliptic Problem Solvers, M. H. Schultz, ed., Academic Press, New York, 1981, pp. 1–16.

[28] ———, *Analysis of a multi-level inverse iteration procedure for eigenvalue problems,* SIAM J. Numer. Anal., 19 (1982), pp. 886–898.

[29] ———, PLTMG *users' guide—Edition* 4.0, Report, Dept. of Mathematics, University of California at San Diego, La Jolla, CA, 1985.

*[30] ———, *A-posteriori error estimates, adaptive local mesh refinement, and multigrid iteration,* Proc. 2nd European Conference on Multigrid Methods, Cologne,

October 1–4, 1985, W. Hackbusch and U. Trottenberg, eds., Lecture Notes in Mathematics, Springer-Verlag, Berlin, to appear.

*[31] R. E. BANK AND T. F. CHAN, PLTMGC: *A multi-grid continuation program for parameterized nonlinear elliptic systems,* SIAM J. Sci. Stat. Comput., 7 (1986), pp. 540–559.

[32] R. E. BANK AND C. C. DOUGLAS, *Sharp estimates for multigrid rates of convergence with general smoothing and acceleration,* SIAM J. Numer. Anal., 22 (1985), pp. 617–633.

[33] R. E. BANK AND T. F. DUPONT, *Analysis of a two-level scheme for solving finite element equations,* Report CNA-159, Center for Numerical Analysis, University of Texas at Austin, Austin, TX, 1980.

[34] ———, *An optimal order process for solving finite element equations,* Math. Comp., 36 (1981), pp. 35–51.

*[35] R. E. BANK AND H. D. MITTELMANN, *Continuation and multi-grid for nonlinear elliptic systems,* Proc. 2nd European Conference on Multigrid Methods, Cologne, October 1–4, 1985, W. Hackbusch and U. Trottenberg, eds., Lecture Notes in Mathematics, Springer-Verlag, Berlin, to appear.

*[36] R. E. BANK AND D. J. ROSE, *Global approximate Newton methods,* Numer. Math., 37 (1981), pp. 279–295.

[37] ———, *Analysis of a multi-level iterative method for nonlinear finite element equations,* Math. Comp., 39 (1982), pp. 453–465.

[38] R. E. BANK AND A. H. SHERMAN, *A comparison of smoothing iterations for multi-level methods,* in Advances in Computer Methods for Partial Differential Equations III, R. Vichnevetsky and R. S. Stepleman, eds., IMACS, New York, 1979, pp. 143–147.

[39] ———, *Algorithmic aspects of the multi-level solution of finite element equations,* Sparse Matrix Proceedings 1978, I. S. Duff and G. W. Stewart, eds., Society for Industrial and Applied Mathematics, Philadelphia, PA, 1979, pp. 62–89.

[40] ———, PLTMG *users' guide—July* 1979 *version,* Report CNA-152, Center for Numerical Analysis, University of Texas at Austin, Austin, TX, 1979.

[41] ———, *The use of adaptive grid refinement for badly behaved elliptic partial differential equations,* Math. Comput. Simulation XXII, 1980, pp. 18–24.

[42] ———, *An adaptive multi-level method for elliptic boundary value problems,* Computing, 26 (1981), pp. 91–105.

*[43] R. E. BANK AND A. WEISER, *Some a posteriori error estimators for elliptic partial differential equations.* Math. Comp., 44 (1985), pp. 283–301.

*[44] R. E. BANK, J. W. JEROME AND D. J. ROSE, *Analytical and numerical aspects of semiconductor device modeling,* in Computing Methods in Applied Sciences and Engineering V, Proc. Fifth International Symposium, Versailles, December 1981, R. Glowinski and J. L. Lions, eds., North-Holland, Amsterdam, 1982, pp. 593–597.

*[45] R. E. BANK, A. H. SHERMAN AND A. WEISER, *Refinement algorithms and data structures for regular local mesh refinement,* in Scientific Computing, R. S. Stepleman, ed., North-Holland, Amsterdam, 1983.

*[46] F. BANNASCH, *Mehrgitterverfahren für die dreidimensionale Poissongleichung,* Diplomarbeit, Institut für Angewandte Mathematik, Universität Bonn, 1983.

[47] D. BARKAI AND A. BRANDT, *Vectorized multigrid poisson solver for the* CDC *Cyber* 205, in Appl. Math. Comp., 13, Proc. Internat. Multigrid Conference, April 6–8, 1983, Copper Mountain, CO, S. F. McCormick and U. Trottenberg, eds., North Holland, 1983, pp. 215–228.

*[48] K. E. BARRETT AND D. M. BUTTERFIELD, *Multigrid finite element method,* Report PC A03/MF A01, Coventry Polytechnic, UK, 1983.

*[49] K. E. BARRETT, D. M. BUTTERFIELD, S. E. ELLIS, C. J. JUDD AND J. H. TABOR, *Multigrid analysis of linear elastic stress problems,* in Multigrid Methods for Integral and Differential Equations, D. J. Paddon and H. Holstein, eds., The Institute of Mathematics and its Applications Conference Series, 3, Clarendon Press, Oxford, 1985, pp. 263–282.

J. BARTON, cf. [165].

*[50] J. R. BAUMGARTNER, *Three-dimensional treatment of convective flow in the earth's mantle,* J. Stat. Phys., 29 (1985), pp. 501–511.

*[51] A. BAYLISS, C. I. GOLDSTEIN AND E. TURKEL, *The numerical solution of the Helmholtz equation for wave propagation problems in underwater acoustics,* ICASE Report 84-49, 1984.

[52] K. BECKER, *Mehrgitterverfahren zur Lösung der Helmholtz-Gleichung im Rechteck mit Neumannschen Randbedingungen,* Diplomarbeit, Institut für Angewandte Mathematik, Universität Bonn, 1981.

[53] ———, *COMFLO- ein Experimentierprogramm zur Mehrgitterbehandlung subsonischer Potentialströmungen um Tragflächenprofile,* Preprint 604, Sonderforschungsbereich 72, Universität Bonn, 1983.

*[54] ———, *Ein Mehrgitterprogramm zur Berechnung subsonischer Potentialströmungen um Tragflächenprofile,* Dissertation, Institut für Angewandte Mathematik, Universität Bonn, 1984.

*[55] ———, *Ein Mehrgitterverfahren zur Lösung der vollen Potentialgleichung im Falle transsonischer Strömungen,* Rechnerarchitekturen für die numerische Simulation auf der Basis superschneller Lösungsverfahren II, U. Trottenberg and P. Wypior, eds., GMD-Studien Nr. 102, Gesellschaft für Mathematik und Datenverarbeitung, St. Augustin, 1985, pp. 199–210.

*[56] ———, *A multigrid solver for two-dimensional transonic full potential flow calculations,* Proc. 2nd European Conference on Multigrid Methods, Cologne, October 1–4, 1985, W. Hackbusch and U. Trottenberg, eds., Lecture Notes in Mathematics, Springer-Verlag, Berlin, to appear.

[57] K. BECKER AND U. TROTTENBERG, *Fast multigrid methods and applications—a short survey and one result on a special nearly singular problem,* Proc. Third Internat. Symposium on Numerical Methods in Engineering (Tome 1), March 14–18, 1983, Paris, P. Lascaux, ed., Pluralis, Paris, 1983, pp. 81–91.

*[58] ———, *Development of multigrid algorithms for problems from fluid dynamics,* Arbeitspapiere der GMD, Nr. 111. Gesellschaft für Mathematik und Datenverarbeitung, St. Augustin, 1984.

H. BEER, cf. [465].

[59] A. BEHIE AND P. A. FORSYTH, JR., *Multi-grid solution of the pressure equation in reservoir simulation,* Proc. 6th Annual Meeting of Reservoir Simulation, Society of Petroleum Engineers, New Orleans, LA, 1982.

*[60] ———, *Comparison of fast iterative methods for symmetric systems,* IMA J. Numer. Anal., 3 (1983), pp. 41–63.

[61] ———, *Multi-grid solution of three-dimensional problems with discontinuous coefficients,* Appl. Math. Comp., 13, Proc. Internat. Multigrid Conference, April 6–8, 1983, Copper Mountain, CO, S. F. McCormick and U. Trottenberg, eds., North-Holland, Amsterdam, 1983, pp. 229–240.

M. BERCOVIER, cf. [544].

*[62] M. J. BERGER AND A. JAMESON, *An adaptive multigrid method for the Euler equation,* Proc. Ninth International Conference on Numerical Methods in Fluid Dynamics,

Soubbaramayer and J. P. Boujot, eds., Lecture Notes in Physics 218, Springer-Verlag, Berlin, 1985.

*[63] ———, *Automatic adaptive grid refinement for the Euler equations*, AIAA J., 23 (1985), pp. 561–568.

*[64] P. BJÖRSTAD, *The direct solution of a generalized biharmonic equation on a disk*, in Efficient Solution of Elliptic Systems, Proc. of a GAMM-Seminar in Kiel, January 27–29, 1984, W. Hackbusch, ed., Notes on Numerical Fluid Mechanics, Vol. 10, Vieweg, Braunschweig, 1984, pp. 1–10.

*[65] A. BODE, *Ein Mehrgitter/Gleitkomma-Zusatz für den Knotenprozessor eines Multiprozessors*, Rechnerarchitekturen für die numerische Simulation auf der Basis superschneller Lösungsverfahren I, Workshop 'Rechnerarchitektur,' Erlangen, June 14–15, 1984, U. Trottenberg and P. Wypior, eds., GMD-Studien Nr. 88, Gesellschaft für Mathematik und Datenverarbeitung, St. Augustin, 1984, pp. 153–160.

*[66] K. BÖHMER, P. W. HEMKER, H. J. STETTER, *The defect correction approach*, in Computing, Suppl. 5, Defect Correction Methods, K. Böhmer and H. J. Stetter, eds., 1984, pp. 1–32.

[67] C. BÖRGERS, *Mehrgitterverfahren für eine Mehrstellendiskretisierung der Poissongleichung und für eine zweidimensionale singulär gestörte Aufgabe*, Diplomarbeit, Institut für Angewandte Mathematik, Universität Bonn, 1981.

*[68] ———, *A Lagrangian fractional step method for the incompressible Navier–Stokes equations*, Technical report, 183, Computer Science Department, New York University, 1985.

[69] J. W. BOERSTOEL, *A fast-solver algorithm for steady transonic potential-flow computations with Newton iteration and multigrid relaxation*, Proc. Fourth GAMM-Conference on Numerical Methods in Fluid Mechanics, H. Viviand, ed., Vieweg, Braunschweig, 1982, pp. 21–41.

[70] ———, *A multigrid algorithm for steady transonic potential flows around aerofoils using Newton iteration*, J. Comput. Phys., 48 (1982), pp. 314–343.

[71] ———, *Numerical modelling and fast-solver calculation of approximately normal shocks*, in Computational and Asymptotic Methods for Boundary and Interior Layers, Proc. BAIL II Conference, Boole Press Conference Series 4, J. J. H. Miller, ed., 1982, pp. 151–159.

[72] J. W. BOERSTOEL AND A. KASSIES, *Integrating multigrid relaxation into a robust fast-solver for transonic potential flows around lifting airfoils*, Proc. AIAA Sixth Computational Fluid Dynamics Conference, July 13–15, 1983, Danvers, MA, submitted.

R. F. BOISVERT, cf. [464].

*[73] C. BOLLRATH, *Two multi-level algorithms for the dam problem*, in Advances in Multi-Grid Methods. Proc. Conference held in Oberwolfach, December 8–13, 1984, D. Braess, W. Hackbusch and U. Trottenberg, eds., Notes on Numerical Fluid Mechanics, Vol. 11, Vieweg, Braunschweig, 1985, pp. 12–23.

*[74] ———, *Zwei Mehrgitterverfahren zur numerischen Berechnung von stationären Strömungen durch poröse Medien mit freiem Rand*, Dissertation, Abteilung für Mathematik, Ruhr-Universität Bochum, 1985.

*[75] J. H. BOLSTAD AND H. B. KELLER, *A multigrid continuation method for elliptic problems with turning points*, SIAM J. Sci. Stat. Comput., 7 (1986), pp. 1081–1104.

*[76] ———, *Computation of anomalous modes in the Taylor experiment*, Report 217–50, California Institute of Technology, Applied Mathematics, Pasadena, CA, 1985.

*[77] R. BOYER AND B. MARTINET, *Multigrid methods in convex optimization*, Proc. 2nd

European Conference on Multigrid Methods, Cologne, October 1–4, 1985, W. Hackbusch and U. Trottenberg, eds., GMD-Studien Nr. 110 Gesellschaft für Mathematik und Datenverarbeitung, St. Augustin, to appear.

*[78] B. J. BRAAMS, *Magnetohydrodynamic equilibrium calculations using multigrid,* Proc. 2nd European Conference on Multigrid Methods, Cologne, October 1–4, 1985, W. Hackbusch and U. Trottenberg, eds., Lecture Notes in Mathematics, Springer-Verlag, Berlin, to appear.

[79] D. BRAESS, *The contraction number of a multigrid method for solving the Poisson equation,* Numer. Math., 37 (1981), pp. 387–404.

[80] ———, *The convergence rate of a multigrid method with Gauss–Seidel relaxation for the Poisson equation,* in Multigrid Methods, Proc. conference held at Köln-Porz, November 23–27, 1981, W. Hackbusch and U. Trottenberg, eds., Lecture Notes in Mathematics 960, Springer-Verlag, Berlin, 1982, pp. 368–386.

[81] ———, *The convergence rate of a multigrid method with Gauss–Seidel relaxation for the Poisson equation (revised),* Math. Comp., 42 (1984), pp. 505–519.

*[82] ———, *On the combination of the multigrid method and conjugate gradients,* Proc. 2nd European Conference on Multigrid Methods, Cologne, October 1–4, 1985, W. Hackbusch and U. Trottenberg, eds., Lecture Notes in Mathematics, Springer-Verlag, Berlin, to appear.

[83] D. BRAESS AND W. HACKBUSCH, *A new convergence proof for the multigrid method including the V-cycle,* SIAM J. Numer. Anal., 20 (1983), pp. 967–975.

*[84] D. BRAESS AND P. PEISKER, *On the numerical solution of the biharmonic equation and the role of squaring matrices for preconditioning,* Bericht Nr. 31, Institut für Mathematik, Ruhr-Universität Bochum, 1984.

*[85] D. BRAESS, W. HACKBUSCH AND U. TROTTENBERG, eds., *Advances in multi-grid methods,* Proc. conference held in Oberwolfach, December 8–13, 1984, Notes on Numerical Fluid Mechanics, Vol. 11, Vieweg, Braunschweig, 1984.

H. W. BRANCA, cf. [404].

K. BRAND, cf. [396].

A. BRANDT, cf. [5], [24], [47], [190], [524].

[86] A. BRANDT, *Multi-level adaptive technique (MLAT) for fast numerical solution to boundary value problems,* Proc. Third International Conference on Numerical Methods in Fluid Mechanics, Paris 1972, H. Cabannes and R. Teman, eds., Lecture Notes in Physics 18, Springer-Verlag, Berlin, 1973, pp. 82–89.

[87] ———, *Multi-level adaptive techniques (MLAT). I. The multi-grid method,* Research Report RC 6026, IBM T. J. Watson Research Center, Yorktown Heights, NY, 1976.

[88] ———, *Multi-level adaptive solutions to boundary-value problems,* Math. Comp., 31 (1977), pp. 333–390.

[89] ———, *Multi-level adaptive techniques (MLAT) for partial differential equations: ideas and software,* in Mathematical Software III, J. R. Rice, ed., Academic Press, New York, 1977, pp. 277–318.

[90] ———, *Multi-level adaptive finite-element methods. I. Variational problems,* in Special Topics of Applied Mathematics, J. Frehse, D. Pallaschke and U. Trottenberg, eds., North-Holland, Amsterdam, 1979, pp. 91–128.

[91] ———, *Multi-level adaptive techniques (MLAT) for singular-perturbation problems,* in Numerical Analysis of Singular Perturbation Problems, P. W. Hemker and J. J. H. Miller, eds., Academic Press, London, 1979, pp. 53–142.

[92] ———, *Multi-level adaptive computations in fluid dynamics,* AIAA J., 18 (1980), pp. 1165–1172.

[93] ———, *Numerical stability and fast solutions to boundary value problems,* in Boundary

and Interior Layers—Computational and Asymptotic Methods, J. J. H. Miller, ed., Boole Press, Dublin, 1980, pp. 29–49.

[94] ———, *Stages in developing multigrid solutions,* in Numerical Methods for Engineering I, E. Absi, R. Glowinski, P. Lascaux, H. Veysseyre, eds., Dunod, Paris, 1980, pp. 23–45.

[95] ———, *Multigrid solvers for non-elliptic and singular-perturbation steady-state problems,* Research Report, Dept. of Applied Mathematics, Weizmann Institute of Science, Rehovot, Israel, 1981.

[96] ———, *Multigrid solvers on parallel computers,* in Elliptic Problem Solvers, M. H. Schultz, ed., Academic Press, New York, 1981, pp. 39–84.

[97] ———, *Guide to multigrid development,* in Multigrid Methods, Proc. conference held at Köln-Porz, November 23–27, 1981, W. Hackbusch and U. Trottenberg, eds., Lecture Notes in Mathematics 960, Springer-Verlag, Berlin, 1982.

[98] ———, *Introductory remarks on multigrid methods,* in Numerical Methods in Fluid Dynamics, Proc. conference held at University of Reading, 1982, K. W. Morton and M. J. Baines, eds., 1982.

[99] ———, *Multigrid solutions to steady-state compressible Navier–Stokes equations.* I. in Computing Methods in Applied Sciences and Engineering V, Proc. Fifth International Symposium, Versailles, December 1981, R. Glowinski and J. L. Lions, eds., 1982, pp. 407–423.

*[100] ———, *Local and multi-level parallel processing mill,* Rechnerarchitekturen für die numerische Simulation auf der Basis superschneller Lösungsverfahren I, Workshop 'Rechnerarchitektur,' Erlangen, June 14–15, 1984, U. Trottenberg and P. Wypior, eds., GMD-Studien Nr. 88, Gesellschaft für Mathematik und Datenverarbeitung, St. Augustin, 1984, pp. 31–40.

*[101] ———, *Multigrid techniques:* 1984 *guide with applications to fluid dynamics,* GMD-Studien Nr. 85. Gesellschaft für Mathematik und Datenverarbeitung, St. Augustin, 1984.

*[102] ———, *Introduction—levels and scales,* in Multigrid Methods for Integral and Differential Equations, D. J. Paddon and H. Holstein, eds., The Institute of Mathematics and its Applications Conference Series 3, Clarendon Press, Oxford, 1985, pp. 1–10.

*[103] ———, *Algebraic multigrid theory: The symmetric case,* in Appl. Math. Comp., Proc. 2nd Internat. Multigrid Conference, April 1985, Copper Mountain, CO, S. F. McCormick, ed., North-Holland, Amsterdam, to appear.

[104] A. Brandt and C. W. Cryer, *Multigrid algorithms for the solution of linear complementarity problems arising from free boundary problems,* SIAM J. Sci. Stat. Comput., 4 (1983), pp. 655–684.

[105] A. Brandt and N. Dinar, *Multi-grid solutions to elliptic flow problems,* in Numerical Methods for Partial Differential Equations, S. V. Parter, ed., Academic Press, New York, 1979, pp. 53–147.

[106] A. Brandt and D. Ophir, *Language for processes of numerical solutions to differential equations. Second annual report,* U.S. Army Contract DAJA37-79-C-0504, 1981.

[107] ———, *Gridpack: Toward unification of general grid programming,* Proc. IFIP-Conference on PDE Software: Modules, Interfaces and Systems, Söderköping, Sweden, August 22–26, 1983, B. Enquist and T. Smedsaas, eds., North-Holland, Amsterdam, 1983.

[108] A. Brandt and S. Ta'asan, *Multi-grid methods for highly oscillatory problems,* Research Report, Dept. of Applied Mathematics, Weizmann Institute of Science, Rehovot, Israel, 1981.

*[109] A. BRANDT AND S. TA'ASAN, *Multigrid solutions to quasi-elliptic schemes,* in Progress and Supercomputing in Computational Fluid Dynamics, E. M. Murmann and S. S. Abarbanely, eds., Birkhäuser Verlag, Boston, 1985, pp. 235–255.

*[110] ———, *Multigrid methods for nearly singular and slightly indefinite problems,* Proc. 2nd European Conference on Multigrid Methods, Cologne, October 1–4, 1985, W. Hackbusch, U. Trottenberg, eds., Lecture Notes in Mathematics, Springer-Verlag, Berlin, to appear.

[111] A. BRANDT, J. E. DENDY, JR. AND H. M. RUPPEL, *The multigrid method for semi-implicit hydrodynamics codes,* J. Comput. Phys., 34 (1980), pp. 348–370.

*[112] A. BRANDT, S. R. FULTON AND G. D. TAYLOR, *Improved spectral multigrid methods for periodic elliptic problems,* J. Comput. Phys., to appear.

[113] A. BRANDT, S. F. MCCORMICK AND J. RUGE, *Algebraic multigrid (AMG) for automatic multigrid solution with application to geodetic computations,* Report, Inst. Comp. Studies, Colorado State University, Ft. Collins, CO, 1982.

[114] ———, *Multigrid methods for differential eigenproblems,* SIAM J. Sci. Stat. Comput., 4 (1983), pp. 244–260.

*[115] ———, *Algebraic multigrid (AMG) for sparse matrix equations,* in Sparsity and Its Applications, D. J. Evans, ed., Cambridge Univ. Press, Cambridge, 1984.

*[116] A. BRANDT, D. RON AND D. AMIT, *Multilevel approaches to discrete-state and stochastic problems,* Proc. 2nd European Conference on Multigrid Methods, Cologne, October 1–4, 1985, W. Hackbusch and U. Trottenberg, eds., Lecture Notes in Mathematics, Springer-Verlag, Berlin, to appear.

[117] M. BREDIF, *Une methode d'elements finis multigrille pour le calcul d'ecoulements potentiels transsoniques,* Proc. Third International Symposium on Numerical Methods in Engineering (Tome 1), March 14–18, 1983, Paris, P. Lascaux, ed., Pluralis, Paris, 1983, pp. 247–254.

F. W. BRINKLEY, cf. [4].

*[118] U. BROCKMEIER, N. K. MITRA AND M. FIEBIG, *Implementation of multigrid in SOLA algorithm,* Proc. 2nd European Conference on Multigrid Methods, Cologne, October 1–4, 1985, W. Hackbusch and U. Trottenberg, eds., GMD-Studien Nr. 110, Gesellschaft für Mathematik und Datenverarbeitung, St. Augustin, to appear.

[119] J. L. BROWN, *A multigrid mesh-embedding technique for three-dimensional transonic potential flow analysis,* in Multigrid Methods, H. Lomax, ed., NASA Conference Publication 2202, Ames Research Center, Moffett Field, CA, 1981, pp. 131–150.

*[120] ———, *An embedded-mesh potential flow analysis,* AIAA J., 22 (1984), pp. 174–178.

D. M. BUTTERFIELD, cf. [48], [49].

R. CAMARERO, cf. [342], [343].

*[121] R. CAMARERO AND M. REGGIO, *Multigrid scheme for three-dimensional body-fitted coordinates in turbomachine applications,* J. Fluids Engng.. Trans. ASME, 105 (1983), pp. 76–82.

[122] R. CAMARERO AND M. YOUNIS, *Efficient generation of body-fitted coordinates for cascades using multigrid,* AIAA J., 18 (1980), pp. 487–488.

D. A. CAUGHEY, cf. [517], [518].

[123] D. A. CAUGHEY, *Multi-grid calculation of three-dimensional transonic potential flows,* in Appl. Math. Comp., 13, Proc. Internat. Multigrid Conference, April 6–8, 1983, Copper Mountain, CO, S. F. McCormick and U. Trottenberg, eds., North-Holland, Amsterdam, 1983, pp. 241–260.

T. F. CHAN , cf. [31].

*[124] T. F. CHAN, *Techniques for large sparse systems arising from continuation methods,* Technical Report no. 291, Computer Science Dept., Yale University, New Haven, CT, 1983.

*[125] T. F. CHAN, *An efficient modular algorithm for coupled nonlinear systems,* Research Report no. 328, Computer Science Dept., Yale University, New Haven, CT, 1984.

*[126] T. F. CHAN AND H. B. KELLER, *Arc-length continuation and multi-grid techniques for nonlinear elliptic eigenvalue problems,* Report, Computer Science Dept., Yale University, New Haven, CT, 1981.

[127] ———, *Arc-length continuation and multi-grid techniques for nonlinear elliptic eigenvalue problems,* SIAM J. Sci. Stat. Comput., 3 (1982), pp. 173–194.

*[128] T. F. CHAN AND Y. SAAD, *Multigrid algorithms on the Hypercube multiprocessor,* Res. Rpt. no. 368, Computer Science Dept., Yale University, New Haven, CT, 1985.

[129] T. F. CHAN AND F. SAIED, *A comparison of elliptic solvers for general two-dimensional regions,* Technical Report no. 238, Computer Science Dept., Yale University, New Haven, CT, 1983.

*[130] T. F. CHAN AND R. SCHREIBER, *Parallel networks for multigrid algorithms: architecture and complexity,* Report no. 262, Department of Computer Science, Yale University, New Haven, CT, 1983.

*[131] T. F. CHAN, Y. SAAD AND M. H. SCHULTZ, *Solving elliptic partial differential equations on the Hypercube multiprocessor,* Research Report no. 373, Computer Science Dept., Yale University, New Haven, CT, 1985.

*[132] M. S. CHARRON AND M. C. MARCHE, *Une generation interactive des donnes pour appliquer la methode des elements finis,* Canad. J. Ci. Engrg., 11 (1984).

J. J. CHATTOT, cf. [335].

*[133] R. V. CHIMA, *Analysis of inviscid and viscous flows in cascades with an explicit multiple-grid algorithm,* NASA-TM-83636. National Aeronautics and Space Administration, Cleveland, OH, 1984.

*[134] R. V. CHIMA AND G. M. JOHNSON, *Efficient solution of the Euler and Navier–Stokes equations with a vectorized multi-grid algorithm,* American Institute of Aeronautics and Astronautics, 83–1893, Danvers, 1983.

P. E. CIESIELSKI, cf. [182].

D. S. CLEMM, cf. [206].

*[135] V. COUAILLIER, *Solution of the Euler equations: Explicit schemes acceleration by a multigrid method,* Proc. 2nd European Conference on Multigrid Methods, Cologne, October 1–4, 1985, W. Hackbusch and U. Trottenberg, eds., GMD-Studien Nr. 110, Gesellschaft für Mathematik und Datenverarbeitung, St. Augustin, to appear.

*[136] A. W. CRAIG AND O. C. ZIENKIEWICZ, *A multigrid algorithm using a hierarchical finite element basis,* in Multigrid Methods for Integral and Differential Equations, D. J. Paddon and H. Holstein, eds., The Institute of Mathematics and its Applications Conference Series 3, Clarendon Press, Oxford, 1985, pp. 301–312.

*[137] A. W. CRAIG, J. Z. ZHU AND O. C. ZIENKIEWICZ, *A-posteriori error estimation, adaptive mesh refinement and multigrid methods using hierarchical finite element bases,* Report C/R/483/84, Institute for Numerical Methods in Engineering, University College of Swansea, Swansea, UK, 1984.

C. W. CRYER, cf. [104].

*[138] R. L. DAVIS, *The prediction of compressible laminar viscous flows using a time marching control-volume and multi grid technique,* American Institute of Aeronautics and Astronautics, 83–1896, Danvers, 1983.

*[139] B. DEBUS, *Ansatz spezieller Mehrgitterkomponenten für ein zweidimensionales, singulär gestörtes Modellproblem: Grobgitter—und Glättungsoperatoren,* Diplomarbeit, Institut für Angewandte Mathematik, Universität Bonn, 1985.

[140] H. DECONINCK AND C. HIRSCH, *A multigrid finite element method for the transonic potential equation,* in Multigrid Methods, Proc. conference held at Köln-Porz,

November 23–27, 1981, W. Hackbusch and U. Trottenberg, eds., Lecture Notes in Mathematics 960, Springer-Verlag, Berlin, 1982, pp. 387–409.

[141] ———, *A multigrid method for the transonic full potential equation discretized with finite elements on an arbitrary body fitted mesh,* J. Comput. Phys., 48 (1982), pp. 344–365.

J. E. DENDY, JR., cf. [5], [111].

[142] J. E. DENDY, JR., *Black box multigrid,* J. Comput. Phys., 48 (1982), pp. 366–386.

*[143] ———, *A priori local grid refinement in the multigrid method,* Proc. Conference on Elliptic Problem Solvers, Monterey, CA, Jan. 1983.

[144] ———, *Black box multigrid for nonsymmetric problems,* in Appl. Math. Comp., 13, Proc. Internat. Multigrid Conference, April 6–8, 1983, Copper Mountain, CO, S. F. McCormick and U. Trottenberg, eds., North-Holland, Amsterdam, 1983, pp. 261–284.

*[145] ———, *Multigrid semi-implicit hydrodynamics revisited,* Proc. Conference on Large Scale Scientific Computation, Madison, WI, May 1983.

*[146] ———, *Black box multigrid for systems,* in Appl. Math. Comp., Proc. 2nd Internat. Multigrid Conference, April 1985, Copper Mountain, CO, S. F. McCormick, ed., North-Holland, Amsterdam, to appear.

[147] J. E. DENDY, JR. AND J. M. HYMAN, *Multi-grid and ICCG for problems with interfaces,* in Elliptic Problem Solvers, M. H. Schultz, ed., Academic Press, New York, 1981, pp. 247–253.

*[148] J. D. DENTON, *An improved time marching method for turbomachinery flow calculation,* in Numerical Methods in Aeronautical Fluid Dynamics, Proc. Conference, P. L. Roe, ed., Reading, UK, March/April, 1981, Academic Press, London, 1982, pp. 189–210.

[149] C. R. DEVORE, *Vectorization and implementation of an efficient multigrid algorithm for the solution of elliptic partial differential equations,* Report, NRL-MR-5504, Naval Research Laboratory, Washington, D.C., 1984.

*[150] E. DICK, *A multigrid method for Cauchy–Riemann and steady Euler equations based on flux-difference splitting,* in Efficient Solution of Elliptic Systems: Proc. GAMM-Seminar in Kiel, January 27–29, 1984, W. Hackbusch, ed., Notes on Numerical Fluid Mechanics, Volume 10, Vieweg, Braunschweig, 1984, pp. 20–37.

*[151] ———, *A multigrid technique for steady Euler equations based on flux-difference splitting,* Proc. Ninth International Conference on Numerical Methods in Fluid Dynamics, Soubbaramayer and J. P. Boujot, eds., Lecture Notes in Physics 218, Springer-Verlag, Berlin, 1985.

N. DINAR, cf. [105].

[152] N. DINAR, *Fast methods for the numerical solution of boundary-value problems,* Ph.D. thesis, Dept. of Applied Mathematics, Weizmann Institute of Science, Rehovot, 1978.

[153] M. DONOVANG, *Defektkorrekturen nach Stetter und Pereyra und MG-Extrapolation nach Brandt: Beziehungen und Anwendung auf elliptische Randwertaufgaben,* Diplomarbeit, Institut für Angewandte Mathematik, Universität Bonn, 1981.

*[154] M. R. DORR, *L3AMG—an algebraic multigrid routine,* Report, UCID-20117, Lawrence Livermore National Lab., Livermore, CA, 1984.

C. C. DOUGLAS, cf. [32].

[155] C. C. DOUGLAS, *Multi-grid algorithms for elliptic boundary-value problems,* Technical Report no. 223, Computer Science Dept., Yale University, New Haven, CT, 1982.

[156] ———, *A multi-grid optimal order solver for elliptic boundary-value problems,* Research Report no. 248, Computer Science Dept., Yale University, New Haven, CT, 1983.

[157] C. C. Douglas, *Abstract multi-grid with applications to elliptic boundary-value problems,* Research Report no. 275, Computer Science Dept., Yale University, New Haven, CT, 1983.

[158] ———, *Multi-grid algorithms with applications to elliptic boundary-value problems,* Research Report no. 247, Computer Science Dept., Yale University, New Haven, CT, 1983.

T. F. Dupont, cf. [33], [34].

S. E. Ellis, cf. [49].

G. Enden, cf. [294].

B. Epstein, cf. [359].

L. E. Eriksson, cf. [472].

*[159] R. Ewing, S. F. McCormick and J. Thomas, *The fast adaptive composite grid method for solving differential boundary value problems,* Proc. Fifth ASCE-EMD Speciality Conference, August 1984.

*[160] B. Favini and G. Guj, *MG techniques for staggered differences,* in Multigrid Methods for Integral and Differential Equations, D. J. Paddon and H. Holstein, eds., The Institute of Mathematics and its Applications Conference Series 3, Clarendon Press, Oxford, 1985, pp. 253–262.

R. P. Fedorenko, cf. [444].

[161] R. P. Fedorenko, *A relaxation method for solving elliptic difference equations,* U.S.S.R. Comput. Math. and Math. Phys., 1 (1962), pp. 1092–1096.

[162] ———, *The speed of convergence of an iterative process,* U.S.S.R. Comput. Math. and Math. Phys., 4 (1964), pp. 227–235.

*[163] G. Fessler, *Überlegungen zu Mehrgitteralgorithmen auf Baumrechnern,* Rechnerarchitekturen für die numerische Simulation auf der Basis superschneller Lösungsverfahren II, U. Trottenberg and P. Wypior, eds., GMD-Studien Nr. 102, Gesellschaft für Mathematik und Datenverarbeitung, St. Augustin, 1985, pp. 7–14.

M. Fiebig, cf. [118].

[164] G. J. Fix and M. D. Gunzburger, *On numerical methods for acoustic problems,* Comput. Math. Appl., 6 (1980), pp. 265–278.

*[165] J. Flores, J. Barton, T. Holst and T. H. Pulliam, *Comparison of the full potential and Euler formulations for computing transonic airfoil flows,* NASA-TM-85983. National Aeronautics and Space Administration, Moffett Field, CA, 1984.

[166] H. Foerster and K. Witsch, *On efficient multigrid software for elliptic problems on rectangular domains,* Math. Comput. Simulation, XXIII, (1981), pp. 293–298.

[167] ———, *Multigrid software for the solution of elliptic problems on rectangular domains: MG00 (Release 1),* in Multigrid Methods, Proc. conference held at Köln-Porz, November 23–27, 1981, W. Hackbusch and U. Trottenberg, eds., Lecture Notes in Mathematics 960, Springer-Verlag, Berlin, 1982.

[168] H. Foerster, K. Stüben and U. Trottenberg, *Nonstandard multigrid techniques using checkered relaxation and intermediate grids,* in Elliptic Problem Solvers, M. H. Schultz, ed., Academic Press, New York, 1981, pp. 285–300.

[169] C. K. Forester, *Advantages of multi-grid methods for certifying the accuracy of PDE modeling,* in Multigrid Methods, H. Lomax, ed., NASA Conference Publication 2202, Ames Research Center, Moffett Field, CA, 1981, pp. 23–45.

*[170] ———, *Error norms for the adaptive solution of the Navier–Stokes equations,* NASA-CR-165828, National Aeronautics and Space Administration, Washington, D.C., 1982.

P. A. Forsyth, Jr., cf. [59], [60], [61].

*[171] B. Franke and H.-O. Leilich, *Berechnungsgrundlagen für das Datenkommunikationssystem in einem Multiprocessor für Mehrgitterverfahren,*

Rechnerarchitekturen für numerische Simulation auf der Basis superschneller Lösungsverfahren II, U. Trottenberg and P. Wypior, eds., GMD-Studien Nr. 102, Gesellschaft für Mathematik und Datenverarbeitung, St. Augustin, 1985, pp. 15–120.

J. M. J. FRAY, cf. [443].

G. FRITSCH, cf. [281].

L. FUCHS, cf. [325], [326], [554].

*[172] L. FUCHS, *Finite-difference methods for plane steady inviscid transonic flows,* Report TRITA-GAD-2, Division of Gasdynamics, Royal Institute of Technology, Stockholm, 1977.

[173] ———, *A Newton-multi-grid method for the solution of non-linear partial differential equations,* in Boundary and Interior Layers—Computational and Asymptotic Methods, J. J. H. Miller, ed., Boole Press, Dublin, 1980, pp. 291–296.

[174] ———, *Multi-grid solution of the Navier–Stokes equations on non-uniform grids,* in Multigrid Methods, H. Lomax, ed., NASA Conference Publication 2202, Ames Research Center, Moffett Field, CA, 1981, pp. 84–101.

[175] ———, *Transonic flow computation by a multi-grid method,* in Numerical Methods for the Computation of Inviscid Transonic Flows with Shock Waves, A. Rizzi and H. Viviand, eds., Vieweg, Braunschweig, 1981, pp. 58–65.

*[176] ———, *New relaxation methods for incompressible flow problems,* in Numerical Methods in Laminar and Turbulent Flow, Proc. Third International Conference held in Seattle, August 8–11, 1983, C. Taylor, J. A. Johnson and W. R. Smith, eds., Pineridge Press, Swansea, 1983, pp. 627–641.

*[177] ———, *Defect corrections and higher numerical accuracy,* in Efficient Solution of Elliptic Systems, Proc. GAMM-Seminar in Kiel, January 27–29, 1984, W. Hackbusch ed., Notes on Numerical Fluid Mechanics, Volume 10, Vieweg, Braunschweig, 1984, pp. 52–66.

*[178] ———, *Multi-grid schemes for incompressible flows,* in Efficient Solution of Elliptic Systems: Proc. GAMM-Seminar in Kiel, January 27–29, 1984, W. Hackbusch, ed., Notes on Numerical Fluid Mechanics, Volume 10, Vieweg, Braunschweig, 1984, pp. 38–51.

*[179] ———, *An adaptive multi-grid scheme for simulation of flows,* in Proc. 2nd European Conference on Multigrid Methods, Cologne, October 1–4, 1985, W. Hackbusch and U. Trottenberg, eds., Lecture Notes in Mathematics, Springer-Verlag, Berlin, to appear.

*[180] L. FUCHS AND N. TILLMARK, *Numerical and experimental study of driven flow in a polar cavity,* Internat. J. Numer. Meth. Fluids, 5 (1985), pp. 311–329.

*[181] L. FUCHS AND H.-S. ZHAO, *Solution of three-dimensional viscous incompressible flows by a multigrid method,* Internat. J. Numer. Meth. Fluids, 4 (1984), pp. 539–555.

S. R. FULTON, cf. [112].

*[182] S. R. FULTON, P. E. CIESIELSKI AND W. H. SCHUBERT, *Multigrid methods for elliptic problems,* Monthly Weather Review, 1985.

*[183] D. B. GANNON, *Self adaptive methods for parabolic partial differential equations,* Report UIUCDCS-R-80-1020, Dept. of Computer Science, University of Illinois at Urbana-Champaign, Urbana, IL, 1980.

*[184] ———, *On the structure of parallelism in a highly concurrent PDE solver,* Proc. 7th Symposium on Computer Arithmetic, H. Kai, ed., Urbana, IL, June 1985, IEEE, Silver Springs, MD, 1985, pp. 252–259.

*[185] D. B. GANNON AND J. R. VAN ROSENDALE, *Highly parallel multigrid solvers for elliptic PDE's: an experimental analysis,* ICASE Report No. 82-36, 1982.

*[186] D. B. GANNON AND J. R. VAN ROSENDALE, *Parallel architectures for iterative methods on adaptive block-structured grids,* Report NASA CR-172195, 1983.

[187] J. GARY, *The multigrid iteration applied to the collocation method,* SIAM J. Numer. Anal., 18 (1981), pp. 211–224.

[188] ———, *On higher order multigrid methods with application to a geothermal reservoir model,* Internat. J. Numer. Meth. Fluids, 2 (1982), pp. 43–60.

[189] J. GARY, S. F. MCCORMICK AND R. A. SWEET, *Successive overrelaxation, multigrid, and preconditioned conjugate gradients algorithms for solving a diffusion problem on a vector computer,* in Appl. Math. Comp., 13, Proc. Internat. Multigrid Conference, April 6–8, 1983, Copper Mountain, CO, S. F. McCormick and U. Trottenberg, eds., North-Holland, Amsterdam, 1983, pp. 285–310.

[190] S. P. GAUR AND A. BRANDT, *Numerical solution of semiconductor transport equations in two dimensions by multi-grid method,* in Advances in Computer Methods for Partial Differential Equations II, R. Vichnevetsky, ed., IMACS (AICA), New Brunswick, NJ, 1977, pp. 327–329.

*[191] W. GENTZSCH, *Vectorization of computer programs with applications to computational fluid dynamics,* Notes on Numerical Fluid Dynamics, Vol. 8, Vieweg, Braunschweig, 1984.

*[192] L. GEUS, *Parallelisierung eines Mehrgitterverfahrens für die Navier–Stokes–Gleichungen auf EGPA-Systemen,* Diplomarbeit, Institut für Mathematische Maschinen und Datenverarbeitung (Informatik), Universität Erlangen/Nürnberg, Arbeitsbericht 18, 3, 1985.

*[193] L. GEUS, W. HENNING, W. SEIDL, AND J. VOLKERT, MG00-*Implementierungen auf EGPA-Multiprozessorsystemen,* Rechnerarchitekturen für die numerische Simulation auf der Basis superschneller Lösungsverfahren II, U. Trottenberg and P. Wypior, eds., GMD-Studien Nr. 102, Gesellschaft für Mathematik und Datenverarbeitung, St. Augustin, 1985, pp. 121–136.

K. N. GHIA, cf. [196].

*[194] K. N. GHIA AND U. GHIA, *Analysis of three-dimensional viscous internal flows,* AFOSR-TR-83-1053, Air Force Office of Scientific Research, Bolling AFB, Washington, D.C., 1983.

[195] K. N. GHIA, U. GHIA, D. R. REDDY AND C. T. SHIN, *Multigrid simulation of asymptotic curved-duct flows using a semi-implicit numerical technique,* Proc. Symposium on Computers in Flow and Fluid Dynamics Experiments, ASME, Winter Annual Meeting, Washington, D.C., 1981.

U. GHIA, cf. [194], [195].

[196] U. GHIA, K. N. GHIA AND C. T. SHIN, *Solution of incompressible Navier–Stokes equations by coupled strongly-implicit multi-grid method,* J. Comput. Phys., 48 (1982), pp. 387–411.

*[197] W. GILOI, *Kritische Betrachtungen und konstruktive Vorschläge zur Frage der Entwicklung eines großen MIMD-Multiprozessorsystems für numerische Anwendungen (schnelle Lösungsverfahren),* Rechnerarchitekturen für die numerische Simulation auf der Basis superschneller Lösungsverfahren I, Workshop 'Rechnerarchitektur,' Erlangen, June 14–15, 1984, U. Trottenberg and P. Wypior, eds., GMD-Studien Nr. 88, Gesellschaft für Mathematik und Datenverarbeitung, St. Augustin, 1984, pp. 161–194.

C. I. GOLDSTEIN, cf. [51].

[198] U. GÖLLNER, *Ein singulär gestörtes parabolisches Anfangsrandwertproblem und seine numerische Behandlung,* Diplomarbeit, Institut für Angewandte Mathematik, Universität Bonn, 1983.

*[199] B. GÖRG AND O. KOLP, *Parallele Rechnerarchitekturen für Mehrgitteralgorithmen,*

Rechnerarchitekturen für die numerische Simulation auf der Basis superschneller Lösungsverfahren II, U. Trottenberg and P. Wypior, eds., GMD-Studien Nr. 102, Gesellschaft für Mathematik und Datenverarbeitung, St. Augustin, 1985, pp. 137–150.

*[200] J. GOZANI, *Conjugate gradient coupled with multigrid for an indefinite problem,* in Advances in Computer Methods for Partial Differential Equations V, R. Vichnevetsky and R. Stepleman, eds., 1984.

[201] A. GREENBAUM, *Analysis of a multigrid method as an iterative technique for solving linear systems,* SIAM J. Numer. Anal., 21 (1984), pp. 473–485.

*[202] ———, *A multigrid method for multiprocessors,* in Appl. Math. Comp., Proc. 2nd Internat. Multigrid Conference, April 1985, Copper Mountain, CO, S. F. McCormick, ed., North-Holland, Amsterdam, to appear.

[203] F. F. GRINSTEIN, H. RABITZ AND A. ASKAR, *The multigrid method for accelerated solution of the discretized Schrödinger equation,* J. Comput. Phys., 49 (1983), pp. 423–512.

*[204] ———, *Steady state reactive kinetics on surfaces exhibiting defect structures,* J. Chem. Phys., 82 (1985), pp. 3434–3441.

[205] C. E. GROSCH, *Performance analysis of Poisson solvers on array computers,* in Supercomputers, 2, Infotech International, Maidenhead, 1979, pp. 147–181.

[206] K. G. GUDERLEY AND D. S. CLEMM, *Eigenvalue and near eigenvalue problems solved by Brandt's multigrid method,* Flight Dyn. Report AFFDL-TR-79-3147, Wright-Patterson Air-Force Base, 1979.

G. GUJ, cf. [160].

M. D. GUNZBURGER, cf. [164].

*[207] K. GUSTAFSON AND R. LEBEN, *Multigrid calculation of subvortices,* in Appl. Math. Comp., Proc. 2nd Internat. Multigrid Conference, April 1985, Copper Mountain, CO, S. F. McCormick, ed., North-Holland, Amsterdam, to appear.

I. GUSTAFSSON, cf. [23].

[208] F. G. GUSTAVSON, *Implementation of the multi-grid method for solving partial differential equations,* Research Report RA 82, IBM T. J. Watson Research Center, Yorktown Heights, NY, 1976, pp. 51–57.

W. HACKBUSCH, cf. [83], [85].

[209] W. HACKBUSCH, *Ein iteratives Verfahren zur schnellen Auflösung elliptischer Randwertprobleme,* Report 76-12, Institut für Angewandte Mathematik, Universität Köln, 1976.

[210] ———, *A fast numerical method for elliptic boundary value problems with variable coefficients,* Proc. Second GAMM-Conference on Numerical Methods in Fluid Mechanics, E. H. Hirschel and W. Geller, eds., DFVLR, Köln, 1977, pp. 50–57.

[211] ———, *A multi-grid method applied to a boundary problem with variable coefficients in a rectangle,* Report 77-17, Institut für Angewandte Mathematik, Universität Köln, 1977.

[212] ———, *On the convergence of a multi-grid iteration applied to finite element equations,* Report 77-8, Institut für Angewandte Mathematik, Universität Köln, 1977.

[213] ———, *A fast iterative method for solving Helmholtz's equation in a general region,* in Fast Elliptic Solvers, U. Schumann, ed., Advance Publications, London, 1978, pp. 112–124.

[214] ———, *A fast iterative method for solving Poisson's equation in a general region,* in Numerical Treatment of Partial Differential Equations, Proc. conference held at Oberwolfach, July 4–10, 1976, R. Bulirsch, R. D. Grigorieff, J. Schröder, eds., Lecture Notes in Mathematics 631, Springer-Verlag, Berlin, 1978, pp. 51–62.

[215] W. Hackbusch, *On the multigrid method applied to difference equations,* Computing, 20 (1978), pp. 291–306.

[216] ———, *On the computation of approximate eigenvalues and eigenfunctions of elliptic operators by means of a multi-grid method,* SIAM J. Numer. Anal., 16 (1979), pp. 201–215.

[217] ———, *On the fast solution of parabolic boundary control problems,* SIAM J. Control Optim., 17 (1979), pp. 231–244.

[218] ———, *On the fast solutions of nonlinear elliptic equations,* Numer. Math., 32 (1979), pp. 83–95.

[219] ———, *Analysis and multigrid solutions of mixed finite element and mixed difference equations,* preprint, Institut für Angewandte Mathematik, Ruhr-Universität Bochum, 1980.

[220] ———, *Convergence of multi-grid iterations applied to difference equations,* Math. Comp., 34 (1980), pp. 425–440.

[221] ———, *Numerical solution of nonlinear equations by the multigrid iteration of the second kind,* in Numerical Methods for Nonlinear Problems, Proc. International Conference held at the University College Swansea, C. Taylor, ed., Pineridge Press, 1980, pp. 1041–1050.

[222] ———, *On the fast solving of elliptic control problems,* J. Optim. Theory Appl., 31 (1980), pp. 565–581.

[223] ———, *Survey of convergence proofs for multigrid iterations,* Special Topics of Applied Mathematics, J. Frehse, D. Pallaschke and U. Trottenberg, eds., North-Holland, Amsterdam, 1980, pp. 151–164.

[224] ———, *The fast numerical solution of very large elliptic difference schemes,* J. Inst. Math. Appl., 26 (1980), pp. 119–132.

[225] ———, *Bemerkungen zur iterierten Defektkorrekture und zu ihrer Kombination mit Mehrgitterverfahren,* Rev. Roumaine Math. Pures Appl., 26 (1981), pp. 1319–1329.

[226] ———, *Die schnelle Auflösung der Fredholmschen Integralgleichung zweiter Art,* Beiträge Numer. Math., 9 (1981), 47–62.

[227] ———, *Error analysis of the nonlinear multigrid method of the second kind,* Appl. Math., 26 (1981), pp. 18–29.

[228] ———, *Numerical solution of linear and nonlinear parabolic control problems,* in Optimization and Optimal Control, A. Auslender, W. Oettli and J. Stoer, eds., Lecture Notes in Control and Information Sciences, 30, Springer-Verlag, Berlin, 1981, pp. 179–185.

[229] ———, *On the convergence of multi-grid iterations,* Beiträge Numer. Math., 9 (1981), pp. 213–239.

[230] ———, *On the regularity of difference schemes,* Ark. Mat., 19 (1981), pp. 71–95.

[231] ———, *Optimal H**p,p/2 error estimates for a parabolic Galerkin method,* SIAM J. Numer. Anal., 18 (1981), pp. 681–692.

[232] ———, *The fast numerical solution of time periodic parabolic problems,* SIAM J. Sci. Stat. Comput., 2 (1981), pp. 198–206.

[233] ———, *Multi-grid convergence theory,* in Multigrid Methods, Proc. conference held at Köln-Porz, November 23–27, 1981, W. Hackbusch and U. Trottenberg, eds., Lecture Notes in Mathematics, 960, Springer-Verlag, Berlin, 1982.

[234] ———, *Multi-grid solution of continuation problems,* in Iterative Solution of Nonlinear Systems of Equations, Proc., Oberwolfach 1982, R. Ansorge, T. Meis and W. Törnig, eds., Lecture Notes in Mathematics 953, Springer-Verlag, Berlin, 1982, pp. 20–45.

[235] ———, *On multi-grid iterations with defect correction,* in Multigrid Methods, Proc. conference held at Köln-Porz, November 23–27, 1981, W. Hackbusch and U.

Trottenberg, eds., Lecture Notes in Mathematics 960, Springer-Verlag, Berlin, 1982, pp. 461–473.

[236] ———, *Introduction to multi-grid methods for the numerical solution of boundary value problems,* in Computational Methods for Turbulent, Transonic and Viscous Flows, J. A. Essers, ed., Hemisphere, Washington, D.C., 1983.

*[237] ———, *On the regularity of difference schemes—part* II: *regularity estimates for linear and nonlinear problems,* Ark. Mat., 21 (1983), pp. 3–28.

*[238] ———, ed., *Efficient solution of elliptic systems,* Proc. GAMM-Seminar in Kiel, January 27–29, 1984, Notes on Numerical Fluid Mechanics, Volume 10, Vieweg, Braunschweig, 1984.

*[239] ———, *Local defect correction method and domain decomposition techniques,* in Computing, Suppl. 5, Defect Correction Methods, K. Böhmer and H. J. Stetter, eds., Springer-Verlag, Wien, 1984, pp. 89–113.

[240] ———, *Multi-grid convergence for a singular perturbation problem,* Linear Algebra Appl., 58 (1984), pp. 125–145.

[241] ———, *Multi-grid solutions to linear and nonlinear eigenvalue problems for integral and differential equations,* Rostock Math. Colloq. 25 (1984), pp. 79–98.

*[242] ———, *Parabolic multi-grid methods,* in Computing Methods in Applied Sciences and Engineering VI: Proc. Sixth International Symposium, Versailles, December 1983, R. Glowinski and J. L. Lions, eds., North-Holland, Amsterdam, 1984.

*[243] ———, *Multi-grid eigenvalue computation,* in Advances in Multi-Grid Methods, Proc. conference held in Oberwolfach, December 8–13, 1984, D. Braess, W. Hackbusch and U. Trottenberg, eds., Notes on Numerical Fluid Mechanics, Volume 11, Vieweg, Braunschweig, 1985, pp. 24–32.

*[244] ———, *Multigrid methods and applications,* Springer Series in Comp. Math. 4, Springer-Verlag, Berlin, 1985.

*[245] ———, *Multigrid methods of the second kind,* in Multigrid Methods for Integral and Differential Equations, D. J. Paddon and H. Holstein, eds., The Institute of Mathematics and Its Applications Conference Series 3, Clarendon Press, Oxford, 1985, pp. 11–84.

[246] W. Hackbusch and G. Hofmann, *Results of the eigenvalue problem for the plate equation,* Z. Angew. Math. Phys., 31 (1980), pp. 730–739.

[247] W. Hackbusch and H. D. Mittelmann, *On multi-grid methods for variational inequalities,* Numer. Math., 42 (1983), pp. 65–76.

*[248] W. Hackbusch and Z. P. Nowak, *Multigrid methods for calculating the lifting potential incompressible flows around three-dimensional bodies,* Proc. 2nd European Conference on Multigrid Methods, Cologne, October 1–4, 1985, W. Hackbusch and U. Trottenberg, eds., Lecture Notes in Mathematics, Springer-Verlag, Berlin, to appear.

[249] W. Hackbusch and U. Trottenberg, eds., *Multigrid methods,* Proc. conference held at Köln-Porz, November 23–27, 1981, Lecture Notes in Mathematics, 960, Springer-Verlag, Berlin, 1982.

*[250] W. Hackbusch and T. Will, *A numerical method for a parabolic bang–bang problem,* Control Cybernet. 12 (1983), pp. 99–110.

W. Händler, cf. [281].

D. Hänel, cf. [504], [505], [506].

*[251] H. Hahn, *Mehrskalen-Methoden der Statistischen Physik: Ausgangspunkte zu Mehrgitter-Verfahren für die statistisch-physikalische Numerik,* Rechnerarchitekturen für die numerische Simulation auf der Basis superschneller Lösungsverfahren II, U. Trottenberg and P. Wypior, eds., GMD-Studien Nr. 102, Gesellschaft für Mathematik und Datenverarbeitung, St. Augustin, 1985, pp. 233–240.

*[252] M. G. Hall, *Advances and shortcomings in the calculation of inviscid flows with shock*

waves, in Numerical Methods in Aeronautical Fluid Dynamics, Proc. Conference, P. L. Roe, ed., Reading, UK, March/April 1981, Academic Press, London, 1982, pp. 33–60.

*[253] L. HART, S. F. MCCORMICK, A. O'GALLAGER AND J. THOMAS, *The fast adaptive composite grid method (FAC): Algorithms for advanced computers,* in Appl. Math. Comp., Proc. 2nd Internat. Multigrid Conference, April 1985, Copper Mountain, CO, S. F. McCormick, ed., North-Holland, Amsterdam, to appear.

*[254] F. K. HEBEKER, *On a multigrid method to solve the integral equations of* 3-D *Stokes flow,* in Efficient Solution of Elliptic Systems: Proc. GAMM-Seminar in Kiel, January 27–29, 1984, W. Hackbusch, ed., Notes on Numerical Fluid Mechanics, Volume 10, Vieweg, Braunschweig, 1984, pp. 67–73.

P. W. HEMKER, cf. [66].

[255] P. W. HEMKER, *Fourier analysis of gridfunctions, prolongations and restrictions,* Report NW 93/80, Dept. of Numerical Mathematics, Mathematical Center, Amsterdam, 1980.

[256] ———, *Multi-grid bibliography,* in Colloquium Numerical Integration of Partial Differential Equations, J. Verwer, ed., Dept. of Numerical Mathematics, Mathematical Center, Amsterdam, 1980.

[257] ———, *On the structure of an adaptive multi-level algorithm,* BIT, 20 (1980), pp. 289–301.

[258] ———, *The incomplete LU-decomposition as a relaxation method in multi-grid algorithms,* in Boundary and Interior Layers—Computational and Asymptotic Methods, J. J. H. Miller, ed., Boole Press, Dublin, 1980, pp. 306–311.

[259] ———, Algol 68 *Fourier analysis program,* Programlisting, Dept. of Numerical Mathematics, Mathematical Centre, Amsterdam, 1981.

[260] ———, Algol 68 *multigrid library,* Programlisting, Dept. of Numerical Mathematics, Mathematical Centre, Amsterdam, 1981.

[261] ———, *Introduction to multigrid methods,* Nieuw Archief voor Wiskunke (3), 29 (1981), pp. 71–101.

[262] ———, *Lecture notes of a seminar on multiple grid methods,* Report NN 24/81, Dept. of Numerical Mathematics, Mathematical Centre, Amsterdam, 1981.

[263] ———, *A note on defect correction processes with an approximate inverse of deficient rank,* Appl. Math. Comp., 8 (1982), pp. 137–139.

*[264] ———, *Extensions of the defect correction principle,* in An Introduction to Computational and Asymptotic Methods for Boundary and Interior Layers, J. J. H. Miller, ed., Short Course, Dublin, June 1982, Boole Press, Dublin, 1982, pp. 33–45.

[265] ———, *Mixed defect correction iteration for the accurate solution of the convection diffusion equation,* in Multigrid Methods, Proc. conference held at Köln-Porz, November 23–27, 1981, W. Hackbusch and U. Trottenberg, eds., Lecture Notes in Mathematics 960, Springer-Verlag, Berlin, 1982, pp. 485–501.

*[266] ———, *Numerical aspects of singular perturbation problems,* in Asymptotic Analysis II, F. Verhulst, ed., Lecture Notes in Mathematics 985, Springer-Verlag, Berlin, 1982, pp. 267–287.

[267] ———, *On the comparison of Line-Gauss-Seidel and ILU relaxation in multigrid algorithms,* in Computational and Asymptotic Methods for Boundary and Interior Layers, Proc. BAIL II Conference, Boole Press Conference Series 4, J. J. H. Miller, ed., Boole Press, Dublin, 1982, pp. 269–277.

*[268] ———, *Multigrid methods for problems with a small parameter in the highest derivative,* in Numerical Analysis, Proc., Dundee, 1983, G. A. Watson, ed., Lecture Notes in Mathematics 1066, Springer-Verlag, Berlin, 1983, pp. 106–121.

*[269] ———, *Mixed defect correction iteration for the solution of a singular perturbation*

problem, Computing, Suppl. 5, Defect Correction Methods, K. Böhmer and H. J. Stetter, eds., 1984, pp. 123–145.

*[270] ———, *Multigrid algorithms run on supercomputers,* Supercomputer, 4 (1984), pp. 44–51.

*[271] ———, *Defect correction and higher order schemes for the multigrid solution of the steady Euler equations,* Proc. 2nd European Conference on Multigrid Methods, Cologne, October 1–4, 1985, W. Hackbusch and U. Trottenberg, eds., Lecture Notes in Mathematics, Springer-Verlag, Berlin, to appear.

*[272] P. W. HEMKER AND G. M. JOHNSON, *Multigrid approaches to the Euler equations,* in Multigrid Methods, S. F. McCormick, ed., Society for Industrial and Applied Mathematics, Philadelphia, 1987.

[273] P. W. HEMKER AND H. SCHIPPERS, *Multiple grid methods for the solution of Fredholm integral equations of the second kind,* Math. Comp., 36 (1981), pp. 215–232.

*[274] P. W. HEMKER AND S. P. SPEKREIJSE, *Multigrid solutions of the steady Euler equations,* in Advances in Multi-Grid Methods, Proc. Conference held in Oberwolfach, December 8–13, 1984, D. Braess, W. Hackbusch and U. Trottenberg, eds., Notes on Numerical Fluid Mechanics, Volume 11, Vieweg, Braunschweig, 1985, pp. 33–44.

*[275] ———, *Multiple grid and Osher's scheme for the efficient solution of the steady Euler equations,* Appl. Numer. Math., to appear.

*[276] P. W. HEMKER AND P. M. DE ZEEUW, *Defect correction for the solution of a singular perturbation problem,* in Scientific Computing, R. S. Stepleman, ed., North-Holland, Amsterdam, 1983, pp. 113–118.

*[277] ———, *Some implementations of multigrid linear system solvers,* in Multigrid Methods for Integral and Differential Equations, D. J. Paddon and H. Holstein, eds., The Institute of Mathematics and its Applications Conference Series 3, Clarendon Press, Oxford, 1985, pp. 85–116.

*[278] P. W. HEMKER, P. WESSELING AND P. M. DE ZEEUW, *A portable vector-code for autonomous multigrid modules,* Proc. IFIP-Conference on PDE Software: Modules, Interfaces and Systems, Söderköping, Sweden, August 22–26, 1983, B. Enquist and T. Smedsass, eds., North-Holland, Amsterdam, 1984.

[279] P. W. HEMKER, R. KETTLER, P. WESSELING AND P. M. DE ZEEUW, *Multigrid methods: development of fast solvers,* in Appl. Math. Comp., 13, Proc. Internat. Multigrid Conference, April 6–8, 1983, Copper Mountain, CO, S. F. McCormick and U. Trottenberg, eds., North-Holland, Amsterdam, 1983, pp. 311–326.

W. HENNING, cf. [193], [281].

*[280] G. T. HERMAN, H. LEVKOWITZ, S. F. MCCORMICK AND H. TUY, *Multigrid image reconstruction,* Proc. Workshop on Multilevel Image Processing and Analysis, Leesbury, July 1982, A. Rosenfeld, ed., Springer-Verlag, Berlin, 1983.

C. HIRSCH, cf. [140], [141].

*[281] F. HOFMANN, W. HÄNDLER, J. VOLKERT, W. HENNING AND G. FRITSCH, *Multiprocessor-Architekturkonzept für Mehrgitterverfahren,* Rechnerarchitekturen für die numerische Simulation auf der Basis superschneller Lösungsverfahren I, Workshop 'Rechnerarchitektur,' Erlangen, 14–15 Juni, 1984, U. Trottenberg and P. Wypior, eds., GMD-Studien Nr. 88, Gesellschaft für Mathematik und Datenverarbeitung, St. Augustin, 1984, pp. 65–76.

G. HOFMANN, cf. [246].

*[282] G. HOFMANN, *Analyse eines Mehrgitterverfahrens zur Berechnung von Eigenwerten elliptischer Differentialoperatoren,* Dissertation, Universität Kiel, 1985.

*[283] ———, *Analysis of a SOR-like multi-grid algorithm for eigenvalue problems,* in Advances in Multi-Grid Methods, Proc. conference held in Oberwolfach, Decem-

ber 8–13, 1984, D. Braess, W. Hackbusch and U. Trottenberg, eds., Notes on Numerical Fluid Mechanics, Volume 11, Vieweg, Braunschweig, 1985, pp. 45–57.

[284] W. HOLLAND, S. F. MCCORMICK AND J. RUGE, *Unigrid methods for boundary value problems with nonrectangular domains,* J. Comput. Phys., 48 (1982), pp. 412–422.

T. HOLST, cf. [165].

H. HOLSTEIN, cf. [445].

*[285] H. HOLSTEIN AND G. PAPAMANOLIS, *A multigrid treatment of stream function normal derivative boundary conditions,* in Advances in Multi-Grid Methods, Proc. conference held in Oberwolfach, December 8–13, 1984, D. Braess, W. Hackbusch and U. Trottenberg, eds., Notes on Numerical Fluid Mechanics, Volume 11, Vieweg, Braunschweig, 1985, pp. 58–63.

*[286] W. H. HOLTER, *A vectorized multigrid solver for the three-dimensional Poisson equation,* in Supercomputer Applications, A. H. L. Emmen, ed., North-Holland, Amsterdam, 1985, pp. 17–32.

*[287] M. H. L. HOUNJET, *Field panel/finite difference method for potential unsteady transonic flow,* AIAA J., 23 (1985).

P. J. VAN DER HOUWEN, cf. [522].

[288] P. J. VAN DER HOUWEN, *On the time integration of parabolic differential equations,* in Numerical Analysis, Proceedings, Dundee, 1981, G. A. Watson, ed., Lecture Notes in Mathematics 912, Springer-Verlag, Berlin, 1982, pp. 157–168.

[289] P. J. VAN DER HOUWEN AND B. P. SOMMEIJER, *Analysis of Richardson iteration in multigrid methods for nonlinear parabolic differential equations,* Report NW 105/81, Dept. of Numerical Mathematics, Mathematical Centre, Amsterdam, 1981.

[290] P. J. VAN DER HOUWEN AND H. B. DE VRIES, *Preconditioning and coarse grid corrections in the solution of the initial value problem for nonlinear partial differential equations,* SIAM J. Sci. Stat. Comput., 3 (1982), pp. 473–485.

M. Y. HUSSAINI, cf. [360], [456], [457], [532], [604], [605], [606].

*[291] M. Y. HUSSAINI AND T. A. ZANG, *Iterative spectral methods and spectral solutions to compressible flows,* NASA-CR-173014, National Aeronautics and Space Administration, Hampton, VA, 1982.

[292] M. Y. HUSSAINI, M. D. SALAS AND T. A. ZANG, *Spectral methods for inviscid, compressible flows,* NASA-CR-172710, National Aeronautics and Space Administration, Hampton, VA, 1983.

Q. HUYNH, cf. [360].

*[293] Q. HUYNH, Y. S. WONG AND L. R. LUSTMAN, *Numerical methods for singular perturbation problems,* Technical Report, Naval Underwater Systems Center, Newport, RI, 1983.

J. M. HYMAN, cf. [147].

*[294] M. ISRAELI AND G. ENDEN, *A two-grid method for fluid dynamic problems with disparate time scales,* Proc. Ninth Internat. Conference on Numerical Methods in Fluid Dynamics, Soubbaramayer and J. P. Boujot, eds., Lecture Notes in Physics 218, Springer-Verlag, Berlin, 1985.

*[295] M. ISRAELI AND M. ROSENFELD, *Marching multigrid solutions to the parabolized Navier–Stokes equations,* Proc. Fifth GAMM-Conference on Numerical Methods in Fluid Mechanics, Rome, October 5–7, 1983, Notes on Numerical Fluid Mechanics, Volume 7, Vieweg, Braunschweig, 1983, pp. 137–145.

*[296] D. A. H. JACOBS, *Some iterative methods: past, present, future,* IEEE Colloquium on Numerical Solution Techniques for Large Sparse Systems of Equations, London, November 1983.

A. JAMESON, cf. [62], [63], [355], [452], [503].

[297] A. JAMESON, *Acceleration of transonic potential flow calculations on arbitrary meshes by the multiple grid method,* AIAA-79-1458, American Institute of Aeronautics and Astronautics Fourth Computational Fluid Dynamics Conference, New York, NY, 1979.

*[298] ———, *Evolution of computational methods in aerodynamics,* J. Appl. Mech., 50, (1983), pp. 1052–1070.

*[299] ———, *Numerical solution of the Euler equations for compressible inviscid fluids,* Proc. Sixth International Conference on Computational Methods in Applied Science and Engineering, Versailles, France, December 1983.

[300] ———, *Solution of the Euler equations for two dimensional transonic flow by a multigrid method,* in Appl. Math. Comp., 13, Proc. International Multigrid Conference, April 6–8, 1983, Copper Mountain, CO, S. F. McCormick and U. Trottenberg, eds., North-Holland, Amsterdam, 1983, pp. 327–356.

*[301] ———, *Transonic flow calculations for aircraft,* in Numerical Methods in Fluid Dynamics, Lecture Notes in Mathematics 1127, F. Brezzi, ed., Springer-Verlag, Berlin, 1985.

*[302] ———, *Multigrid algorithms for compressible flow calculations,* Proc. 2nd European Conference on Multigrid Methods, Cologne, October 1–4, 1985, W. Hackbusch and U. Trottenberg, eds., Lecture Notes in Mathematics, Springer-Verlag, Berlin, to appear.

*[303] A. JAMESON AND T. J. BAKER, *Multigrid solution of the Euler equations for aircraft configurations,* American Institute of Aeronautics and Astronautics, AIAA-84-0093, 1984.

*[304] A. JAMESON AND S. YOON, *Multigrid solution of the Euler equations using implicit schemes,* American Institute of Aeronautics and Astronautics, AIAA-85-0293, 1985.

*[305] A. JAMESON, W. SCHMIDT AND E. TURKEL, *Numerical solutions of the Euler equations by finite volume methods using Runge–Kutta time-stepping schemes,* American Institute of Aeronautics and Astronautics, AIAA-81-1259, 1981.

J. W. JEROME, cf. [44].

D. C. JESPERSEN, cf. [354].

[306] D. C. JESPERSEN, *Multilevel techniques for nonelliptic problems,* in Multigrid Methods, H. Lomax, ed., NASA Conference Publication 2202, Ames Research Center, Moffett Field, CA, 1981, pp. 1–22.

[307] ———, *Multigrid methods for the Euler equations: the relaxation technique,* Report MS 202a-1, NASA Ames Research Center, Moffett Field, CA, 1982.

[308] ———, *A multigrid method for the Euler equations,* AIAA-83-0124, Proc. American Institute of Aeronautics and Astronautics 21st Aerospace Science Meeting, January 10–13, 1983, Reno, NV, 1983.

[309] ———, *Design and implementation of a multigrid code for the Euler equations,* in Appl. Math. Comp., 13, Proc. Internat. Multigrid Conference, April 6–8, 1983, Copper Mountain, CO, S. F. McCormick and U. Trottenberg, eds., North-Holland, Amsterdam, pp. 357–374.

*[310] ———, *Recent developments in multigrid methods for the steady Euler equations,* Lecture Notes for Lecture Series on Computational Fluid Dynamics, March 12–16, von Karman Institute for Fluid Dynamics, Rhode-St.-Genese, Belgium, 1984.

*[311] ———, *A time-accurate multiple-grid algorithm,* American Institute of Aeronautics and Astronautics, AIAA 85-1493-CP, 1985.

[312] ———, *Multigrid methods for partial differential equations,* in Studies in Numerical Analysis, G. Golub, ed., to appear.

G. M. JOHNSON, cf. [134], [272], [541], [542].

[313] G. M. JOHNSON, *Convergence acceleration of viscous flow computations,* National Aeronautics and Space Administration, NASA-TM-83039, 1982.

[314] ———, *Multiple-grid acceleration of Lax–Wendroff algorithms,* National Aeronautics and Space Administration, NASA-TM-82843, 1982.

*[315] ———, *Flux based acceleration of the Euler equations,* National Aeronautics and Space Administration, Report NASA TM 83453, 1983.

[316] ———, *Multiple-grid convergence acceleration of viscous and inviscid flow computation,* in Appl. Math. Comp., 13, Proc. Internat. Multigrid Conference, April 6–8, 1983, Copper Mountain, CO, S. F. McCormick and U. Trottenberg, eds., North-Holland, Amsterdam, 1983, pp. 375–398.

*[317] ———, *Multiple grid acceleration of Lax–Wendroff algorithm,* National Aeronautics and Space Administration, NASA-TM-82843, 1984.

*[318] ———, *Accelerated solution of the steady Euler equations,* in Recent Advances in Numerical Methods in Fluids, Vol. 4, W. G. Habashi, ed., Pineridge Press, to appear.

*[319] G. M. JOHNSON AND J. M. SWISSHELM, *Multiple-grid solution of the three dimension Euler and Navier–Stokes equations,* Proc. Ninth International Conference on Numerical Methods in Fluid Dynamics, Soubbaramayer and J. P. Boujot, eds., Lecture Notes in Physics 218, Springer-Verlag, Berlin, 1985.

*[320] G. M. JOHNSON, J. M. SWISSHELM AND S. P. KUMAR, *Concurrent processing adaptation of a multiple-grid algorithm,* AIAA J., 1508-CP (1985).

*[321] W. JOPPICH, *Ein Mehrgitteransatz zur Lösung der Diffusionsgleichung auf einem zeitabhängigen Gebiet: Diffusion unter einer oxidierenden Oberfläche,* Arbeitspapiere der GMD, Nr. 178, Gesellschaft für Mathematik und Datenverarbeitung, St. Augustin, 1985.

*[322] K. E. JORDAN, G. C. PAPANICOLAOU AND R. SPIGLER, *On the numerical solution of a nonlinear stochastic Helmholtz equation with a multigrid preconditioner,* in Appl. Math. Comp., Proc. 2nd Internat. Multigrid Conference, April 1985, Copper Mountain, CO, S. F. McCormick, ed., North-Holland, Amsterdam, to appear.

C. J. JUDD, cf. [49].

*[323] D. KAMOWITZ AND S. V. PARTER, *A study of some multigrid ideas,* Appl. Math. Comp., 17 (1985), pp. 153–184.

*[324] ———, *MGR (Upsilon) multigrid methods,* AFOSR-TR-85-0351. Air Force Office of Scientific Research, Bolling AFB, Washington, D.C., 1985.

*[325] A. KARLSSON AND L. FUCHS, *Fast and accurate solution of time-dependent incompressible flow,* in Numerical Methods in Laminar and Turbulent Flow, Proc. Third International Conference held in Seattle, August 8–11, 1983, C. Taylor, J. A. Johnson and W. R. Smith, eds., Pineridge Press, Swansea, 1983, pp. 606–616.

*[326] ———, *Multigrid solution of time-dependent incompressible flow,* Proc. Fifth GAMM-Conference on Numerical Methods in Fluid Mechanics, Rome, October 5–7, 1983, Notes on Numerical Fluid Mechanics, Volume 7, Vieweg, Braunschweig, 1983.

A. KASSIES, cf. [72].

H. B. KELLER, cf. [75], [76], [126], [127].

*[327] H. B. KELLER AND H. O. KREISS, *Mathematical software for hyperbolic equations and two point boundary value problems,* AFOSR-TR-85-0272. Air Force Office of Scientific Research, Bolling AFB, Washington, D.C., 1985.

R. KETTLER, cf. [279].

[328] R. KETTLER, *Analysis and comparison of relaxation schemes in robust multigrid and preconditioned conjugate gradient methods,* in Multigrid Methods, Proc. conference held at Köln-Porz, November 23–27, 1981, W. Hackbusch and U. Trottenberg,

eds., Lecture Notes in Mathematics 960, Springer-Verlag, Berlin, 1982, pp. 502–534.

[329] R. KETTLER AND J. A. MEIJERINK, *A multigrid method and a combined multigrid-conjugate gradient method for elliptic problems with strongly discontinuous coefficients in general domains,* Shell Publication 604, KSEPL, Rijswijk, 1981.

*[330] R. KETTLER AND P. WESSELING, *Aspects of multigrid methods for problems in three dimensions,* in Appl. Math. Comp., Proc. 2nd Internat. Multigrid Conference, April 1985, Copper Mountain, CO, S. F. McCormick, ed., North-Holland, Amsterdam, to appear.

[331] K. KNEILE, *Accelerated convergence of structured banded systems using constrained corrections,* in Multigrid Methods, H. Lomax, ed., NASA Conference Publication 2202, Ames Research Center, Moffett Field, CA, 1981, pp. 285–303.

*[332] R. KNURA, *Lösung von Navier–Stokes–Gleichungen in L-förmigen Gebieten mit Mehrgittermethoden,* Diplomarbeit, Institut für Angewandte Mathematik, Universität Düsseldorf, 1985.

*[333] C. KOECK, *Calcul de l'ecoulement dans une entree d'air par resolution numerique des equations d'Euler,* AAAF-NT-84-24. Association Aeronautique et Astronomique de France, Paris, 1984.

*[334] ———, *Computation of three-dimensional flow using the Euler equations and a multiple-grid scheme,* Internat. J. Numer. Meth. Fluids, 5 (1985), pp. 483–500.

*[335] C. KOECK AND J. J. CHATTOT, *Computation of three dimensional vortex flows past wings using the Euler equations and a multiple-grid scheme,* Proc. Ninth International Conference on Numerical Methods in Fluid Dynamics, Soubbaramayer and J. P. Boujot, eds., Lecture Notes in Physics, 218, Springer-Verlag, Berlin, 1985.

O. KOLP, cf. [199].

*[336] O. KOLP, *Parallelisierung eines Mehrgitterverfahrens für einen Baumrechner,* Arbeitspapiere der GMD, Nr. 82, Gesellschaft für Mathematik und Datenverarbeitung, St. Augustin, 1984.

*[337] O. KOLP AND H. MIERENDORFF, *Systemunabhängige Organisation von Mehrgitterverfahren auf Parallelrechnern,* GI-14, Jahrestagung, Okt. 1984, Informatik Fachberichte, Nr. 88, H.-D. Ehrich, ed., Springer-Verlag, Berlin, 1985.

*[338] ———, *Efficient multigrid algorithms for locally constrained parallel systems,* in Appl. Math. Comp., Proc. 2nd Internat. Multigrid Conference, April 1985, Copper Mountain, CO, S. F. McCormick, ed., North-Holland, Amsterdam, to appear.

*[339] ———, *Bus coupled systems for multigrid algorithms,* Proc. 2nd European Conference on Multigrid Methods, Cologne, October 1–4, 1985, W. Hackbusch and U. Trottenberg, eds., Lecture Notes in Mathematics, Springer-Verlag, Berlin, to appear.

H. O. KREISS, cf. [327].

[340] N. KROLL, *Direkte Anwendungen von Mehrgittertechniken auf parabolische Anfangsrandwertaufgaben,* Diplomarbeit, Institut für Angewandte Mathematik, Universität Bonn, 1981.

*[341] ———, *Anforderungen an Rechnerleistungen aus der Sicht der numerischen Aerodynamik,* Rechnerarchitekturen für die numerische Simulation auf der Basis superschneller Lösungsverfahren II, U. Trottenberg and P. Wypior, eds., GMD-Studien Nr. 102, Gesellschaft für Mathematik und Datenverarbeitung, St. Augustin, 1985, pp. 241–252.

S. P. KUMAR, cf. [320], [542].

*[342] M. LACROIX, R. CAMARERO AND A. TAPUCU, *A multigrid scheme for the thermal-hydraulics of a blocked channel,* Numer. Heat Transfer, 7, pp. 375–393.

*[343] M. LACROIX, R. CAMARERO AND A. TAPUCU, *Multigrid scheme for thermohydraulic flow,* Numer. Heat Transfer, 7 (1984), pp. 375–393.

*[344] U. LANGER, *On the choice of iterative parameters in the relaxation method on a sequence of meshes,* U.S.S.R. Comput. Math. and Math. Phys., 22, no. 5 (1982), pp. 98–114.

G. K. LEAF, cf. [566].

R. LEBEN, cf. [207].

[345] H. N. LEE, *Multi-grid and step technique for atmospheric chemical pollutant transport and diffusion,* Preliminary Proc. for Internat. Multigrid Conference, April 6–8, 1983, Copper Mountain, CO, Institute for Computational Studies at Colorado State University, Ft. Collins, CO, 1983.

[346] H. N. LEE AND R. E. MEYERS, *On time dependent multi-grid numerical techniques,* Comput. Math. Appl., 6 (1980), pp. 61–65.

H. LEHMANN, cf. [405].

S. LEICHER, cf. [580].

H.-O. LEILICH, cf. [171].

*[347] M. LEMKE, *Experiments with a vectorized multigrid Poisson solver on the CDC Cyber 205, Cray* X-MP *and Fujitsu* VP 200, Arbeitspapiere der GMD, Nr. 179, Gesellschaft für Mathematik und Datenverarbeitung, St. Augustin, 1985.

H. LEVKOWITZ, cf. [280].

E. LEWIS, cf. [450].

*[348] C. P. LI, *A multigrid factorization technique for the flux-split Euler equations,* Proc. Ninth International Conference on Numerical Methods in Fluid Dynamics, Soubbaramayer and J. P. Boujot, eds., Lecture Notes in Physics 218, Springer-Verlag, Berlin, 1985.

Q. LIN, cf. [508].

J. LINDEN, cf. [537].

[349] J. LINDEN, *Mehrgitterverfahren für die Poisson-Gleichung in Kreis und Ringgebiet unter Verwendung lokaler Koordinaten,* Diplomarbeit, Institut für Angewandte Mathematik, Universität Bonn, 1981.

*[350] ———, *A multigrid method for solving the biharmonic equation on rectangular domains,* in Advances in Multi-Grid Methods, Proc. conference held in Oberwolfach, December 8–13, 1984, D. Braess, W. Hackbusch and U. Trottenberg, eds., Notes on Numerical Fluid Mechanics, Volume 11, Vieweg, Braunschweig, 1985, pp. 64–76.

*[351] ———, *Mehrgitterverfahren für das erste Randwertproblem der biharmonischen Gleichung und Anwendung auf ein inkompressibles Strömungsproblem,* Dissertation, Institut für Angewandte Mathematik, Universität Bonn, 1985.

[352] J. LINDEN, U. TROTTENBERG AND K. WITSCH, *Multigrid computation of the pressure of an incompressible fluid in a rotating spherical gap,* Proc. Fourth GAMM-Conference on Numerical Methods in Fluid Mechanics, H. Viviand, ed., Vieweg, Braunschweig, 1982, pp. 183–193.

*[353] R. LÖHNER AND K. MORGAN, *Unstructured multigrid methods,* Proc. 2nd European Conference on Multigrid Methods, Cologne, October 1–4, 1985, W. Hackbusch and U. Trottenberg, eds., GMD-Studien Nr. 110, Gesellschaft für Mathematik und Datenverarbeitung, St. Augustin, to appear.

[354] H. LOMAX, J. PULLIAM AND D. C. JESPERSEN, *Eigensystem analysis techniques for finite-difference equations.* I. *Multigrid techniques,* AIAA-81-1027, American Institute of Aeronautics and Astronautics Fifth Computational Fluid Dynamics Conference, New York, NY, 1981, pp. 55–80.

*[355] J. M. LONGO, W. SCHMIDT AND A. JAMESON, *Viscous transonic airfoil flow simulation,* Z. Flugw, Weltraumforsch., 7 (1983), pp. 47–56.

*[356] G. LONSDALE, *Multigrid methods for the solution of the Navier–Stokes equations,* Ph.D. Thesis, University of Manchester, 1985.

*[357] ———, *Solution of a rotating Navier–Stokes problem by a nonlinear multigrid algorithm,* Proc. 2nd European Conference on Multigrid Methods, Cologne, October 1–4, 1985, W. Hackbusch and U. Trottenberg, eds., GMD-Studien Nr. 110, Gesellschaft für Mathematik und Datenverarbeitung, St. Augustin, to appear.

*[358] G. LONSDALE AND J. E. WASH, *The pressure correction method, and the use of a multigrid technique, for laminar source-sink flow between corotating discs,* Technical Report, Mathematics Department, University of Manchester, 1984.

*[359] A. L. LUNTZ AND B. EPSTEIN, *A multigrid full potential transonic code for arbitrary configuration,* Proc. 2nd European Conference on Multigrid Methods, Cologne, October 1–4, 1985, W. Hackbusch and U. Trottenberg, eds., GMD-Studien Nr. 110, Gesellschaft für Mathematik und Datenverarbeitung, St. Augustin, to appear.

L. R. LUSTMAN, cf. [293].

[360] L. R. LUSTMAN, Q. HUYNH AND M. Y. HUSSAINI, *Multigrid method with weighted mean scheme,* in Multigrid Methods, H. Lomax, ed., NASA Conference Publication 2202, Ames Research Center, Moffett Field, CA, 1981, pp. 47–59.

J. F. MAITRE, cf. [1].

[361] J. F. MAITRE AND F. MUSY, *The contraction number of a class of two-level methods; an exact evaluation for some finite element subspaces and model problems,* in Multigrid Methods, Proc. conference held at Köln-Porz, November 23–27, 1981, W. Hackbusch and U. Trottenberg, eds., Lecture Notes in Mathematics 960, Springer-Verlag, Berlin, 1982, pp. 535–544.

[362] J. F. MAITRE AND F. MUSY, *Analyse numerique, Methodes multigrilles: operateur associe et estimations du facteur de convergence; le cas du V-cycle,* C.R. Acad. Sc. Paris, Ser. I, 296 (1983), pp. 521–524.

*[363] ———, *Multigrid methods: convergence theorie in a variational framework,* SIAM J. Numer. Anal., 21 (1984), pp. 657–671.

*[364] ———, *Algebraic formalisation of the multigrid method in the symmetric and positive definite case—a convergence estimation for the V-cycle,* in Multigrid Methods for Integral and Differential Equations, D. J. Paddon and H. Holstein, eds., The Institute of Mathematics and its Applications Conference Series 3, Clarendon Press, Oxford, 1985, pp. 213–224.

*[365] ———, *Multigrid methods for symmetric variational problems: A general theory and convergence estimates for usual smoothers,* Appl. Math. Comp., to appear.

*[366] J. F. MAITRE, F. MUSY AND P. NIGON, *A fast solver for the Stokes equations using multigrid with a UZAWA smoother,* in Advances in Multi-Grid Methods, Proc. conference held in Oberwolfach, December 8–13, 1984, D. Braess, W. Hackbusch and U. Trottenberg, eds., Notes on Numerical Fluid Mechanics, Volume 11, Vieweg, Braunschweig, 1985, pp. 77–83.

*[367] J. MANDEL, *Convergence of an iterative method for the system $Ax + y = x$ using aggregation,* Ekonom.-Mat. Obzor., 17 (1981), pp. 287–291.

*[368] ———, *A convergence analysis of the iterative aggregation method with one parameter,* Linear Algebra Appl., 59 (1984), pp. 159–169.

[369] ———, *A multi-level iterative method for symmetric, positive definite linear complementarity problems,* Appl. Math. Optim., 11 (1984), pp. 77–95.

*[370] ———, *Algebraic study of a multigrid method for some free boundary problem,* C.R. Acad. Sc. Paris, Ser. I, 298 (1984), pp. 469–472.

*[371] ———, *On some two-level iterative methods,* in Computing Suppl. 5, Defect Correction

Methods, K. Böhmer and H. J. Stetter, eds., Springer-Verlag, Wien, 1984, pp. 75–88.

*[372] ———, *On multilevel iterative methods for integral equations of the second kind and related problems,* Numer. Math., 46 (1985), pp. 147–157.

*[373] ———, *Multigrid convergence for nonsymmetric, indefinite variational problems and one smoothing step,* in Appl. Math. Comp., Proc. 2nd Internat. Multigrid Conference, April 1985, Copper Mountain, CO, S. F. McCormick, ed., North-Holland, Amsterdam, 19 (1986), pp. 201–206.

*[374] ———, *On multigrid and iterative aggregation methods for nonsymmetric problems,* Proc. 2nd European Conference on Multigrid Methods, Cologne, October 1–4, 1985, W. Hackbusch and U. Trottenberg, eds, Lecture Notes in Mathematics, Springer-Verlag, Berlin, 1987, pp. 219–231.

*[375] ———, *Algebraic study of multigrid methods for symmetric, definite problems,* Appl. Math. Comp., to appear.

*[376] J. Mandel and B. Sekerka, *A local convergence proof for the iterative aggregation method,* Linear Algebra Appl., 51 (1983), 163–172.

*[377] J. Mandel, S. F. McCormick and J. Ruge, *An algebraic theory for multigrid methods for variational problems,* SIAM J. Numer. Anal., to appear.

[378] L. Mansfield, *On the solution of nonlinear finite element systems,* SIAM J. Numer. Anal., 17 (1980), pp. 752–765.

[379] ———, *On the multi-grid solution of finite element equations with isoparametric elements,* Numer. Math., 37 (1981), pp. 423–432.

M. C. Marche, cf. [132].

D. R. Marr, cf. [4].

B. Martinet, cf. [77].

*[380] O. A. McBryan and E. F. van de Velde, *The multigrid method on parallel processors,* Proc. 2nd Conference on Multigrid Methods, Cologne, October 1–4, 1985, W. Hackbusch and U. Trottenberg, eds., Lecture Notes in Mathematics, Springer-Verlag, Berlin, to appear.

[381] D. R. McCarthy, *Embedded mesh multigrid treatment of two-dimensional transonic flows,* in Appl. Math. Comp., 13, Proc. Internat. Multigrid Conference, April 6–8, 1983, Copper Mountain, CO, S. F. McCormick and U. Trottenberg, eds., North-Holland, Amsterdam, 1983, pp. 399–418.

[382] D. R. McCarthy and T. A. Reyhner, *A multigrid code for the three-dimensional transonic potential flow about inlets,* AIAA J., 20 (1982), pp. 45–50.

S. F. McCormick, cf. [113], [114], [115], [159], [189], [253], [280], [284], [377].

[383] S. F. McCormick, *Multigrid methods: An alternate viewpoint,* Report UCID-18487, Lawrence Livermore Laboratory, Livermore, CA, 1979.

[384] ———, *Mesh refinement methods for integral equations,* in Numerical Treatment of Integral Equations, J. Albrecht and L. Collatz, eds., Birkhäuser-Verlag, Basel, 1980, pp. 183–190.

[385] ———, *A mesh refinement method for* $A*x = Lambda*B*x$, Math. Comp., 36 (1981), pp. 485–498.

[386] ———, *Multigrid short course notes,* Report, Lockheed Palo Alto Research Center, 1981.

[387] ———, *Numerical software for fixed point microprocessor applications and for fast implementation of multigrid techniques,* ARO Report 81-3, Proc. Army Numerical Analysis and Computers Conference, 1981.

[388] ———, *An algebraic interpretation of multigrid methods,* SIAM J. Numer. Anal., 19 (1982), pp. 548–560.

[389] S. F. McCormick, *Multigrid methods for variation problems: The V-cycle,* in Math. Comput., Simulation XXV, North-Holland, Amsterdam, 1983, pp. 63–65.

*[390] ———, *Fast adaptive composite grid (FAC) methods: Theory for the variational case,* in Computing, Suppl. 5, Defect Correction Methods, K. Böhmer and H. J. Stetter, eds., Springer-Verlag, Vienna, 1984.

[391] ———, *Multigrid methods for variational problems: further results,* SIAM J. Numer. Anal., 21 (1984), pp. 255–263.

*[392] ———, *A variational theory for multi-level adaptive techniques (MLAT),* in Multigrid Methods for Integral and Differential Equations, D. J. Paddon and H. Holstein, eds., The Institute of Mathematics and its Applications Conference Series 3, Clarendon Press, Oxford, 1985, pp. 225–230.

*[393] ———, ed., *Multigrid Methods,* Frontiers in Applied Mathematics, Vol. 3, Society for Industrial and Applied Mathematics, Philadelphia, PA, this Volume 1987.

*[394] ———, *A variational method for $Au = LBu$ on composite grids,* submitted.

*[395] ———, *Multigrid methods for variational problems: general theory for the V-cycle,* SIAM J. Numer. Anal., 22 (1985), pp. 634–643.

[396] S. F. McCormick and K. Brand, eds., *Multigrid Newsletter,* Dept. of Mathematics, University of Colorado, Denver, CO; Institut für methodische Grundlagen (F1/T), Gesellschaft für Mathematik und Datenverarbeitung, 1986.

[397] S. F. McCormick and G. H. Rodrigue, *Multigrid methods for multiprocessor computers,* Technical Report, Lawrence Livermore Laboratory, Livermore, CA, 1979.

[398] S. F. McCormick and J. Ruge, *Multigrid methods for variational problems,* SIAM J. Numer. Anal., 19 (1982), pp. 924–929.

*[399] ———, *Unigrid for multigrid simulation,* Math. Comp., 41 (1983), pp. 43–62.

[400] S. F. McCormick and J. Thomas, *Multigrid methods applied to water wave problems,* Proc. Third Internat. Conference Ship Hydrodynamics, Paris, June 1981.

*[401] S. F. McCormick and J. Thomas, *The fast adaptive composite grid method (FAC) for elliptic boundary value problems,* Math. Comp., to appear.

*[402] S. F. McCormick, J. Ruge, S. Schaffer and J. Thomas, *Multigrid methods and adaptive techniques for oil reservoir simulation,* Report, Inst. Comp. Studies, Colorado State University, Ft. Collins, CO, 1983.

J. H. Meelker, cf. [587].

J. A. Meijerink, cf. [329].

*[403] J. A. Meijerink, *Incomplete block LU-factorisation,* IEEE Colloquium on Numerical Solution Techniques for Large Sparse Systems of Equations, London, November, 1983.

[404] T. Meis and H. W. Branca, *Schnelle Lösung von Randwertaufgaben,* Z. Angew. Math. Mech., 62 (1982), pp. T263–T270.

[405] T. Meis, H. Lehmann and H. Michael, *Application of the multigrid method to a nonlinear indefinite problem,* in Multigrid Methods, Proc. conference held at Köln-Porz, November 23–27, 1981, W. Hackbusch and U. Trottenberg, eds., Lecture Notes in Mathematics 960, Springer-Verlag, Berlin, 1982, pp. 545–557.

*[406] N. D. Melson, *Vectorizable multigrid algorithms for transonic flow calculations,* in Appl. Math. Comp., Proc. 2nd Internat. Multigrid Conference, April 1985, Copper Mountain, CO, S. F. McCormick, ed., North-Holland, Amsterdam, to appear.

[407] M. L. Merriam, *Formal analysis of multigrid techniques applied to Poisson's equation in three dimensions,* AIAA-81-1028, American Institute of Aeronautics and Astronautics Fifth Computational Fluid Dynamics Conference. New York. 1981.

*[408] ———, *Application of data flow concepts of a multigrid solver for the Euler equations,* in Appl. Math. Comp., Proc. 2nd Internat. Multigrid Conference, April 1985,

Copper Mountain, CO, S. F. McCormick, ed., North-Holland, Amsterdam, to appear.

*[409] G. H. MEYER, *Hele–Shaw flow with a cusping free boundary,* J. Comput. Phys., 44 (1981), pp. 262–276.

R. E. MEYERS, cf. [346].

H. MICHAEL, cf. [405].

H. MIERENDORFF, cf. [337], [338], [339].

*[410] H. MIERENDORFF, *Ein Konzept zur Nutzung eng gekoppelter Mehrrechnersysteme,* Mitteilungen—Gesellschaft für Informatik e.V., Parallel-Algorithmen und Rechnerstrukturen Heft 2, 1984, pp. 116–123.

*[411] ———, *Transportleistung und Größe paralleler Systeme bei speziellen Mehrgitteralgorithmen,* Rechnerarchitekturen für die numerische Simulation auf der Basis superschneller Lösungsverfahren I, Workshop 'Rechnerarchitektur,' Erlangen, June 14–15, 1984, U. Trottenberg and P. Wypior, eds., GMD-Studien Nr. 88, Gesellschaft für Mathematik und Datenverarbeitung, St. Augustin, 1984, pp. 41–54.

N. K. MITRA, cf. [118].

H. D. MITTELMANN, cf. [35], [247].

[412] H. D. MITTELMANN, *A fast solver for nonlinear eigenvalue problems,* in Iterative Solution of Nonlinear Systems of Equations, Proceedings, Oberwolfach 1982, R. Ansorge, T. Meis and W. Törnig, eds., Lecture Notes in Mathematics 953, Springer-Verlag, Berlin, 1982, pp. 46–67.

[413] ———, *Multi-grid methods for simple bifurcation problems,* in Multigrid Methods, Proc. conference held at Köln-Porz, November 23–27, 1981, W. Hackbusch and U. Trottenberg, eds., Lecture Notes in Mathematics 960, Springer-Verlag, Berlin, 1982, pp. 558–575.

*[414] ———, *Multilevel continuation techniques for nonlinear boundary value problems with parameter dependence,* in Appl. Math. Comp., Proc. 2nd Internat. Multigrid Conference, April 1985, Copper Mountain, CO, S. F. McCormick, ed., North-Holland, Amsterdam, to appear.

[415] H. D. MITTELMANN AND H. WEBER, *Multi-grid solution of bifurcation problems,* in Preliminary Proc. Internat. Multigrid Conference, April 6–8, 1983, Copper Mountain, CO; Institute for Computational Studies at Colorado State University, Ft. Collins, CO, 1983.

[416] W. J. A. MOL, *A multigrid method applied to some simple problems,* Memorandum no. 287, Dept. of Numerical Mathematics, Twente University of Technology, Twente, 1979.

[417] ———, *Computation of flows around a Karman–Trefftz profile,* Preprint NW 114/81, Dept. of Numerical Mathematics, Mathematical Centre, Amsterdam, 1981.

[418] ———, *Numerical solution of the Navier–Stokes equations by means of a multigrid method and Newton-iteration,* Proc. Seventh Internat. Conference on Numerical Methods in Fluid Dynamics, Stanford 1980, W. C. Reynolds and R. W. MacCormack, eds., Lecture Notes in Physics 141, Springer-Verlag, Berlin, 1981, pp. 285–291.

[419] ———, *On the choice of suitable operators and parameters in multigrid methods,* Report NW 107/81, Dept. of Numerical Mathematics, Mathematical Centre, Amsterdam, 1981.

[420] ———, *Smoothing and coarse grid approximation properties of multigrid methods,* Report NW 110/81, Dept. of Numerical Mathematics, Mathematical Centre, Amsterdam, 1981.

K. MORGAN, cf. [353].

*[421] H. MÜHLENBEIN, *Modellierung von parallelen Mehrgitteralgorithmen und Rechnerstrukturen,* Rechnerarchitekturen für die numerische Simulation auf der Basis superschneller Lösungsverfahren I, Workshop 'Rechnerarchitektur,' Erlangen, June 14–15, 1984, U. Trottenberg and P. Wypior, eds., GMD-Studien Nr. 88, Gesellschaft für Mathematik und Datenverarbeitung, St. Augustin, 1984, pp. 55–64.

*[422] H. MÜHLENBEIN AND S. WARHAUT, *Concurrent multigrid methods in an object oriented environment—A case study,* Rechnerarchitekturen für die numerische Simulation auf der Basis superschneller Lösungsverfahren II, U. Trottenberg and P. Wypior, eds., GMD-Studien Nr. 102, Gesellschaft für Mathematik und Datenverarbeitung, St. Augustin, 1985, pp. 151–156.

*[423] W. A. MULDER, *Multigrid relaxation for the Euler equations,* J. Comput. Phys., 60 (1985), pp. 235–252.

F. MUSY, cf. [1], [361], [362], [363], [364], [365], [366].

[424] F. MUSY, *Analyse Numerique—Sur les methodes multigrilles: formalisation algebraique et demonstration de convergence,* C.R. Acad. Sc. Paris, Ser. I, 295 (1982), pp. 471–474.

*[425] ———, *Methodes multigrilles: demonstration de convergence incluant le V-cycle et le W-cycle; applications au lissage de Gauss–Seidel,* C.R. Acad. Sc. Paris, Ser. I, 298 (1984), pp. 369–372.

*[426] ———, *Etude d'une classe de methodes multigrilles pour les problemes variationel: theorie generale et estimations sur taux de convergence,* Ph.D. Thesis, Universite Claude Bernard Lyon, 1985.

*[427] M. NAPOLITANO, *Incremental multigrid strategy for the fluid dynamics equations,* Research Report no. 357, Computer Science Dept., Yale University, New Haven, CT, 1985.

[428] R. H. NI, *A multiple grid scheme for solving Euler equations,* AIAA J., 20 (1982), pp. 1565–1571.

[429] R. A. NICOLAIDES, *On multiple grid and related techniques for solving discrete elliptic systems,* J. Comput. Phys., 19 (1975), pp. 418–431.

[430] ———, *On the 1**2 convergence of an algorithm for solving finite element equations,* Math. Comp., 31 (1977), pp. 892–906.

[431] ———, *On multi-grid convergence in the indefinite case,* Math. Comp., 32 (1978), pp. 1082–1086.

[432] ———, *On the observed rate of convergence of an iterative method applied to a model elliptic difference equation,* Math. Comp., 32 (1978), pp. 127–133.

[433] ———, *On finite element multigrid algorithms and their use,* in The Mathematics of Finite Elements and Applications III, MAFELAP 1978, J. R. Whiteman, ed., Academic Press, London, 1979, pp. 459–466.

[434] ———, *On some theoretical and practical aspects of multigrid methods,* Math. Comp., 33 (1979), pp. 933–952.

*[435] A. NIESTEGGE, *Untersuchungen von Mehrgitterverfahren für die Stokes–Gleichungen,* Diplomarbeit, Institut für Angewandte Mathematik, Universität Düsseldorf, 1985.

P. NIGON, cf. [366].

Z. P. NOWAK, cf. [248].

[436] Z. P. NOWAK, *Use of the multigrid method for Laplacian problem in three dimensions,* in Multigrid Methods, Proc. conference held at Köln-Porz, November 23–27, 1981, W. Hackbusch and U. Trottenberg, eds., Lecture Notes in Mathematics 960, Springer-Verlag, Berlin, 1982, pp. 576–598.

*[437] ———, *A quasi-Newton multigrid method for determining the transonic lifting flows*

around airfoils, Report 84-11, Delft University of Technology, Department of Mathematics and Informatics, Delft, 1984.

*[438] ———, *Calculations of transonic flows around single and multi-element airfoils on a smaller computer,* in Advances in Multi-Grid Methods, Proc. conference held in Oberwolfach, December 8–13, 1984, D. Braess, W. Hackbusch and U. Trottenberg, eds., Notes on Numerical Fluid Mechanics, Volume 11, Vieweg, Braunschweig, 1985, pp. 84–101.

*[439] Z. P. NOWAK AND P. WESSELING, *Multigrid acceleration of an iterative method with applications to transonic potential flow,* in Computing Methods in Applied Sciences and Engineering VI: Proc. Sixth Internat. Symposium, Versailles, December 1983, R. Glowinski and J. L. Lions, eds., North-Holland, Amsterdam, 1984, pp. 199–217.

A. O'GALLAGER, cf. [253].

*[440] K.-D. OERTEL, *Praktische und theoretische Aspekte der ILU-Glättung bei Mehrgitterverfahren,* Diplomarbeit, Institut für Angewandte Mathematik, Universität Bonn, to appear.

*[441] S. OHRING, *Application of the multigrid method to Poisson's equation in boundary-fitted coordinates,* J. Comput. Phys., 50 (1983), pp. 307–315.

D. OPHIR, cf. [106], [107].

[442] D. OPHIR, *Language for processes of numerical solutions to differential equations,* Ph.D. Thesis, Dept. of Applied Mathematics, Weizmann Institute of Science, Rehovot, Israel, 1978.

[443] B. OSKAM AND J. M. J. FRAY, *General relaxation schemes in multigrid algorithms for higher order singularity methods,* J. Comput. Phys., 48 (1982), pp. 423–440.

*[444] I. V. OTROSHCENKO AND R. P. FEDORENKO, *A relaxation method for solving a biharmonic difference equation,* U.S.S.R. Computational Math. and Math. Phys., 23 (1983), pp. 57–63.

*[445] D. J. PADDON AND H. HOLSTEIN, eds., *Multigrid methods for integral and differential equations,* in The Institute of Mathematics and its Applications Conference Series 3, Clarendon Press, Oxford, 1985.

J. W. PAINTER, cf. [5].

*[446] J. W. PAINTER, *Grid coupling methods for the multigrid solution of the neutron diffusion equation,* Trans. Amer. Nucl. Soc., 33 (1979), pp. 345–346.

[447] ———, *Multigrid experience with the neutron diffusion equation,* Report LA-UR 79-1634, Los Alamos Scientific Laboratory, Los Alamos, NM, 1979.

*[448] A. PAPAMANOLIS, *Multigrid methods in fluid dynamics,* Master's Thesis, Dept. of Computer Science, University of Wales, Aberystwyth, 1984.

G. PAPAMANOLIS, cf. [285].

G. C. PAPANICOLAOU, cf. [322].

S. V. PARTER, cf. [323], [324].

*[449] S. V. PARTER, *A note on convergence of the multigrid V-cycle,* Appl. Math. Comp., 17 (1985), pp. 137–152.

*[450] V. PAU AND E. LEWIS, *Application of the multigrid technique to the pressure-correction equation for the SIMPLE algorithm,* Proc. 2nd European Conference on Multigrid Methods, Cologne, October, 1–4, 1985, W. Hackbusch and U. Trottenberg, eds., GMD-Studien Nr. 110, Gesellschaft für Mathematik und Datenverarbeitung, St. Augustin, to appear.

P. PEISKER, cf. [84].

*[451] P. PEISKER, *A multilevel algorithm for the biharmonic problem,* Numer. Math., 46 (1985), pp. 623–634.

*[452] R. B. PELZ AND A. JAMESON, *Transonic flow calculations using triangular finite elements,* AIAA J., 23 (1985), pp. 569–576.

*[453] R. E. PHILLIPS AND F. W. SCHMIDT, *Multigrid techniques for the numerical solution of the diffusion equation,* Numer. Heat Transfer, 7 (1984), pp. 251–268.

*[454] ———, *Multigrid techniques for the solution of the passive scalar advection-diffusion equation,* Numer. Heat Transfer, 8 (1985), pp. 25–43.

*[455] T. N. PHILLIPS, *Numerical solution of a coupled pair of elliptic equations from solid state electronics,* J. Comput. Phys., 53 (1984), pp. 472–483.

*[456] T. N. PHILLIPS, T. A. ZANG AND M. Y. HUSSAINI, *Preconditioners for the spectral multigrid methods,* NASA-CR-172202, ICASE, NASA Langley Research Center, 1983.

*[457] ———, *Spectral multigrid methods for Dirichlet problems,* in Multigrid Methods for Integral and Differential Equations, D. J. Paddon and H. Holstein, eds., The Institute of Mathematics and its Applications Conference Series 3, Clarendon Press, Oxford, 1985, pp. 231–252.

*[458] J. PITKÄRANTA AND T. SAARINEN, *A multigrid version of a simple finite element method for the Stokes problem,* Math. Comp., 45 (1985), pp. 1–14.

[459] T. C. POLING, *Numerical experiments with multi-grid methods,* M.A. Thesis, Dept. of Numerical Mathematics, College of William and Mary, Williamsburg, VA, 1978.

U. PROJAHN, cf. [465].

J. PULLIAM, cf. [354].

T. H. PULLIAM, cf. [165].

*[460] R. RABENSTEIN, *A signal processing approach to the numerical solution of parabolic differential equations,* Proc. 2nd European Conference on Multigrid Methods, Cologne, October 1–4, 1985, W. Hackbusch and U. Trottenberg, eds., GMD-Studien Nr. 110, Gesellschaft für Mathematik und Datenverarbeitung, St. Augustin, to appear.

H. RABITZ, cf. [203], [204].

*[461] P. RAJ, *Multi-grid method for transonic wing analysis and design,* J. Aircr., 21 (1984), pp. 143–150.

D. R. REBBY, cf. [478].

D. R. REDDY, cf. [195].

M. REGGIO, cf. [121].

*[462] M. REIMERS, *On local mode analysis in multi-grid methods,* Technical Report No. 83-14, Institute of Applied Mathematics and Statistics, University of British Columbia, Vancouver, Canada, 1983.

T. A. REYHNER, cf. [382].

*[463] T. A. REYHNER, *Three dimensional transonic potential flow about complex 3-dimensional configurations,* NASA-CR-3814, National Aeronautics and Space Administration, Washington, D.C., 1984.

*[464] J. R. RICE AND R. F. BOISVERT, *Solving elliptic problems using ELLPACK,* Springer Series in Comp. Math., 2, Springer-Verlag, Berlin, 1985.

[465] H. RIEGER, U. PROJAHN AND H. BEER, *Fast iterative solution of Poisson equation with Neumann boundary conditions in nonorthogonal curvilinear coordinate systems by a multiple grid method,* Numer. Heat Transfer, 6 (1983), pp. 1–15.

[466] M. RIES, *Lösung elliptischer Randwertaufgaben mit approximativen und iterativen Reduktionsverfahren,* Dissertation, Institut für Angewandte Mathematik, Universität Bonn, 1981.

[467] M. RIES AND U. TROTTENBERG, *MGR—eine blitzschneller elliptischer Löser,* Preprint no. 277, Sonderforschungsbereich 72, Universität Bonn, 1979.

[468] M. RIES, U. TROTTENBERG AND G. WINTER, *A note on MGR methods,* Linear Algebra Appl., 49 (1983), pp. 1–26.

*[469] H. RITZDORF, *Lokal verfeinerte Mehrgitter-Methoden für Gebiete mit einspringenden Ecken,* Diplomarbeit, Institut für Angewandte Mathematik, Universität Bonn, 1984.

*[470] M. C. RIVARA, *Algorithms for refining triangular grids suitable for adaptive and multigrid techniques,* J. Numer. Meth. Engrg., 20 (1984), pp. 745–756.

*[471] ———, *Design and data structure of fully adaptive, multigrid, finite element software,* ACM Trans. Math. Software, 10 (1984), pp. 242–264.

[472] A. RIZZI AND L. E. ERIKSSON, *Unigrid projection method to solve the Euler equations for steady transonic flow,* Eighth Internat. Conference on Numerical Methods in Fluid Dynamics, Proc. Conference, TH Aachen, June 28–July 2, 1982, E. Krause, ed., Lecture Notes in Physics 170, Springer-Verlag, Berlin, 1982, pp. 427–432.

*[473] P. J. ROACHE AND S. STEINBERG, *Application of a single equation MG-FAS solver to elliptic grid generation equations (sub-grid and super-grid coefficient generation),* Appl. Math. Comp., Proc. 2nd Internat. Multigrid Conference, April 1985, Copper Mountain, CO, S. F. McCormick, ed., North-Holland, Amsterdam, to appear.

G. H. RODRIGUE, cf. [397].

*[474] NI RON-HO, *A multiple grid scheme for solving the Euler equations,* AIAA J., 20 (1982), 1565–1571.

D. RON, cf. [116].

D. J. ROSE, cf. [36], [37], [44].

J. R. VAN ROSENDALE, cf. [185], [186].

[475] J. R. VAN ROSENDALE, *Rapid solution of finite element equations on locally refined grids by multi-level methods,* Report UIUCDS-R-80-1021, Dept. of Computer Science, University of Illinois at Urbana-Champaign, Urbana, IL, 1980.

[476] ———, *Algorithms and data structures for adaptive multigrid elliptic solvers,* Appl. Math. Comp., 13, Proc. Internat. Multigrid Conference, April 6–8, 1983, Copper Mountain, CO, S. F. McCormick and U. Trottenberg, eds., North-Holland, Amsterdam, 1983, pp. 453–470.

M. ROSENFELD, cf. [295].

[477] S. G. RUBIN, *Incompressible Navier–Stokes and parabolized Navier–Stokes procedures and computational techniques,* in Computational Fluid Dynamics, Lecture Series 1982-04, von Karman Institute for Fluid Dynamics, Rhode-St.-Genese, Belgium, 1982.

*[478] S. G. RUBIN AND D. R. REBBY, *Global parabolized Navier–Stokes solutions for laminar and turbulent flow,* AFOSR-TR-84-0028, Air Force Office of Scientific Research, Bolling AFB, Washington, D.C., 1983.

*[479] U. RÜDE, *Anwendung der Mehrgittermethode zur Berechnung von digitalen Höhenmodellen in der Photogrammetrie,* Report, Institut für Informatik, Technische Universität München, TUMI8525, 1985.

*[480] ———, *Discretizations for multigrid methods,* Proc. 2nd European Conference on Multigrid Methods, Cologne, October 1–4, 1985, W. Hackbusch and U. Trottenberg, eds., GMD-Studien Nr. 110, Gesellschaft für Mathematik und Datenverarbeitung, St. Augustin, to appear.

*[481] U. RÜDE AND C. ZENGER, *On the treatment of singularities in the multigrid method,* Proc. 2nd European Conference on Multigrid Methods, Cologne, October 1–4, 1985, W. Hackbusch and U. Trottenberg, eds., Lecture Notes in Mathematics, Springer-Verlag, Berlin, to appear.

J. RUGE, cf. [113], [114], [115], [284], [377], [398], [399], [402].

[482] J. RUGE, *Multigrid methods for differential eigenvalue and variational problems and multigrid simulation,* Ph.D. Thesis, Dept. of Mathematics, Colorado State University, Ft. Collins, CO, 1981.

[483] ———, *Algebraic multigrid (AMG) for geodetic survey problems,* in Preliminary Proc. Internat. Multigrid Conference, April 6–8, 1983, Copper Mountain, CO, Institute for Computational Studies at Colorado State University, Ft. Collins, CO, 1983.

*[484] J. RUGE, *Final Report on* AMG02, Report, GMD, Gesellschaft für Mathematik und Datenverarbeitung, St. Augustin, 1985.

*[485] ———, *AMG for problems of elasticity,* in Appl. Math. Comp., Proc. 2nd Internat. Multigrid Conference, April 1985, Copper Mountain, CO, S. F. McCormick, ed., North-Holland, Amsterdam, to appear.

*[486] J. RUGE AND K. STÜBEN, *Efficient solution of finite difference and finite element equations by algebraic multigrid (AMG),* in Multigrid Methods for Integral and Differential Equations, D. J. Paddon and H. Holstein, eds., The Institute of Mathematics and its Applications Conference Series, Clarendon Press, Oxford, 1985, pp. 169–212.

*[487] ———, *Algebraic multigrid (AMG),* Frontiers in Appl. Math., Vol. 3. Society for Industrial and Applied Mathematics, S. F. McCormick, ed., Philadelphia, PA, this Volume 1987.

L. A. RUKHOVETS, cf. [16], [17].

H. M. RUPPEL, cf. [111].

*[488] B. RUTTMANN AND K. SOLCHENBACH, *A multigrid solver for the in-cylinder turbulent flows in engines,* in Efficient Solution of Elliptic Systems: Proc. GAMM-Seminar in Kiel, January 27–29, 1984, W. Hackbusch, ed., Notes on Numerical Fluid Mechanics, Volume 10, Vieweg, Braunschweig, 1984, pp. 87–108.

*[489] ———, *Numerische Simulation der Strömungs- und Verbrennungsvorgänge im Motorzylinder,* Rechnerarchitekturen für die numerische Simulation auf der Basis superschneller Lösungsverfahren II, U. Trottenberg and P. Wypior, eds., GMD-Studien Nr. 102, Gesellschaft für Mathematik und Datenverarbeitung, St. Augustin, 1985, pp. 259–272.

Y. SAAD, cf. [128], [131].

T. SAARINEN, cf. [458].

F. SAIED, cf. [129].

*[490] V. V. SAJDUROV, *On the solution of eigenvalue problems of variational difference equations in a sequence of grids,* in Variational Difference Methods in Mathematical Physics, Proc. May 1983, Moskow, N. S. Bachvalow and J. A. Kuznecov, eds., AN SSSR, Moskow, 1984, pp. 149–160.

M. D. SALAS, cf. [292].

*[491] N. L. SANKAR, *A multigrid strongly implicit procedure for transonic potential flow problems,* AIAA J., 21 (1983).

S. SCHAFFER, cf. [402].

[492] S. SCHAFFER, *High order multi-grid methods to solve the Poisson equation,* in Multigrid Methods, H. Lomax, ed., NASA Conference Publication 2202, Ames Research Center, Moffett Field, CA, 1981, pp. 275–284.

*[493] ———, *Higher order multi-grid methods,* Math. Comp., 43 (1984), pp. 89–115.

*[494] S. SCHAFFER AND F. STENGER, *Multigrid-sinc methods,* in Appl. Math. Comp., Proc. 2nd Internat. Multigrid Conference, April 1985, Copper Mountain, CO, S. F. McCormick, ed., North-Holland, Amsterdam, to appear.

H. SCHIPPERS, cf. [273].

[495] H. SCHIPPERS, *Multigrid techniques for the solution of Fredholm integral equations of the second kind,* MCS 41, Colloquium on the Numerical Treatment of Integral Equations, H. J. J. Te Riele, ed., Dept. of Numerical Mathematics, Mathematical Centre, Amsterdam, 1979, pp. 29–49.

[496] ———, *Multiple grid methods for oscillating disk flow,* in Boundary and Interior Layers—Computational and Asymptotic Methods, J. J. H. Miller, ed., Boole Press, Dublin, 1980, pp. 410–414.

[497] ———, *The automatic solution of Fredholm integral equations of the second kind,*

Report NW 99/80, Dept. of Numerical Mathematics, Mathematical Centre, Amsterdam, 1980.

[498] ———, *Application of multigrid methods for integral equations to two problems from fluid dynamics,* J. Comput. Phys., 48 (1982), pp. 441–461.

[499] ———, *Multigrid methods for boundary integral equations,* Report NLR MP 82059 U, National Aerospace Laboratory NLR, Amsterdam, 1982.

[500] ———, *On the regularity of the principal value of the double layer potential,* J. Engrg. Math., 16 (1982), pp. 59–76.

[501] ———, *Multiple grid methods for equations of the second kind with applications in fluid mechanics,* Report, Contract no. 163, Dept. of Numerical Mathematics, Mathematical Centre, Amsterdam, 1984.

*[502] ———, *Theoretical and practical aspects of multigrid methods in boundary element calculations,* in Progress in Boundary Element Research, Vol. 6, C. A. Brebbia, ed., Springer-Verlag, Berlin, submitted.

F. W. SCHMIDT, cf. [453], [454].

W. SCHMIDT, cf. [305], [355], [580].

[503] W. SCHMIDT AND A. JAMESON, *Applications of multi-grid methods for transonic flow calculations,* in Multigrid Methods, Proc. conference held at Köln-Porz, November 23–27, 1981, W. Hackbusch and U. Trottenberg, eds., Lecture Notes in Mathematics 960, Springer-Verlag, Berlin, 1982, pp. 599–613.

R. SCHREIBER, cf. [130].

*[504] W. SCHRÖDER AND D. HÄNEL, *Multigrid solution of the Navier–Stokes equations for the flow in a rapidly rotating cylinder,* Proc. Ninth Internat. Conference on Numerical Methods in Fluid Dynamics, Soubbaramayer and J. P. Boujot, eds., Lecture Notes in Physics 218, Springer-Verlag, Berlin, 1985.

*[505] ———, *Applications of the multigrid method to the solution of parabolic differential equations,* Notes on Numerical Fluid Dynamics, Vol. 14, Vieweg, Braunschweig, 1986.

*[506] ———, *A comparison of several MG-methods for the solution of the time-dependent Navier–Stokes equations,* Proc. 2nd European Conference on Multigrid Methods, Cologne, October 1–4, 1985, W. Hackbusch and U. Trottenberg, eds., Lecture Notes in Mathematics, Springer-Verlag, Berlin, to appear.

W. H. SCHUBERT, cf. [182].

[507] A. SCHÜLLER, *Anwendung von Mehrgittermethoden auf das Problem einer ebenen, reibungsfreien und kompressiblen Unterschallströmung am Beispiel des Kreisprofils,* Diplomarbeit, Institut für Angewandte Mathematik, Universität Bonn, 1983.

*[508] A. SCHÜLLER AND Q. LIN, *Efficient high order algorithms for elliptic boundary value problems combining full multigrid techniques and extrapolation methods,* Arbeitspapiere der GMD, Nr. 192, Gesellschaft für Mathematik und Datenverarbeitung, St. Augustin, 1985.

M. H. SCHULTZ, cf. [131].

*[509] U. SCHUMANN AND H. VOLKERT, *Three-dimensional mass—and momentum—consistent Helmholtz-equation in terrain-following coordinates,* in Efficient Solution of Elliptic Systems, Proc. GAMM-Seminar in Kiel, January 27–29, 1984, W. Hackbusch, ed., Notes on Numerical Fluid Mechanics, Volume 10, Vieweg, Braunschweig, 1984, pp. 109–131.

*[510] H. SCHWICHTENBERG, *Erweiterungsmöglichkeiten des Programmpaketes* MG01 *auf nichtlineare Aufgaben,* Diplomarbeit, Institut für Angewandte Mathematik, Universität Bonn, 1985.

*[511] T. SCOTT, *Multigrid methods for oil reservoir simulation in three dimensions,* in Multigrid Methods for Integral and Differential Equations, D. J. Paddon and H.

Holstein, eds., The Institute of Mathematics and its Applications Conference Series 3, Clarendon Press, Oxford, 1985, pp. 283–300.

*[512] A. SEIDL, *A multigrid method for solution of the diffusion equation in VLSI process modeling,* IEEE Trans. on Electron Devices, Vol. ED-30, 9 (1983), pp. 999–1004.

W. SEIDL, cf. [193].

B. SEKERKA, cf. [376].

*[513] G. W. SHAW, *Multigrid methods in Fluid Dynamics,* Ph.D. thesis, University of Oxford, 1986.

*[514] Q. M. SHEIKH, *Systems of nonlinear algebraic equations arising in simulation of semiconductor devices,* Report no. 1133, Dept. of Computer Science, University of Ullinois, Urbana, IL, 1983.

A. H. SHERMAN, cf. [38], [39], [40], [41], [42], [45].

*[515] A. S. L. SHIEH, *Solution of coupled system of PDE by the transistorized multi-grid method,* Proc. Conference on Numerical Simulation of VLSI devices, Boston, MA, Nov. 1984.

[516] Y. SHIFTAN, *Multi-grid methods for solving elliptic difference equations,* M.Sc. thesis, Dept. of Applied Mathematics, Weizmann Institute of Science, Rehovot, Israel, 1972.

C. T. SHIN, cf. [195], [196].

[517] A. SHMILOVICH AND D. A. CAUGHEY, *Application of the multi-grid method to calculations of transonic potential flow about wing-fuselage combinations,* J. Comput. Phys., 48 (1982), pp. 462–484.

*[518] ———, *Calculation of transonic potential flow past wing-tail-fuselage combinations using multigrid techniques,* Proc. Ninth Internat. Conference on Numerical Methods in Fluid Dynamics, Soubbaramayer and J. P. Boujot, eds., Lecture Notes in Physics 218, Springer-Verlag, Berlin, 1985.

*[519] S. SLAMET, *Quasi–Newton and multigrid methods for semiconductor device simulation,* Report no. 1154, Dept. of Computer Science, University of Illinois, Urbana, IL, 1983.

K. SOLCHENBACH, cf. [488], [489].

*[520] K. SOLCHENBACH AND B. STECKEL, geb. Ruttmann, *Numerical simulation of the flow in 3D-cylindrical combustion chambers using multigrid methods,* Arbeitspapiere der GMD, Gesellschaft für Mathematik und Datenverarbeitung, St. Augustin, to appear.

[521] K. SOLCHENBACH, K. STÜBEN, U. TROTTENBERG AND K. WITSCH, *Efficient solution of a nonlinear heat conduction problem by use of fast reduction and multigrid methods,* in Numerical Integration of Differential Equations and Large Linear Systems, Proc. two workshops held at the University of Bielefeld, J. Hinze, ed., Lecture Notes in Mathematics 968, Springer-Verlag, Berlin, 1982, pp. 114–148.

B. P. SOMMEIJER, cf. [289].

*[522] B. P. SOMMEIJER AND P. J. VAN DER HOUWEN, *Algorithm 621; Software with low storage requirements for two-dimensional, nonlinear, parabolic differential equations,* ACM Trans. Math. Software, 10 (1984), pp. 378–396.

P. SONNEVELD, cf. [595].

*[523] P. SONNEVELD AND P. WESSELING, *Multigrid and conjugate gradient methods as convergence acceleration techniques,* in Multigrid Methods for Integral and Differential Equations, D. J. Paddon and H. Holstein, eds., The Institute of Mathematics and its Applications Conference Series 3, Clarendon Press, Oxford, 1985, pp. 117–168.

[524] J. C. SOUTH AND A. BRANDT, *Application of a multi-level grid method to transonic flow*

calculations, Transonic Flow Problems in Turbomachinery, T. C. Adamson and M. F. Platzer, eds., Hemisphere, Washington, D.C., 1977.

S. P. SPEKREIJSE, cf. [274], [275].

*[525] S. P. SPEKREIJSE, *Second order accurate upwind solutions of the 2D steady Euler equations by the use of a defect correction method,* Proc. 2nd European Conference on Multigrid Methods, Cologne, October 1–4, 1985, W. Hackbusch and U. Trottenberg, eds., Lecture Notes in Mathematics, Springer-Verlag, Berlin, to appear.

R. SPIGLER, cf. [322].

B. STECKEL, geb. Ruttmann, cf. [520].

B. STEFFEN, cf. [588].

*[526] B. STEFFEN, *Überlegungen zur Berechnung von Hohlraumresonantoren,* Rechnerarchitekturen für die numerische Simulation auf der Basis superschneller Lösungsverfahren II, U. Trottenberg and P. Wypior, eds., GMD-Studien Nr. 102, Gesellschaft für Mathematik und Datenverarbeitung, St. Augustin, 1985, pp. 287–296.

*[527] ———, *Incorporation of multigrid methods in accelerator software,* Proc. 2nd European Conference on Multigrid Methods, Cologne, October 1–4, 1985, W. Hackbusch and U. Trottenberg, eds., GMD-Studien Nr. 110, Gesellschaft für Mathematik und Datenverarbeitung, St. Augustin, to appear.

*[528] J. L. STEGER, *A preliminary study of relaxation methods for the inviscid conservative gasdynamics equations using flux splitting,* National Aeronautics and Space Administration, NASA-CR-3415, 1981.

*[529] ———, *Algorithms for zonal methods and developments of three dimensional mesh generation procedures,* Report, Dept. of Aeronautics and Astronautics, Stanford University, Palo Alto, CA, 1984.

S. STEINBERG, cf. [473].

F. STENGER, cf. [494].

*[530] P. STEPHANY, *Eine Erweiterung des Mehrgitterprogramms* MG01: *Einbeziehung von Ableitungen erster Ordnung, variable Koeffizienten,* Diplomarbeit, Institut für Angewandte Mathematik, Universität Bonn, 1983.

H. J. STETTER, cf. [19], [66].

[531] L. G. STRAKHOVSKAYA, *An iterative method for evaluating the first eigenvalue of an elliptic operator,* U.S.S.R. Computational Math. and Math. Phys., 17 no. 3 (1977), pp. 88–101.

[532] C. L. STREETT, T. A. ZANG AND M. Y. HUSSAINI, *Spectral multigrid methods with applications to transonic potential flow,* in Preliminary Proc. Internat. Multigrid Conference, April 6–8, 1983, Copper Mountain, CO, Institute for Computational Studies at Colorado State University, Ft. Collins, CO, 1983.

*[533] R. M. STUBBS, *Multiple gridding of the Euler equations with an implicit scheme,* American Institute of Aeronautics and Astronautics, AIAA-83-1945, Danvers, 1983.

[534] ———, *Multiple-grid strategies for accelerating the convergence of the Euler equations,* Proc. Fifth GAMM-Conference on Numerical Methods in Fluid Mechanics, Rome, October 5–7, 1983, Notes on Numerical Fluid Mechanics, Volume 7, Vieweg, Braunschweig, 1983.

K. STÜBEN, cf. [168], [486], [487], [521].

[535] K. STÜBEN, MG01: *A multi-grid program to solve Delta* $U - c(x, y)U = f(x, y)$ *(on Omega),* $U = g(x, y)$ *(on dOmega), on nonrectangular bounded domains Omega,* IMA-Report Nr. 82.02.02, Gesellschaft für Mathematik und Datenverarbeitung, St. Augustin, 1982.

[536] ———, *Algebraic multigrid (AMG): experiences and comparisons,* in Appl. Math.

Comp., 13, Proc. Internat. Multigrid Conference, April 6–8, 1983, Copper Mountain, CO, S. F. McCormick and U. Trottenberg, eds., North-Holland, Amsterdam, 1983, pp. 419–452.

*[537] K. Stüben and J. Linden, *Multigrid methods: An overview with emphasis on grid generation processes,* Proc. First Internat. Conference on Numerical Grid Generations in Computational Fluid Dynamics, J. Häuser, ed., Pineridge Press, Swansea, 1986.

[538] K. Stüben and U. Trottenberg, *Multigrid methods: Fundamental algorithms, model problem analysis and applications,* in Multigrid Methods, Proc. conference held at Köln-Porz, November 23–27, 1981, W. Hackbusch and U. Trottenberg, eds., Lecture Notes in Mathematics 960, Springer-Verlag, Berlin, 1982, pp. 1–176.

[539] ———, *On the construction of fast solvers for elliptic equations,* in Computational Fluid Dynamics, Lecture Series 1982-04, von Karman Institute for Fluid Dynamics, Rhode-St.-Genese, Belgium, 1982.

*[540] K. Stüben, U. Trottenberg and K. Witsch, *Software development based on multigrid techniques,* Proc. IFIP-Conference on PDE Software, Modules, Interfaces and Systems, Söderköping, Sweden, August 22–26, 1983, B. Enquist and T. Smedsaas, eds., 1983.

R. A. Sweet, cf. [189].

J. M. Swisshelm, cf. [319], [320].

*[541] J. M. Swisshelm and G. M. Johnson, *Numerical simulation of three-dimensional flowfields using the Cyber* 205, in Supercomputer Applications, R. W. Numrich, ed., Lecture Notes in Physics, Plenum, New York, 1984.

*[542] J. M. Swisshelm, G. M. Johnson and S. P. Kumar, *Parallel computation of Euler and Navier–Stokes flows,* in Appl. Math. Comp., Proc. 2nd Internat. Multigrid Conference, April 1985, Copper Mountain, CO, S. F. McCormick, ed., North-Holland, Amsterdam, to appear.

S. Ta'asan, cf. [108], [109], [110].

*[543] S. Ta'asan, *Multigrid methods for highly oscillatory problems,* Research Report, Dept. of Applied Mathematics, Weizmann Institute of Science, Rehovot, Israel, 1981.

J. H. Tabor, cf. [49].

[544] A. Tal-Nir and M. Bercovier, *Implementation of the multigrid method "MLAT" in the finite element method,* Report no. 54, Institut National de Recherche en Informatique et en Automatique (INRIA), Le Chesnay, France, 1981.

A. Tapucu, cf. [342], [343].

G. D. Taylor, cf. [112].

*[545] D. Terzopoulos, *Multi-level reconstruction of visual surfaces: Variational principles and finite element representation,* Memo no. 671, Massachusetts Institute of Technology, Cambridge, MA, 1982.

[546] F. C. Thames, *Multigrid applications to three-dimensional elliptic coordinate generation,* Preliminary Proc. Internat. Multigrid Conference, April 6–8, 1983, Copper Mountain, CO; Institute for Computational Studies at Colorado State University, Ft. Collins, CO, 1983.

[547] H. J. Thiebes, *Mehrgittermethoden und Reduktionsverfahren für indefinite elliptische Randwertaufgaben,* Dissertation, Institut für Angewandte Mathematik, Universität Bonn, 1983.

*[548] C.-A. Thole, *Beiträge zur Fourieranalyse von Mehrgittermethoden: V-Cycle, ILU-Glättung, anisotrope Operatoren,* Diplomarbeit, Institut für Angewandte Mathematik, Universität Bonn, 1983.

*[549] ———, *Mehrgitterverfahren für anisotrope* 3D-*Aufgaben: Anforderungen an MIMD-*

Rechnerarchitekturen, Rechnerarchitekturen für die numerische Simulation auf der Basis superschneller Lösungsverfahren II, U. Trottenberg and P. Wypior, eds., GMD-Studien Nr. 102, Gesellschaft für Mathematik und Datenverarbeitung, St. Augustin, 1985, pp. 297–312.

*[550] ———, *Experiments with multigrid methods on the Caltech Hypercube*, GMD-Studien Nr. 103, Gesellschaft für Mathematik und Datenverarbeitung, St. Augustin, 1986.

*[551] ———, *Performance of a multigrid method on a parallel architecture*, Proc. 2nd European Conference on Multigrid Methods, Cologne, October 1–4, 1985, W. Hackbusch and U. Trottenberg, eds., Lecture Notes in Mathematics, Springer-Verlag, Berlin, to appear.

*[552] C.-A. THOLE AND U. TROTTENBERG, *Basic smoothing procedures for the multigrid treatment of elliptic* 3D-*operators*, in Advances in Multi-Grid methods, Proc. conference held in Oberwolfach, December 8–13, 1984, D. Braess, W. Hackbusch and U. Trottenberg, eds., Notes on Numerical Fluid Mechanics, Volume 11, Vieweg, Braunschweig, 1985, pp. 102–111.

J. THOMAS, cf. [159], [253], [400], [401], [402].

*[553] C. P. THOMPSON, *A preliminary investigation into techniques for the adaptation of algebraic multigrid algorithms to advanced computer architectures*, Report, GMD, Contract no. 01/022079, Gesellschaft für Mathematik und Datenverarbeitung, St. Augustin, 1985.

*[554] T. THUNELL AND L. FUCHS, *Numerical solution of the Navier–Stokes equations by multi-grid techniques*, in Numerical Methods in Laminar and Turbulent Flow, Proc. Second Internat. Conference held in Venice, July 13–16, 1981, C. Taylor and A. B. Schrefler, eds., Pineridge Press, Swansea, 1981, pp. 141–152.

N. TILLMARK, cf. [180].

U. TROTTENBERG, cf. [57], [58], [85], [168], [249], [352], [467], [468], [521], [538], [539], [540], [552].

[555] U. TROTTENBERG, *Schnelle Lösung partieller Differentialgleichungen—Idee und Bedeutung des Mehrgitterprinzips*, Jahresbericht 1980/81, Gesellschaft für Mathematik und Datenverarbeitung, 85–95, St. Augustin, 1981.

*[556] ———, *Mehrgitterprinzip—moderne Denk—und Arbeitsweise der Numerik*, GI-13, Jahrestagung, Okt. 1983, Informatik Fachberichte, Nr. 73, Springer-Verlag, Berlin, 1953

*[557] ———, *Mehrgitterprinzip und Rechnerarchitektur*, Rechnerarchitekturen für die numerische Simulation auf der Basis superschneller Lösungsverfahren I, Workshop 'Rechnerarchitektur,' Erlangen, June 14–15, 1984, U. Trottenberg and P. Wypior, eds., GMD-Studien Nr. 88, Gesellschaft für Mathematik und Datenverarbeitung, St. Augustin, 1984, pp. 7–30.

[558] ———, *Schnelle Lösung elliptischer Differentialgleichungen nach dem Mehrgitterprinzip*, Mitteilungen der GAMM, 1 (1984), pp. 23–41.

*[559] ———, *Zur SUPRENUM-Konzeption*, Rechnerarchitekturen für die numerische Simulation auf der Basis superschneller Lösungsverfahren II, U. Trottenberg and P. Wypior, eds., GMD-Studien Nr. 102, Gesellschaft für Mathematik und Datenverarbeitung, St. Augustin, 1985, pp. 313–328.

*[560] U. TROTTENBERG AND P. WYPIOR, eds., *Rechnerarchitekturen für die numerische Simulation auf der Basis superschneller Lösungsverfahren* I, GMD-Studien Nr. 88, Proc. Workshop 'Rechnerarchitektur,' Erlangen, June 14–15, 1984, Gesellschaft für Mathematik und Datenverarbeitung, St. Augustin, 1984.

*[561] ———, *Rechnerarchitekturen für die numerische Simulation auf der Basis superschneller Lösungsverfahren* II, GMD-Studien Nr. 102, Gesellschaft für Mathematik und Datenverarbeitung, St. Augustin, 1985.

E. TURKEL, cf. [51], [305].

*[562] E. TURKEL AND B. VANLEER, *Flux-vector splitting and Runge–Kutta methods for the Euler equations,* NASA-CR-172415, National Aeronautics and Space Administration, Hampton, VA, 1984.

H. TUY, cf. [280].

*[563] K.-H. UNTIET, *Mehrgitterverfahren für eine Finite Element Approximation des Stokes–Problems,* Diplomarbeit, Abteilung für Mathematik, Ruhr Universität Bochum, 1985.

*[564] S. P. VANKA, *Block implicit multigrid calculation of three dimensional recirculating flows,* Proc. Internat. Conference on Numerical Methods in Laminar and Turbulent Flow, July 1985, Swansea, U.K.

*[565] ———, *Study of second order upwind differencing in a recirculating flow,* NASA-CR-174939, National Aeronautics and Space Administration, Washington, D.C., 1985.

*[566] S. P. VANKA AND G. K. LEAF, *Block-implicit multigrid solution of Navier–Stokes equations in primitive variables,* Report, DOE/NBN-5004071, Argonne National Lab., Argonne, IL, 1985.

B. VANLEER, cf. [562].

*[567] P. VASSILEVSKI, *Multigrid method in subspace and domain partitioning in the discrete solution of elliptic problems,* Proc. 2nd European Conference on Multigrid Methods, Cologne, October 1–4, 1985, W. Hackbusch and U. Trottenberg, eds., Lecture Notes in Mathematics, Springer-Verlag, Berlin, to appear.

E. F. VAN DE VELDE, cf. [380].

[568] R. VERFÜRTH, *Two algorithms for mixed problems,* Preliminary Proc. for Internat. Multigrid Conference, April 6–8, 1983, Copper Mountain, CO; Institute for Computational Studies at Colorado State University, Ft. Collins, CO, 1983.

[569] ———, *A combined conjugate gradient-multigrid algorithm for the numerical solution of the Stokes problem,* IMA J. Numer. Anal., 4 (1984), pp. 441–455.

*[570] ———, *A multilevel algorithm for mixed problems,* SIAM J. Numer. Anal., 21 (1984), pp. 264–271.

[571] ———, *Numerical solution of mixed finite element problems,* in Efficient Solution of Elliptic Systems, Proc. GAMM-Seminar in Kiel, January 27–29, 1984, W. Hackbusch ed., Notes on Numerical Fluid Mechanics, Volume 10, Vieweg, Braunschweig, 1984, pp. 132–144.

*[572] ———, *The contraction number of a multigrid method with mesh ratio* 2 *for solving Poisson's equation,* Linear Algebra Appl., 60 (1984), pp. 332–348.

*[573] ———, *A preconditioned conjugate residual algorithm for the Stokes problem,* in Advances in Multi-Grid Methods, Proc. conference held in Oberwolfach, December 8–13, 1984, D. Braess, W. Hackbusch and U. Trottenberg, eds., Notes on Numerical Fluid Mechanics, Volume 11, Vieweg, Braunschweig, 1985, pp. 112–119.

*[574] T. VERHUVEN, *Numerische Behandlung von Template-Matching in der automatischen Bildverarbeitung,* Diplomarbeit, Institut für Angewandte Mathematik, Universität Bonn, 1984.

H. VOLKERT, cf. [509].

J. VOLKERT, cf. [193], [281].

*[575] D. R. VONDY, *Multigrid iteration solution procedure for solving two-dimensional sets of coupled equations (high temperature gas cooled reactors),* Report ORNL/TM-9030, Oak Ridge National Lab., Oak Ridge, TN, 1984.

J. VAN DER VOOREN, cf. [587].

*[576] M. H. VOYTKO, *Comparison of the multigrid and ICCG methods in solving diffusion equations,* Master Thesis, Lawrence Livermore National Laboratory, UC, Livermore, CA, 1984.

H. B. DE VRIES, cf. [290].

[577] H. B. DE VRIES, *The two-level algorithm in the solution of the initial value problem for partial differential equations,* Report NW 136/82, Dept. of Numerical Mathematics, Mathematical Centre, Amsterdam, 1982.

*[578] ———, *The multigrid method in the solution of time-dependent nonlinear partial differential equations,* Report, Department of Numerical Mathematics, Mathematical Centre, Amsterdam, 1983.

*[579] E. L. WACHPRESS, *Split-level iteration,* Comput. Math. Appl., 10 (1984).

*[580] B. WAGNER, S. LEICHER AND W. SCHMIDT, *Applications of a multigrid finite volume method with Runge–Kutta time integration for solving the Euler and Navier–Stokes equations,* Proc. 2nd European Conference on Multigrid Methods, Cologne, October 1–4, 1985, W. Hackbusch and U. Trottenberg, eds., Lecture Notes in Mathematics, Springer-Verlag, Berlin, to appear.

S. WARHAUT, cf. [422].

J. E. WASH, cf. [358].

H. WEBER, cf. [415].

[581] H. WEBER, *A multi-grid technique for the computation of stable bifurcation branches,* Report, Rechenzentrum und Fachbereich Mathematik der Johannes Gutenberg Universität, Mainz, 1982.

[582] ———, *Multi-grid bifurcation iteration,* Preprint no. 3, Rechenzentrum, Universität Mainz, 1983.

*[583] ———, *A singular multi-grid iteration method for bifurcation problems,* Numerical Methods for Bifurcation Problems, T. Küpper, H. D. Mittelmann and H. Weber, eds., Birkhäuser-Verlag, Basel, 1984.

[584] ———, *Efficient computation of stable bifurcating branches of nonlinear eigenvalue problems,* Proc. EQUADIFF-Conference, Würzburg, 1982, to appear.

*[585] ———, *Multigrid bifurcation iteration,* SIAM J. Numer. Anal., submitted.

*[586] A. J. VAN DER WEES, *FAS multigrid employing ILU/SIP smoothing: A robust fast solver for 3D transonic potential flow,* Proc. 2nd European Conference on Multigrid Methods, Cologne, October 1–4, 1985, W. Hackbusch and U. Trottenberg, eds., Lecture Notes in Mathematics, Springer-Verlag, Berlin, to appear.

[587] A. J. VAN DER WEES, J. VAN DER VOOREN AND J. H. MEELKER, *Robust calculation of 3D transonic potential flow based on the nonlinear FAS multi-grid method and incomplete LU decomposition,* Proc. American Institute of Aeronautics and Astronautics Sixth Computational Fluid Dynamics Conference, July 13–15, 1983, Danvers, MA, submitted.

*[588] P. WEIDNER AND B. STEFFEN, *Vektorisierung von Mehrgitterverfahren: Tests auf einer Cray X-MP,* Rechnerarchitekturen für die numerische Simulation auf der Basis superschneller Lösungsverfahren I, Workshop 'Rechnerarchitektur,' Erlangen, June 14–15, 1984, U. Trottenberg and P. Wypior, eds., GMD-Studien Nr. 88, Gesellschaft für Mathematik und Datenverarbeitung, St. Augustin, 1984, pp. 195–198.

A. WEISER, cf. [43], [45].

P. WESSELING, cf. [278], [279], [330], [439], [523].

[589] P. WESSELING, *Numerical solution of stationary Navier–Stokes equations by means of a multiple grid method and Newton iteration,* Report NA-18, Dept. of Mathematics, Delft University of Technology, Delft, 1977.

[590] ———, *A convergence proof for a multiple grid method,* Report NA-21, Dept. of Mathematics, Delft University of Technology, Delft, 1978.

[591] ———, *The rate of convergence of a multiple grid method,* in Numerical Analysis, Proc., Dundee, 1979, G. A. Watson, ed., Lecture Notes in Mathematics 773, Springer-Verlag, Berlin, 1980, pp. 164–184.

[592] P. WESSELING, *A robust and efficient multigrid method,* in Multigrid Methods, Proc. conference held at Köln-Porz, November 23–27, 1981, W. Hackbusch and U. Trottenberg, eds., Lecture Notes in Mathematics 960, Springer-Verlag, Berlin, 1982, pp. 614–630.

[593] ———, *Theoretical and practical aspects of a multigrid method,* SIAM J. Sci. Stat. Comput., 3 (1982), pp. 387–407.

*[594] ———, *Multigrid solution of the Navier–Stokes equations in the vorticity-streamfunction formulation,* in Efficient Solution of Elliptic Systems: Proc. GAMM-Seminar in Kiel, January 27–29, 1984, W. Hackbusch, ed., Notes on Numerical Fluid Mechanics, Volume 10, Vieweg, Braunschweig, 1984, pp. 145–154.

[595] P. WESSELING AND P. SONNEVELD, *Numerical experiments with a multiple grid and a preconditioned Lanczos method,* in Approximation Methods for Navier–Stokes Problems, Proc., Paderborn 1979, R. Rautmann, ed., Lecture Notes in Mathematics 771, Springer-Verlag, Berlin, 1980, pp. 543–562.

T. WILL, cf. [250].

G. WINTER, cf. [468].

[596] G. WINTER, *Fourieranalyse zur Konstruktion schneller MGR-Verfahren,* Dissertation, Institut für Angewandte Mathematik, Universität Bonn, 1983.

K. WITSCH, cf. [166], [167], [352], [521], [540].

*[597] C. J. WOAN, *Euler solution of axisymmetric flows about bodies of revolution using a multigrid method,* American Institute of Aeronautics and Astronautics, AIAA-85-0017, 1985.

[598] H. WOLFF, *Multi-grid techniek voor het oplossen van fredholm-integraalvergelijkingen van de tweede soort,* Report NN 19/79, Dept. of Numerical Mathematics, Mathematical Centre, Amsterdam, 1979.

[599] ———, *Multiple grid methods for the calculation of potential flow around* 3-d *bodies,* Preprint NW 119/82, Dept. of Numerical Mathematics, Mathematical Centre, Amsterdam, 1982.

Y. S. WONG, cf. [293], [605], [606].

P. WYPIOR, cf. [560], [561].

S. YOON, cf. [304].

M. YOUNIS, cf. [122].

[600] H. YSERENTANT, *The convergence of multi-level methods for solving finite element equations in the presence of singularities,* Bericht Nr. 14, Institut für Geometrie und Praktische Mathematik, Technische Hochschule Aachen, 1982.

[601] ———, *The convergence of multi-level methods for strongly nonuniform families of grids and any number of smoothing steps per level,* Computing, 30 (1983), pp. 305–313.

*[602] ———, *On the multi-level splitting of finite element spaces for indefinite elliptic boundary value problems,* Bericht Nr. 26, Institut für Geometrie und Praktische Mathematik, Technische Hochschule Aachen, 1984.

*[603] ———, *Hierarchical bases give conjugate gradient-type methods: A multigrid speed of convergence,* in Appl. Math. Comp., Proc. 2nd Internat. Multigrid Conference, April 1985, Copper Mountain, CO, S. F. McCormick, ed., North-Holland, Amsterdam, to appear.

T. A. ZANG, cf. [291], [292], [456], [457], [532].

*[604] T. A. ZANG AND M. Y. HUSSAINI, *On special multigrid methods for the time-dependent Navier–Stokes equations,* in Appl. Math. Comp., Proc. 2nd Internat. Multigrid Conference, April 1985, Copper Mountain, CO, S. F. McCormick, ed., North-Holland, Amsterdam, to appear.

[605] T. A. ZANG, Y. S. WONG AND M. Y. HUSSAINI, *Spectral multi-grid methods for elliptic equations,* J. Comput. Phys., 48 (1982), pp. 485–501.

*[606] T. A. ZANG, Y. S. WONG AND M. Y. HUSSAINI, *Spectral multigrid methods for elliptic equations* II, NASA-CR-172131, ICASE, NASA Langley Research Center, 1983.

P. M. DE ZEEUW, cf. [276], [277], [278], [279].

*[607] P. M. DE ZEEUW AND E. J. VAN ASSELT, *The convergence rate of multi-level algorithms applied to the convection-diffusion equation,* SIAM J. Sci. Stat. Comput., 6 (1985), pp. 492–503.

C. ZENGER, cf. [481].

H.-S. ZHAO, cf. [181].

J. Z. ZHU, cf. [137].

O. C. ZIENKIEWICZ, cf. [136], [137].

Mehrgitterprogramms MG01: Einbeziehung von	**Ableitungen** erster Ordnung, variable Koeffizienten	Stephany, P.
	Accelerated convergence of structured banded systems u	Kneile, K.
	Accelerated solution of the steady Euler equations	Johnson, G.M
The multigrid method for	**accelerated** solution of the discretized Schrödinger eq	Grinstein, F
code package for two-dimensional, Diffusion-	**accelerated**, Neutral-Particle Transport, Rev. 1	Alcouffe, R.
Multiple-grid strategies for	**accelerating** the convergence of the Euler equations	Stubbs, R.M.
	Acceleration of transonic potential flow calculations	Jameson, A.
Multigrid	**acceleration** of an iterative method with applications	Nowak, Z.P.;
Flux based	**acceleration** of the Euler equations	Johnson, G.M
Convergence	**acceleration** of viscous flow computations	Johnson, G.M
Multiple grid	**acceleration** of Lax Wendroff algorithm	Johnson, G.M
Multiple-grid	**acceleration** of Lax-Wendroff algorithms	Johnson, G.M
Multiple-grid convergence	**acceleration** of viscous and inviscid flow computation	Johnson, G.M
d conjugate gradient methods as convergence	**acceleration** techniques	Sonneveld, P
on of the Euler equations: Explicit schemes	**acceleration** by a multigrid method	Couaillier,
s of convergence with general smoothing and	**acceleration**	Bank, R.E.;
Incorporation of multigrid methods in	**accelerator** software	Steffen, B.
Defect corrections and higher numerical	**accuracy**	Fuchs, L.
es of multi-grid methods for certifying the	**accuracy** of PDE modeling	Forester, C.
rectangular bounded domains Omega with high	**accuracy**	Auzinger, W.
A time-	**accurate** multiple-grid algorithm	Jespersen, D
Fast and	**accurate** solution of time-dependent incompressible flo	Karlsson, A.
Second order	**accurate** upwind solutions of the 2D steady Euler equat	Spekreijse,
Mixed defect correction iteration for the	**accurate** solution of the convection diffusion equation	Hemker, P.W.
l-differential equations using fourth-order-	**accurate** compact differencing	Agarwal, R.K
On numerical methods for	**acoustic** problems	Fix, G.J.; G
for wave propagation problems in underwater	**acoustics**	Bayliss, A.;
Concurrent processing	**adaptation** of a multiple-grid algorithm	Johnson, G.M
inary investigation into techniques for the	**adaptation** of algebraic multigrid algorithms to advanc	Thompson, C.
An	**adaptive** multi-grid scheme for simulation of flows	Fuchs, L.
An	**adaptive** multi-level method for elliptic boundary valu	Bank, R.E.;
An	**adaptive** multigrid method for the Euler equation	Berger, M.J.
Fast	**adaptive** composite grid (FAC) methods: Theory for the	McCormick, S
Self	**adaptive** methods for parabolic partial differential eq	Gannon, D.B.
The fast	**adaptive** composite grid method (FAC) for elliptic boun	McCormick, S
The fast	**adaptive** composite grid method (FAC): Algorithms for a	Hart, L.; Mc
The fast	**adaptive** composite grid method for solving differentia	Ewing, R.; M
Automatic	**adaptive** grid refinement for the Euler equations	Berger, M.J.
The use of	**adaptive** grid refinement for badly behaved elliptic pa	Bank, R.E.;
Multi-level	**adaptive** computations in fluid dynamics	Brandt, A.
Multi-level	**adaptive** finite-element methods. I. Variational proble	Brandt, A.
Multi-level	**adaptive** solutions to boundary-value problems	Brandt, A.
Multi-level	**adaptive** technique (MLAT) for fast numerical solution	Brandt, A.
Multi-level	**adaptive** techniques (MLAT) for partial differential eq	Brandt, A.
Multi-level	**adaptive** techniques (MLAT) for singular-perturbation p	Brandt, A.
Multi-level	**adaptive** techniques (MLAT). I. The multi-grid method	Brandt, A.
Error norms for the	**adaptive** solution of the Navier-Stokes equations	Forester, C.
Multigrid methods and	**adaptive** techniques for oil reservoir simulation	McCormick, S
On the structure of an	**adaptive** multi-level algorithm	Hemker, P.W.
A-posteriori error estimates,	**adaptive** local mesh refinement, and multigrid iteratio	Bank, R.E.
A-posteriori error estimation,	**adaptive** mesh refinement and multigrid methods using h	Craig, A.W.;
Algorithms and data structures for	**adaptive** multigrid elliptic solvers	Rosendale, J
Design and data structure of fully	**adaptive**, multigrid, finite element software	Rivara, M.C.

compressible flows around three-dimensional	**bodies**	Hackbusch, W
he calculation of potential flow around 3-d	**bodies**	Wolff, H.
Efficient generation of	**body**-fitted coordinates for cascades using multigrid	Camarero, R.
Multigrid scheme for three-dimensional	**body**-fitted coordinates in turbomachine applications	Camarero, R.
etized with finite elements on an arbitrary	**body** fitted mesh	Deconinck, H
Unigrid methods for	**boundary** value problems with nonrectangular domains	Holland, W.;
Multigrid methods for	**boundary** integral equations	Schippers, H
A multi-grid method applied to a	**boundary** problem with variable coefficients in a recta	Hackbusch, W
Multi-grid methods for non-linear	**boundary**-value problems	Amini, S.
Multi-level adaptive solutions to	**boundary**-value problems	Brandt, A.
On the fast solution of parabolic	**boundary** control problems	Hackbusch, W
Hele-Shaw flow with a cusping free	**boundary**	Meyer, G.H.
Multi-grid algorithms for elliptic	**boundary**-value problems	Douglas, C.C
A fast numerical method for elliptic	**boundary** value problems with variable coefficents	Hackbusch, W
Numerical stability and fast solutions to	**boundary** value problems	Brandt, A.
Fast methods for the numerical solution of	**boundary**-value problems	Dinar, N.
complementarity problems arising from free	**boundary** problems	Brandt, A.;
-grid methods for the numerical solution of	**boundary** value problems	Hackbusch, W
An adaptive multi-level method for elliptic	**boundary** value problems	Bank, R.E.;
atment of stream function normal derivative	**boundary** conditions	Holstein, H.
c study of a multigrid method for some free	**boundary** problem	Mandel, J.
ct multi-grid with applications to elliptic	**boundary**-value problems	Douglas, C.C
d practical aspects of multigrid methods in	**boundary** element calculations	Schippers, H
e multigrid method to Poisson's equation in	**boundary**-fitted coordinates	Ohring, S.
e solution of Poisson equation with Neumann	**boundary** conditions in nonorthogonal curvilinear coord	Rieger, H.;
f grids for elliptic equations with natural	**boundary** condition	Astrakhantse
fficient high order algorithms for elliptic	**boundary** value problems combining full multigrid techn	Schüller, A.
id algorithms with applications to elliptic	**boundary**-value problems	Douglas, C.C
level continuation techniques for nonlinear	**boundary** value problems with parameter dependence	Mittelmann,
nique (MLAT) for fast numerical solution to	**boundary** value problems	Brandt, A.
nite element spaces for indefinite elliptic	**boundary** value problems	Yserentant,
posite grid method for solving differential	**boundary** value problems	Ewing, R.; M
ulti-grid optimal order solver for elliptic	**boundary**-value problems	Douglas, C.C
ve composite grid method (FAC) for elliptic	**boundary** value problems	McCormick, S
ware for hyperbolic equations and two point	**boundary** value problems	Keller, H.B.
, U = g(x,y) (on D-Omega) on nonrectangular	**bounded** domains Omega with high accuracy	Auzinger, W.
a), U=g(x,y) (on dOmega), on nonrectangular	**bounded** domains Omega	Stüben, K.
Black	**box** multigrid	Dendy, J.E.
Black	**box** multigrid for nonsymmetric problems	Dendy, J.E.
Black	**box** multigrid for systems	Dendy, J.E.
e for the computation of stable bifurcation	**branches**	Weber, H.
Efficient computation of stable bifurcating	**branches** of nonlinear eigenvalue problems	Weber, H.
alue and near eigenvalue problems solved by	**Brandt's** multigrid method	Guderley, K.
etter und Pereyra und MG-Extrapolation nach	**Brandt**: Beziehungen und Anwendung auf elliptische Rand	Donovang, M.
	Bus coupled systems for multigrid algorithms	Kolp, O.; Mi
	Calcul de l'ecoulement dans une entree d'air par resol	Koeck, C.
ethode d'elements finis multigrille pour le	**calcul** d'ecoulements potentiels transsoniques	Bredif, M.
Multigrid methods for	**calculating** the lifting potential incompressible flows	Hackbusch, W
	Calculation of transonic potential flow past wing-tail	Shmilovich,
Robust	**calculation** of 3d transonic potential flow based on th	Wees, A.J. v
Multigrid	**calculation** of subvortices	Gustafson, K
Multi-grid	**calculation** of three-dimensional transonic potential f	Caughey, D.A
Block implicit multigrid	**calculation** of three dimensional recirculating flows	Vanka, S.P.

	Comparison of the multigrid and ICCG method in solving	Voytko, M.H.
A	**comparison** of elliptic solvers for general two-dimensi	Chan, T.F.;
A	**comparison** of several MG-methods for the solution of t	Schröder, W.
A	**comparison** of smoothing iterations for multi-level met	Bank, R.E.;
A	**comparison** of two multi-level iterative methods for no	Bank, R.E.
On the	**comparison** of Line-Gauss-Seidel and ILU relaxation in	Hemker, P.W.
Analysis and	**comparison** of relaxation schemes in robust multigrid a	Kettler, R.
Algebraic multigrid (AMG): experiences and	**comparisons**	Stüben, K.
hod for symmetric, positive definite linear	**complementarity** problems	Mandel, J.
igrid algorithms for the solution of linear	**complementarity** problems arising from free boundary pr	Brandt, A.;
edimensional transonic potential flow about	**complex** 3-dimensional configurations	Reyhner, T.A
for multigrid algorithms: architecture and	**complexity**	Chan, T.F.;
Fast adaptive	**composite** grid (FAC) methods: Theory for the variation	McCormick, S
The fast adaptive	**composite** grid method (FAC) for elliptic boundary valu	McCormick, S
The fast adaptive	**composite** grid method (FAC): Algorithms for advanced c	Hart, L.; Mc
The fast adaptive	**composite** grid method for solving differential boundar	Ewing, R.; M
A variational method for Au = LBu on	**composite** grids	McCormick, S
The prediction of	**compressible** laminar viscous flows using a time marchi	Davis, R.L.
Multigrid algorithms for	**compressible** flow calculations	Jameson, A.
Spectral methods for inviscid,	**compressible** flows	Hussaini, M.
Multigrid solutions to steady-state	**compressible** Navier-Stokes equations. I	Brandt, A.
spectral methods and spectral solutions to	**compressible** flows	Hussaini, M.
merical solution of the Euler equations for	**compressible** inviscid fluids	Jameson, A.
Evolution of	**computational** methods in aerodynamics	Jameson, A.
n of computer programs with applications to	**computational** fluid dynamics	Gentzsch, W.
nd parabolized Navier-Stokes procedures and	**computational** techniques	Rubin, S.G.
Vectorization of	**computer** programs with applications to computational f	Gentzsch, W.
algebraic multigrid algorithms to advanced	**computer** architectures	Thompson, C.
for solving a diffusion problem on a vector	**computer**	Gary, J.; Mc
gle and multi-element airfoils on a smaller	**computer**	Nowak, Z.P.
Multigrid solvers on parallel	**computers**	Brandt, A.
Multigrid methods for multiprocessor	**computers**	McCormick, S
grid method (FAC): Algorithms for advanced	**computers**	Hart, L.; Mc
multigrid smoothers for vector and parallel	**computers**	Axelsson, O.
rmance analysis of Poisson solvers on array	**computers**	Grosch, C.E.
e full potential and Euler formulations for	**computing** transonic airfoil flows	Flores, J.;
Application of data flow	**concepts** of a multigrid solver for the Euler equations	Merriam, M.L
	Concurrent multigrid methods in an object oriented env	Mühlenbein,
	Concurrent processing adaptation of a multiple-grid al	Johnson, G.M
On the structure of parallelism in a highly	**concurrent** PDE solver	Gannon, D.B.
or elliptic equations with natural boundary	**condition**	Astrakhantse
stream function normal derivative boundary	**conditions**	Holstein, H.
n of Poisson equation with Neumann boundary	**conditions** in nonorthogonal curvilinear coordinate sys	Rieger, H.;
Efficient solution of a nonlinear heat	**conduction** problem by use of fast reduction and multig	Solchenbach,
potential flow about complex 3-dimensional	**configurations**	Reyhner, T.A
full potential transonic code for arbitrary	**configurations**	Luntz, A.L.;
olution of the Euler equations for aircraft	**configurations**	Jameson, A.;
	Conjugate gradient coupled with multigrid for an indef	Gozani, J.
A combined	**conjugate** gradient - multigrid algorithm for the numer	Verfürth, R.
Multigrid and	**conjugate** gradient methods as convergence acceleration	Sonneveld, P
A preconditioned	**conjugate** residual algorithm for the Stokes problem	Verfürth, R.
Hierarchical bases give	**conjugate** gradient-type methods: A multigrid speed of	Yserentant,
A multigrid method and a combined multigrid-	**conjugate** gradient method for elliptic problems with s	Kettler, R.;

A note on convergence of the multigrid V-	**cycle**	Parter, S.V.
de convergence incluant le V-cycle et le W-	**cycle**; applications au lissage de Gauss-Seidel	Musy, F.
demonstration de convergence incluant le V-	**cycle** et le W-cycle; applications au lissage de Gauss-	Musy, F.
e case - a convergence estimation for the V-	**cycle**	Maitre, J.F.
iational problems: general theory for the V-	**cycle**	McCormick, S
igrid methods for variation problems: The V-	**cycle**	McCormick, S
ions du facteur de convergence; le cas du V-	**cycle**	Maitre, J.F.
of for the multigrid method including the V-	**cycle**	Braess, D.;
ur Fourieranalyse von Mehrgittermethoden: V-	**Cycle**, ILU-Glättung, anisotrope Operatoren	Thole, C.-A.
A multigrid solver for the in-	**cylinder** turbulent flows in engines	Ruttmann, B.
quations for the flow in a rapidly rotating	**cylinder**	Schröder, W.
Numerical simulation of the flow in 3D-	**cylindrical** combustion chambers using multigrid method	Solchenbach,
Two multi-level algorithms for the	**dam** problem	Bollrath, C.
Design and	**data** structure of fully adaptive, multigrid, finite el	Rivara, M.C.
Algorithms and	**data** structures for adaptive multigrid elliptic solver	Rosendale, J
Application of	**data** flow concepts of a multigrid solver for the Euler	Merriam, M.L
Refinement algorithms and	**data** structures for regular local mesh refinement	Bank, R.E.;
Berechnungsgrundlagen für das	**Datenkommunikationssystem** in einem Multiprocessor für	Franke, B.;
	DCMG01: A multigrid code with defect correction to sol	Auzinger, W.
The incomplete LU-	**decomposition** as a relaxation method in multi-grid alg	Hemker, P.W.
Local defect correction method and domain	**decomposition** techniques	Hackbusch, W
ear FAS multi-grid method and incomplete LU	**decomposition**	Wees, A.J. v
	Defect correction and higher order schemes for the mul	Hemker, P.W.
	Defect correction and multigrid iterations	Auzinger, W.
	Defect correction for the solution of a singular pertu	Hemker, P.W.
	Defect corrections and higher numerical accuracy	Fuchs, L.
The	**defect** correction approach	Böhmer, K.;
Local	**defect** correction method and domain decomposition tech	Hackbusch, W
Mixed	**defect** correction iteration for the accurate solution	Hemker, P.W.
Mixed	**defect** correction iteration for the solution of a sing	Hemker, P.W.
A note on	**defect** correction processes with an approximate invers	Hemker, P.W.
Extensions of the	**defect** correction principle	Hemker, P.W.
DCMG01: A multigrid code with	**defect** correction to solve Delta U - c(x,y) U = f(x,y)	Auzinger, W.
On multi-grid iterations with	**defect** correction	Hackbusch, W
e 2D steady Euler equations by the use of a	**defect** correction method	Spekreijse,
te reactive kinetics on surfaces exhibiting	**defect** structures	Grinstein, F
Bemerkungen zur iterierten	**Defektkorrektur** und zu ihrer Kombination mit Mehrgitte	Hackbusch, W
	Defektkorrekturen nach Stetter und Pereyra und MG-Extr	Donovang, M.
on processes with an approximate inverse of	**deficient** rank	Hemker, P.W.
c study of multigrid methods for symmetric,	**definite** problems	Mandel, J.
el iterative method for symmetric, positive	**definite** linear complementarity problems	Mandel, J.
tigrid method in the symmetric and positive	**definite** case - a convergence estimation for the V-cyc	Maitre, J.F.
nd two-level multigrid methods of arbitrary	**degree** of approximation	Axelsson, O.
MG01: A multi-grid program to solve	**Delta** U - c(x,y)U = f(x,y) (on Omega), U=g(x,y) (on dO	Stüben, K.
tigrid code with defect correction to solve	**Delta** U - c(x,y) U = f(x,y) (on Omega), U = g(x,y) (on	Auzinger, W.
Mehrgitterprinzip - moderne	**Denk**- und Arbeitsweise der Numerik	Trottenberg,
near boundary value problems with parameter	**dependence**	Mittelmann,
On time	**dependent** multi-grid numerical techniques	Lee, H.N.; M
Multigrid solution of time-	**dependent** incompressible flow	Karlsson, A.
Fast and accurate solution of time-	**dependent** incompressible flow	Karlsson, A.
On special multigrid methods for the time-	**dependent** Navier-Stokes equations	Zang, T.A.;
he multigrid method in the solution of time-	**dependent** nonlinear partial differential equations	Vries, H.B.
ral MG-methods for the solution of the time-	**dependent** Navier-Stokes equations	Schröder, W.

blems with a small parameter in the highest	**derivative**	Hemker, P.W.
ltigrid treatment of stream function normal	**derivative** boundary conditions	Holstein, H.
	Design and data structure of fully adaptive, multigrid	Rivara, M.C.
	Design and implementation of a multigrid code for the	Jespersen, D
grid method for transonic wing analysis and	**design**	Raj, P.
Stages in	**developing** multigrid solutions	Brandt, A.
	Development of multigrid algorithms for problems from	Becker, K.;
Software	**development** based on multigrid techniques	Stüben, K.;
Guide to multigrid	**development**	Brandt, A.
Multigrid methods:	**development** of fast solvers	Hemker, P.W.
Recent	**developments** in multigrid methods for the steady Euler	Jespersen, D
Algorithms for zonal methods and	**developments** of three dimensional mesh generation proc	Steger, J.L.
ical and numerical aspects of semiconductor	**device** modeling	Bank, R.E.;
ton and multigrid methods for semiconductor	**device** simulation	Slamet, S.
ions arising in simulation of semiconductor	**devices**	Sheikh, Q.M.
Finite-	**difference** methods for plane steady inviscid transonic	Fuchs, L.
Field panel/finite	**difference** method for potential unsteady transonic flo	Hounjet, M.H
On the regularity of	**difference** schemes	Hackbusch, W
On the regularity of	**difference** schemes - part II: regularity estimates fo	Hackbusch, W
Efficient solution of finite	**difference** and finite element equations by algebraic m	Ruge, J.; St
Application of the Osher-Engquist	**difference** scheme and the full multi-grid method to a	Asselt, E.J.
Fedorenkos method for variational	**difference** schemes with extrapolation	Astrakhantse
On the multigrid method applied to	**difference** equations	Hackbusch, W
Multi-grid methods for solving elliptic	**difference** equations	Shiftan, Y.
A relaxation method for solving elliptic	**difference** equations	Fedorenko, R
Eigensystem analysis techniques for finite-	**difference** equations. I. Multigrid techniques	Lomax, H.; P
relaxation method for solving a biharmonic	**difference** equation	Otroshcenko,
ergence of multi-grid iterations applied to	**difference** equations	Hackbusch, W
nn and steady Euler equations based on flux-	**difference** splitting	Dick, E.
solutions of mixed finite element and mixed	**difference** equations	Hackbusch, W
t numerical solution of very large elliptic	**difference** schemes	Hackbusch, W
terative method applied to a model elliptic	**difference** equation	Nicolaides,
ue for steady Euler equations based on flux-	**difference** splitting	Dick, E.
ution of eigenvalue problems of variational	**difference** equations in a sequence of grids	Sajdurov, V.
MG techniques for staggered	**differences**	Favini, B.;
Study of second order upwind	**differencing** in a recirculating flow	Vanka, S.P.
uations using fourth-order-accurate compact	**differencing**	Agarwal, R.K
Multigrid methods for	**differential** eigenproblems	Brandt, A.;
Multigrid methods for	**differential** eigenvalue and variational problems and m	Ruge, J.
Solving elliptic partial	**differential** equations on the Hypercube multiprocessor	Chan, T.F.;
Multigrid methods for partial	**differential** equations	Jespersen, D
Multigrid methods for integral and	**differential** equations	Paddon, D.J.
On the time integration of parabolic	**differential** equations	Houwen, P.J.
adaptive composite grid method for solving	**differential** boundary value problems	Ewing, R.; M
age for processes of numerical solutions to	**differential** equations	Ophir, D.
age for processes of numerical solutions to	**differential** equations. Second annual report	Brandt, A.;
evel adaptive techniques (MLAT) for partial	**differential** equations: ideas and software	Brandt, A.
f the multi-grid method for solving partial	**differential** equations	Gustavson, F
finement for badly behaved elliptic partial	**differential** equations	Bank, R.E.;
initial value problem for nonlinear partial	**differential** equations	Houwen, P.J.
linear eigenvalue problems for integral and	**differential** equations	Hackbusch, W
ltigrid method to the solution of parabolic	**differential** equations	Schröder, W.
n multigrid methods for nonlinear parabolic	**differential** equations	Houwen, P.J.

ulti-grid techniques for nonlinear elliptic	**eigenvalue** problems	Chan, T.F.;
ysis of a SOR-like multi-grid algorithm for	**eigenvalue** problems	Hofmann, G.
On the computation of approximate	**eigenvalues** and eigenfunctions of elliptic operators b	Hackbusch, W
nes Mehrgitterverfahrens zur Berechnung von	**Eigenwerten** elliptischer Differentialoperatoren	Hofmann, G.
e Erweiterung des Mehrgitterprogramms MG01:	**Einbeziehung** von Ableitungen erster Ordnung, variable	Stephany, P.
einerte Mehrgitter-Methoden für Gebiete mit	**einspringenden** Ecken	Ritzdorf, H.
Multigrid analysis of linear	**elastic** stress problems	Barrett, K.E
AMG for problems of	**elasticity**	Ruge, J.
pair of elliptic equations from solid state	**electronics**	Phillips, T.
On finite	**element** multigrid algorithms and their use	Nicolaides,
Multigrid finite	**element** method	Barrett, K.E
A multigrid finite	**element** method for the transonic potential equation	Deconinck, H
A mixed variable finite	**element** method for the efficient solution of nonlinear	Axelsson, O.
Rapid solution of finite	**element** equations on locally refined grids by multi-le	Rosendale, J
Multi-level adaptive finite-	**element** methods. I. Variational problems	Brandt, A.
Numerical solution of mixed finite	**element** problems	Verfürth, R.
Mehrgitterverfahren für eine Finite	**Element** Approximation des Stokes-Problems	Untiet, K.-H
On the solution of nonlinear finite	**element** systems	Mansfield, L
On the multi-grid solution of finite	**element** equations with isoparametric elements	Mansfield, L
A multigrid version of a simple finite	**element** method for the Stokes problem	Pitkäranta,
On the multi-level splitting of finite	**element** spaces for indefinite elliptic boundary value	Yserentant,
A multigrid solver for a stabilized finite	**element** discretization of the Stokes problem	Abdalass, E.
of transonic flows around single and multi-	**element** airfoils on a smaller computer	Nowak, Z.P.
al aspects of multigrid methods in boundary	**element** calculations	Schippers, H
An optimal order process for solving finite	**element** equations	Bank, R.E.;
e of multi-level methods for solving finite	**element** equations in the presence of singularities	Yserentant,
ethods; an exact evaluation for some finite	**element** subspaces and model problems	Maitre, J.F.
f the multigrid method "MLAT" in the finite	**element** method	Tal-Nir, A.;
igrid algorithm using a hierarchical finite	**element** basis	Craig, A.W.;
is of a two-level scheme for solving finite	**element** equations	Bank, R.E.;
level iterative method for nonlinear finite	**element** equations	Bank, R.E.;
multigrid methods using hierarchical finite	**element** bases	Craig, A.W.;
nonsymmetric and indefinite elliptic finite	**element** equations	Bank, R.E.
nt solution of finite difference and finite	**element** equations by algebraic multigrid (AMG)	Ruge, J.; St
of a multi-grid iteration applied to finite	**element** equations	Hackbusch, W
pects of the multi-level solution of finite	**element** equations	Bank, R.E.;
sis and multigrid solutions of mixed finite	**element** and mixed difference equations	Hackbusch, W
surfaces: Variational principles and finite	**element** representation	Terzopoulos,
ucture of fully adaptive, multigrid, finite	**element** software	Rivara, M.C.
vergence of an algorithm for solving finite	**element** equation	Nicolaides,
Une methode d'	**elements** finis multigrille pour le calcul d'ecoulement	Bredif, M.
potential equation discretized with finite	**elements** on an arbitrary body fitted mesh	Deconinck, H
c flow calculations using triangular finite	**elements**	Pelz, R.B.;
finite element equations with isoparametric	**elements**	Mansfield, L
ve des donnes pour appliquer la methode des	**elements** finis	Charron, M.S
Solving	**elliptic** partial differential equations on the Hypercu	Chan, T.F.;
Solving	**elliptic** problems using ELLPACK	Rice, J.R.;
A comparison of	**elliptic** solvers for general two-dimensional regions	Chan, T.F.;
Efficient solution of	**elliptic** systems	Hackbusch, W
Multigrid methods for	**elliptic** problems	Fulton, S.R.
On the fast solving of	**elliptic** control problems	Hackbusch, W
Multi-grid solutions to	**elliptic** flow problems	Brandt, A.;
Multi-grid algorithms for	**elliptic** boundary-value problems	Douglas, C.C

Solving elliptic problems using	**ELLPACK**	Rice, J.R.;
	Embedded mesh multigrid treatment of two-dimensional t	McCarthy, D.
An	**embedded**-mesh potential flow analysis	Brown, J.L.
A multigrid mesh-	**embedding** technique for three-dimensional transonic po	Brown, J.L.
Multigrid methods: An overview with	**emphasis** on grid generation processes	Stüben, K.;
Ein Konzept zur Nutzung	**eng** gekoppelter Mehrrechnersysteme	Mierendorff,
lver for the in-cylinder turbulent flows in	**engines**	Ruttmann, B.
Application of the Osher-	**Engquist** difference scheme and the full multi-grid met	Asselt, E.J.
ent multigrid methods in an object oriented	**environment** -A case study-	Mühlenbein,
Magnetohydrodynamic	**equilibrium** calculations using **multigrid**	Braams, B.J.
	Error analysis of the nonlinear **multigrid method** of th	Hackbusch, W
	Error norms for the adaptive solution of the Navier-St	Forester, C.
A-posteriori	**error** estimates, adaptive local **mesh refinement**, and m	Bank, R.E.
A-posteriori	**error** estimation, adaptive mesh refinement and multigr	Craig, A.W.;
Optimal H**p,p/2	**error** estimates for a parabolic Galerkin method	Hackbusch, W
Some a posteriori	**error** estimators for **elliptic** partial differential equ	Bank, R.E.;
Mehrgitterverfahren für das	**erste** Randwertproblem der biharmonischen Gleichung und	Linden, J.
rogramms MG01: Einbeziehung von Ableitungen	**erster** Ordnung, variable Koeffizienten	Stephany, P.
	Erweiterungsmöglichkeiten des Programmpaketes MG01 auf	Schwichtenbe
Sharp	**estimates** for multigrid rates of convergence with gene	Bank, R.E.;
A-posteriori error	**estimates**, adaptive local mesh refinement, and multigr	Bank, R.E.
Optimal H**p,p/2 error	**estimates** for a parabolic Galerkin method	Hackbusch, W
problems: A general theory and convergence	**estimates** for usual smoothers	Maitre, J.F.
of difference schemes - part II: regularity	**estimates** for linear and nonlinear problems	Hackbusch, W
A-posteriori error	**estimation**, adaptive mesh refinement and multigrid met	Craig, A.W.;
and positive definite case - a convergence	**estimation** for the V-cycle	Maitre, J.F.
problemes variationel: theorie generale et	**estimations** sur taux de convergence	Musy, F.
Methodes multigrilles: operateur associe et	**estimations** du facteur de convergence; le cas du V-cyc	Maitre, J.F.
Some a posteriori error	**estimators** for elliptic partial differential equations	Bank, R.E.;
	Etude d'une classe de methodes multigrilles pour les p	Musy, F.
	Euler solution of axisymmetric flows about bodies of r	Woan, C.J.
Solution of the	**Euler** equations for two dimensional transonic flow by	Jameson, A.
Solution of the	**Euler** equations: Explicit schemes acceleration by a mu	Couaillier,
Parallel computation of	**Euler** and Navier-Stokes flows	Swisshelm, J
Multiple gridding of the	**Euler** equations with an implicit scheme	Stubbs, R.M.
Efficient solution of the	**Euler** and Navier-Stokes equations with a vectorized mu	Chima, R.V.;
Multigrid methods for the	**Euler** equations: the relaxation technique	Jespersen, D
Multigrid solution of the	**Euler** equations for aircraft configurations	Jameson, A.;
Multigrid solution of the	**Euler** equations using implicit schemes	Jameson, A.;
Numerical solution of the	**Euler** equations for compressible inviscid fluids	Jameson, A.
A multigrid method for the	**Euler** equations	Jespersen, D
Numerical solutions of the	**Euler** equations by finite volume methods using Runge-K	Jameson, A.;
Multigrid approaches to the	**Euler** equations	Hemker, P.W.
Multigrid relaxation for the	**Euler** equations	Mulder, W.A.
Flux based acceleration of the	**Euler** equations	Johnson, G.M
A multigrid technique for steady	**Euler** equations based on flux-difference splitting	Dick, E.
Multigrid solutions of the steady	**Euler** equations	Hemker, P.W.
A multiple grid scheme for solving	**Euler** equations	Ni, R.H.
Accelerated solution of the steady	**Euler** equations	Johnson, G.M
An adaptive multigrid method for the	**Euler** equation	Berger, M.J.
Comparison of the full potential and	**Euler** formulations for computing transonic airfoil flo	Flores, J.;
A multiple grid scheme for solving the	**Euler** equations	Ron-Ho, Ni
Unigrid projection method to solve the	**Euler** equations for steady transonic flow	Rizzi, A.; E

Automatic adaptive grid refinement for the	**Euler** equations	Berger, M.J.
accurate upwind solutions of the 2D steady	**Euler** equations by the use of a defect correction meth	Spekreijse,
factorization technique for the flux-split	**Euler** equations	Li, C.P.
implementation of a multigrid code for the	**Euler** equations	Jespersen, D
flow concepts of a multigrid solver for the	**Euler** equations	Merriam, M.L
ies for accelerating the convergence of the	**Euler** equations	Stubbs, R.M.
iple-grid solution of the three dimensional	**Euler** and Navier-Stokes equations	Johnson, G.M
ir par resolution numerique des equations d'	**Euler**	Koeck, C.
me for the efficient solution of the steady	**Euler** equations	Hemker, P.W.
mensional vortex flows past wings using the	**Euler** equations and a multiple-grid scheme	Koeck, C.; C
opments in multigrid methods for the steady	**Euler** equations	Jespersen, D
r splitting and Runge-Kutta methods for the	**Euler** equations	Turkel, E.;
ralisation of Hall's scheme for solving the	**Euler** equations for two-dimensional flows	Arthur, M.T.
s for the multi grid solution of the steady	**Euler** equations	Hemker, P.W.
tigrid method for Cauchy-Riemann and steady	**Euler** equations based on flux-difference splitting	Dick, E.
unge-Kutta time integration for solving the	**Euler** and Navier-Stokes equations	Wagner, B.;
utation of three-dimensional flow using the	**Euler** equations and a multiple-grid scheme	Koeck, C.
An iterative method for	**evaluating** the first eigenvalue of an elliptic operato	Strakhovskay
r of a class of two-level methods; an exact	**evaluation** for some finite element subspaces and model	Maitre, J.F.
	Evolution of computational methods in aerodynamics	Jameson, A.
number of a class of two-level methods; an	**exact** evaluation for some finite element subspaces and	Maitre, J.F.
Steady state reactive kinetics on surfaces	**exhibiting** defect structures	Grinstein, F
Multigrid	**experience** with the neutron diffusion equation	Painter, J.W
Algebraic multigrid (AMG):	**experiences** and comparisons	Stüben, K.
omputation of anomalous modes in the Taylor	**experiment**	Bolstad, J.H
Numerical and	**experimental** study of driven flow in a polar cavity	Fuchs, L.; T
el multigrid solvers for elliptic PDE's: an	**experimental** analysis	Gannon, D.B.
COMFLO- ein	**Experimentierprogramm** zur Mehrgitterbehandlung subsoni	Becker, K.
	Experiments with a vectorized multigrid Poisson solver	Lemke, M.
	Experiments with multigrid methods on the Caltech Hype	Thole, C.-A.
Numerical	**experiments** with a multiple grid and a preconditioned	Wesseling, P
Numerical	**experiments** with multi-grid methods	Poling, T.C.
Solution of the Euler equations:	**Explicit** schemes acceleration by a multigrid method	Couaillier,
iscid and viscous flows in cascades with an	**explicit** multiple-grid algorithm	Chima, R.V.
	Extensions of the defect correction principle	Hemker, P.W.
ems combining full multigrid techniques and	**extrapolation** methods	Schüller, A.
hod for variational difference schemes with	**extrapolation**	Astrakhantse
korrekturen nach Stetter und Pereyra und MG-	**Extrapolation** nach Brandt: Beziehungen und Anwendung a	Donovang, M.
Fast adaptive composite grid (	**FAC**) methods: Theory for the variational case	McCormick, S
The fast adaptive composite grid method (	**FAC**) for elliptic boundary value problems	McCormick, S
The fast adaptive composite grid method (	**FAC**): Algorithms for advanced computers	Hart, L.; Mc
rilles: operateur associe et estimations du	**facteur** de convergence; le cas du V-cycle	Maitre, J.F.
Incomplete block LU-	**factorisation**	Meijerink, J
A multigrid	**factorization** technique for the flux-split Euler equat	Li, C.P.
Incomplete matrix	**factorizations** as multigrid smoothers for vector and p	Axelsson, O.
multi-level methods for strongly nonuniform	**families** of grids and any number of smoothing steps pe	Yserentant,
	FAS multigrid employing ILU/SIP smoothing: A robust fa	Wees, A.J. v
Application of a single equation MG-	**FAS** solver to elliptic grid generation equations (sub-	Roache, P.J.
sonic potential flow based on the nonlinear	**FAS** multi-grid method and incomplete LU decomposition	Wees, A.J. v
	Fast adaptive composite grid (FAC) methods: Theory for	McCormick, S
	Fast and accurate solution of time-dependent incompres	Karlsson, A.
	Fast iterative solution of Poisson equation with Neuma	Rieger, H.;
	Fast methods for the numerical solution of boundary-va	Dinar, N.

	Fast multigrid methods and applications - a short surv	Becker, K.;
A	**fast** iterative method for solving Helmholtz's equation	Hackbusch, W
A	**fast** iterative method for solving Poisson's equation i	Hackbusch, W
A	**fast** numerical method for elliptic boundary value prob	Hackbusch, W
A	**fast** solver for nonlinear eigenvalue problems	Mittelmann,
A	**fast** solver for the Stokes equations using multigrid w	Maitre, J.F.
A	**fast**-solver algorithm for steady transonic potential-f	Boerstoel, J
The	**fast** adaptive composite grid method (FAC) for elliptic	McCormick, S
The	**fast** adaptive composite grid method (FAC): Algorithms	Hart, L.; Mc
The	**fast** adaptive composite grid method for solving differ	Ewing, R.; M
The	**fast** numerical solution of time periodic parabolic pro	Hackbusch, W
The	**fast** numerical solution of very large elliptic differe	Hackbusch, W
On the	**fast** solution of parabolic boundary control problems	Hackbusch, W
On the	**fast** solutions of nonlinear elliptic equations	Hackbusch, W
On the	**fast** solving of elliptic control problems	Hackbusch, W
Comparison of	**fast** iterative methods for symmetric systems	Behie, A.; F
On the construction of	**fast** solvers for elliptic equations	Stüben, K.;
Numerical modelling and	**fast**-solver calculation of approximately normal shocks	Boerstoel, J
Numerical stability and	**fast** solutions to boundary value problems	Brandt, A.
Multigrid methods: development of	**fast** solvers	Hemker, P.W.
Multi-level adaptive technique (MLAT) for	**fast** numerical solution to boundary value problems	Brandt, A.
d point microprocessor applications and for	**fast** implementation of multigrid techniques	McCormick, S
egrating multigrid relaxation into a robust	**fast**-solver for transonic potential flows around lifti	Boerstoel, J
igrid employing ILU/SIP smoothing: A robust	**fast** solver for 3D transonic potential flow	Wees, A.J. v
nonlinear heat conduction problem by use of	**fast** reduction and multigrid methods	Solchenbach,
	Fedorenkos method for variational difference schemes w	Astrakhantse
	Field panel/finite difference method for potential uns	Hounjet, M.H
Une methode d'elements	**finis** multigrille pour le calcul d'ecoulements potenti	Bredif, M.
nnes pour appliquer la methode des elements	**finis**	Charron, M.S
	Finite-difference methods for plane steady inviscid tr	Fuchs, L.
On	**finite** element multigrid algorithms and their use	Nicolaides,
Multigrid	**finite** element method	Barrett, K.E
A multigrid	**finite** element method for the transonic potential equa	Deconinck, H
Field panel/	**finite** difference method for potential unsteady transo	Hounjet, M.H
A mixed variable	**finite** element method for the efficient solution of no	Axelsson, O.
Rapid solution of	**finite** element equations on locally refined grids by m	Rosendale, J
Multi-level adaptive	**finite**-element methods. I. Variational problems	Brandt, A.
Efficient solution of	**finite** difference and finite element equations by alge	Ruge, J.; St
Applications of a multigrid	**finite** volume method with Runge-Kutta time integration	Wagner, B.;
Numerical solution of mixed	**finite** element problems	Verfürth, R.
Mehrgitterverfahren für eine	**Finite** Element Approximation des Stokes-Problems	Untiet, K.-H
On the solution of nonlinear	**finite** element systems	Mansfield, L
On the multi-grid solution of	**finite** element equations with isoparametric elements	Mansfield, L
A multigrid version of a simple	**finite** element method for the Stokes problem	Pitkäranta,
On the multi-level splitting of	**finite** element spaces for indefinite elliptic boundary	Yserentant,
A multigrid solver for a stabilized	**finite** element discretization of the Stokes problem	Abdalass, E.
Eigensystem analysis techniques for	**finite**-difference equations. I. Multigrid techniques	Lomax, H.; P
An optimal order process for solving	**finite** element equations	Bank, R.E.;
Analysis and multigrid solutions of mixed	**finite** element and mixed difference equations	Hackbusch, W
A multigrid algorithm using a hierarchical	**finite** element basis	Craig, A.W.;
Analysis of a two-level scheme for solving	**finite** element equations	Bank, R.E.;
multi-level iterative method for nonlinear	**finite** element equations	Bank, R.E.;
**2 convergence of an algorithm for solving	**finite** element equation	Nicolaides,

ata structure of fully adaptive, multigrid,	**finite** element software	Rivara, M.C.
ation of the multigrid method "MLAT" in the	**finite** element method	Tal-Nir, A.;
ds for nonsymmetric and indefinite elliptic	**finite** element equations	Bank, R.E.
Efficient solution of finite difference and	**finite** element equations by algebraic multigrid (AMG)	Ruge, J.; St
hmic aspects of the multi-level solution of	**finite** element equations	Bank, R.E.;
ic full potential equation discretized with	**finite** elements on an arbitrary body fitted mesh	Deconinck, H
level methods; an exact evaluation for some	**finite** element subspaces and model problems	Maitre, J.F.
merical solutions of the Euler equations by	**finite** volume methods using Runge-Kutta time-stepping	Jameson, A.;
nt and multigrid methods using hierarchical	**finite** element bases	Craig, A.W.;
ransonic flow calculations using triangular	**finite** elements	Pelz, R.B.;
rgence of a multi-grid iteration applied to	**finite** element equations	Hackbusch, W
vergence of multi-level methods for solving	**finite** element equations in the presence of singularit	Yserentant,
visual surfaces: Variational principles and	**finite** element representation	Terzopoulos,
An iterative method for evaluating the	**first** eigenvalue of an elliptic operator	Strakhovskay
Efficient generation of body-	**fitted** coordinates for cascades using multigrid	Camarero, R.
d with finite elements on an arbitrary body	**fitted** mesh	Deconinck, H
id method to Poisson's equation in boundary-	**fitted** coordinates	Ohring, S.
Multigrid scheme for three-dimensional body-	**fitted** coordinates in turbomachine applications	Camarero, R.
Numerical software for	**fixed** point microprocessor applications and for fast i	McCormick, S
Hele-Shaw	**flow** with a cusping free boundary	Meyer, G.H.
Transonic	**flow** calculations for aircraft	Jameson, A.
Transonic	**flow** calculations using triangular finite elements	Pelz, R.B.;
Transonic	**flow** computation by a multi-grid method	Fuchs, L.
Application of data	**flow** concepts of a multigrid solver for the Euler equa	Merriam, M.L
Axisymmetric transonic	**flow** computations using a multigrid method	Arlinger, B.
Viscous transonic airfoil	**flow** simulation	Longo, J.M.;
An embedded-mesh potential	**flow** analysis	Brown, J.L.
Numerical simulation of the	**flow** in 3D-cylindrical combustion chambers using multi	Solchenbach,
Computation of three-dimensional	**flow** using the Euler equations and a multiple-grid sch	Koeck, C.
Multi-grid solutions to elliptic	**flow** problems	Brandt, A.;
Calculation of transonic potential	**flow** past wing-tail-fuselage combinations using multig	Shmilovich,
Acceleration of transonic potential	**flow** calculations on arbitrary meshes by the multiple	Jameson, A.
Convergence acceleration of viscous	**flow** computations	Johnson, G.M
Multigrid scheme for thermohydraulic	**flow**	Lacroix, M.;
Threedimensional transonic potential	**flow** about complex 3-dimensional configurations	Reyhner, T.A
Multigrid algorithms for compressible	**flow** calculations	Jameson, A.
New Relaxation methods for incompressible	**flow** problems	Fuchs, L.
Three-dimensional treatment of convective	**flow** in the earth's mantle	Baumgartner,
Multiple grid methods for oscillating disk	**flow**	Schippers, H
Numerical and experimental study of driven	**flow** in a polar cavity	Fuchs, L.; T
implicit procedure for transonic potential	**flow** problems	Sankar, N.L.
solve the integral equations of 3-D Stokes	**flow**	Hebeker, F.K
-Stokes solutions for laminar and turbulent	**flow**	Rubin, S.G.;
bust fast solver for 3D transonic potential	**flow**	Wees, A.J. v
cations of multi-grid methods for transonic	**flow** calculations	Schmidt, W.;
d solution of time-dependent incompressible	**flow**	Karlsson, A.
ds with applications to transonic potential	**flow**	Streett, C.L
e for three-dimensional transonic potential	**flow** analysis	Brown, J.L.
e solution of time-dependent incompressible	**flow**	Karlsson, A.
er algorithm for steady transonic potential-	**flow** computations with Newton iteration and multigrid	Boerstoel, J
grid technique applied to lifting transonic	**flow** using full potential equation	Arlinger, B.
id methods for the calculation of potential	**flow** around 3-d bodies	Wolff, H.
ler equations for two dimensional transonic	**flow** by a multigrid method	Jameson, A.

Vorschläge zur Frage der Entwicklung eines	**großen** MIMD-Multiprozessorsystems für numerische Anwen	Giloi, W.
	Guide to multigrid development	Brandt, A.
User's	**guide** for Twodant: A code package for two-dimensional,	Alcouffe, R.
PLTMG user's	**guide** - July 1979 version	Bank, R.E.;
PLTMG users'	**guide** - Edition 4.0	Bank, R.E.
Multigrid techniques: 1984	**guide** with applications to fluid dynamics	Brandt, A.
A generalisation of	**Hall's** scheme for solving the Euler equations for two-	Arthur, M.T.
Efficient solution of a nonlinear	**heat** conduction problem by use of fast reduction and m	Solchenbach,
	Hele-Shaw flow with a cusping free boundary	Meyer, G.H.
The numerical solution of the	**Helmholtz** equation for wave propagation problems in un	Bayliss, A.;
Mehrgitterverfahren zur Lösung der	**Helmholtz**-Gleichung im Rechteck mit Neumannschen Randb	Becker, K.
A fast iterative method for solving	**Helmholtz's** equation in a general region	Hackbusch, W
umerical solution of a nonlinear stochastic	**Helmholtz** equation with a multigrid preconditioner	Jordan, K.E.
-dimensional mass- and momentum- consistent	**Helmoltz**-equation in terrain-following coordinates	Schumann, U.
	Hierarchical bases give conjugate gradient-type method	Yserentant,
A multigrid algorithm using a	**hierarchical** finite element basis	Craig, A.W.;
mesh refinement and multigrid methods using	**hierarchical** finite element bases	Craig, A.W.;
	Higher order multi-grid methods	Schaffer, S.
On	**higher** order multigrid methods with application to a g	Gary, J.
Defect correction and	**higher** order schemes for the multi grid solution of th	Hemker, P.W.
Defect corrections and	**higher** numerical accuracy	Fuchs, L.
axation schemes in multigrid algorithms for	**higher** order singularity methods	Oskam, B.; F
for problems with a small parameter in the	**highest** derivative	Hemker, P.W.
	Highly parallel multigrid solvers for elliptic PDE's:	Gannon, D.B.
Multigrid methods for	**highly** oscillatory problems	Ta'asan, S.
Multi-grid methods for	**highly** oscillatory problems	Brandt, A.;
On the structure of parallelism in a	**highly** concurrent PDE solver	Gannon, D.B.
rgittermethode zur Berechnung von digitalen	**Höhenmodellen** in der Photogrammetrie	Rüde, U.
Mehrgittermethoden: Diskretisierungen	**höherer** Ordnungen für Dirichlet-Standardaufgaben in Re	Alef, M.
Überlegungen zur Berechnung von	**Hohlraumresonatoren**	Steffen, B.
A multigrid scheme for the thermal-	**hydraulics** of a blocked channel	Lacroix, M.;
Multigrid semi-implicit	**hydrodynamics** revisited	Dendy, J.E.
The multigrid method for semi-implicit	**hydrodynamics** codes	Brandt, A.;
Mathematical software for	**hyperbolic** equations and two point boundary value prob	Keller, H.B.
Multigrid algorithms on the	**Hypercube** multiprocessor	Chan, T.F.;
iptic partial differential equations on the	**Hypercube** multiprocessor	Chan, T.F.;
ments with multigrid methods on the Caltech	**Hypercube**	Thole, C.-A.
Multi-grid and	**ICCG** for problems with interfaces	Dendy, J.E.
Comparison of the multigrid and	**ICCG** method in solving diffusion equation	Voytko, M.H.
A study of some multigrid	**ideas**	Kamowitz, D.
(MLAT) for partial differential equations:	**ideas** and software	Brandt, A.
Lösung partieller Differentialgleichungen -	**Idee** und Bedeutung des Mehrgitterprinzips	Trottenberg,
FAS multigrid employing	**ILU**/SIP smoothing: A robust fast solver for 3D transon	Wees, A.J. v
Praktische und theoretische Aspekte der	**ILU**-Glättung bei Mehrgitterverfahren	Oertel, K.-D
On the comparison of Line-Gauss-Seidel and	**ILU** relaxation in multigrid algorithms	Hemker, P.W.
ieranalyse von Mehrgittermethoden: V-Cycle,	**ILU**-Glättung, anisotrope Operatoren	Thole, C.-A.
Multigrid	**image** reconstruction	Herman, G.T.
	Implementation of multigrid in SOLA algorithm	Brockmeier,
	Implementation of the multi-grid method for solving pa	Gustavson, F
	Implementation of the multigrid method "MLAT" in the f	Tal-Nir, A.;
Design and	**implementation** of a multigrid code for the Euler equat	Jespersen, D
Vectorization and	**implementation** of an efficient multigrid algorithm for	DeVore, C.R.
nt microprocessor applications and for fast	**implementation** of multigrid techniques	McCormick, S

Some	**implementations** of multigrid linear system solvers	Hemker, P.W.
MGOO-	**Implementierungen** auf EGPA-Multiprozessorsystemen	Geus, L.; He
Block	**implicit** multigrid calculation of three dimensional re	Vanka, S.P.
Block-	**implicit** multigrid solution of Navier-Stokes equations	Vanka, S.P.;
Multigrid semi-	**implicit** hydrodynamics revisited	Dendy, J.E.
A multigrid strongly	**implicit** procedure for transonic potential flow proble	Sankar, N.L.
The multigrid method for semi-	**implicit** hydrodynamics codes	Brandt, A.;
f asymptotic curved-duct flows using a semi-	**implicit** numerical technique	Ghia, K.N.;
igrid solution of the Euler equations using	**implicit** schemes	Jameson, A.;
Navier-Stokes equations by coupled strongly-	**implicit** multi-grid method	Ghia, U.; Gh
ple gridding of the Euler equations with an	**implicit** scheme	Stubbs, R.M.
	Improved spectral multigrid methods for periodic ellip	Brandt, A.;
An	**improved** time marching method for turbomachinery flow	Denton, J.D.
	Incomplete block LU-factorisation	Meijerink, J
	Incomplete matrix factorizations as multigrid smoother	Axelsson, O.
The	**incomplete** LU-decomposition as a relaxation method in	Hemker, P.W.
on the nonlinear FAS multi-grid method and	**incomplete** LU decomposition	Wees, A.J. v
	Incompressible Navier-Stokes and parabolized Navier-St	Rubin, S.G.
Solution of	**incompressible** Navier-Stokes equations by coupled stro	Ghia, U.; Gh
Multi-grid schemes for	**incompressible** flows	Fuchs, L.
New Relaxation methods for	**incompressible** flow problems	Fuchs, L.
Multigrid solution of time-dependent	**incompressible** flow	Karlsson, A.
Solution of three-dimensional viscous	**incompressible** flows by a multigrid method	Fuchs, L.; Z
A Lagrangian fractional step method for the	**incompressible** Navier-Stokes equations	Börgers, C.
ast and accurate solution of time-dependent	**incompressible** flow	Karlsson, A.
Multigrid computation of the pressure of an	**incompressible** fluid in a rotating spherical gap	Linden, J.;
thods for calculating the lifting potential	**incompressible** flows around three-dimensional bodies	Hackbusch, W
	Incremental multigrid strategy for the fluid dynamics	Napolitano,
On multi-grid convergence in the	**indefinite** case	Nicolaides,
Multigrid convergence for nonsymmetric,	**indefinite** variational problems and one smoothing step	Mandel, J.
evel iterative methods for nonsymmetric and	**indefinite** elliptic finite element equations	Bank, R.E.
evel splitting of finite element spaces for	**indefinite** elliptic boundary value problems	Yserentant,
gate gradient coupled with multigrid for an	**indefinite** problem	Gozani, J.
id methods for nearly singular and slightly	**indefinite** problems	Brandt, A.;
rgittermethoden und Reduktionsverfahren für	**indefinite** elliptische Randwertaufgaben	Thiebes, H.J
tion of the multigrid method to a nonlinear	**indefinite** problem	Meis, T.; Le
On multi-grid methods for variational	**inequalities**	Hackbusch, W
two-level algorithm in the solution of the	**initial** value problem for partial differential equatio	Vries, H.B.
rse grid corrections in the solution of the	**initial** value problem for nonlinear partial differenti	Houwen, P.J.
armonischen Gleichung und Anwendung auf ein	**inkompressibles** Strömungsproblem	Linden, J.
-dimensional transonic potential flow about	**inlets**	McCarthy, D.
rid techniek voor het oplossen van fredholm-	**integraalvergelijkingen** van de tweede soort	Wolff, H.
Multigrid methods for	**integral** and differential equations	Paddon, D.J.
Mesh refinement methods for	**integral** equations	McCormick, S
Multigrid methods for boundary	**integral** equations	Schippers, H
On a multigrid method to solve the	**integral** equations of 3-D Stokes flow	Hebeker, F.K
The automatic solution of Fredholm	**integral** equations of the second kind	Schippers, H
On multilevel iterative methods for	**integral** equations of the second kind and related prob	Mandel, J.
Application of multigrid methods for	**integral** equations to two problems from fluid dynamics	Schippers, H
e grid methods for the solution of Fredholm	**integral** equations of the second kind	Hemker, P.W.
inear and nonlinear eigenvalue problems for	**integral** and differential equations	Hackbusch, W
rid techniques for the solution of Fredholm	**integral** equations of the second kind	Schippers, H
Die schnelle Auflösung der Fredholmschen	**Integralgleichung** zweiter Art	Hackbusch, W

A fast	**iterative** method for solving Poisson's equation in a g	Hackbusch, W
A multi-level	**iterative** method for nonlinear elliptic equations	Bank, R.E.
A multi-level	**iterative** method for symmetric, positive definite line	Mandel, J.
On multilevel	**iterative** methods for integral equations of the second	Mandel, J.
On multigrid and	**iterative** aggregation methods for nonsymmetric problem	Mandel, J.
On the choice of	**iterative** parameters in the relaxation method on a seq	Langer, U.
Convergence of an	**iterative** method for the system Ax + y = x using aggre	Mandel, J.
On some two-level	**iterative** methods	Mandel, J.
Comparison of fast	**iterative** methods for symmetric systems	Behie, A.; F
Analysis of a multi-level	**iterative** method for nonlinear finite element equation	Bank, R.E.;
Parallel architectures for	**iterative** methods on adaptive block-structured grids	Gannon, D.B.
Multigrid acceleration of an	**iterative** method with applications to transonic potent	Nowak, Z.P.;
A convergence analysis of the	**iterative** aggregation method with one parameter	Mandel, J.
The speed of convergence of an	**iterative** process	Fedorenko, R
A comparison of two multi-level	**iterative** methods for nonsymmetric and indefinite elli	Bank, R.E.
A local convergence proof for the	**iterative** aggregation method	Mandel, J.;
Analysis of a multigrid method as an	**iterative** technique for solving linear systems	Greenbaum, A
On the observed rate of convergence of an	**iterative** method applied to a model elliptic differenc	Nicolaides,
her Randwertaufgaben mit approximativen und	**iterativen** Reduktionsverfahren	Ries, M.
Ein	**iteratives** Verfahren zur schnellen Auflösung elliptisc	Hackbusch, W
Bemerkungen zur	**iterierten** Defektkorrektur und zu ihrer Kombination mi	Hackbusch, W
Computation of flows around a	**Karman**-Trefftz profile	Mol, W.J.A.
Multigrid methods of the second	**kind**	Hackbusch, W
ethods for integral equations of the second	**kind** and related problems	Mandel, J.
f Fredholm integral equations of the second	**kind**	Hemker, P.W.
f Fredholm integral equations of the second	**kind**	Schippers, H
f Fredholm integral equations of the second	**kind**	Schippers, H
he nonlinear multigrid method of the second	**kind**	Hackbusch, W
le grid methods for equations of the second	**kind** with applications in fluid mechanics	Schippers, H
ns by the multigrid iteration of the second	**kind**	Hackbusch, W
Steady state reactive	**kinetics** on surfaces exhibiting defect structures	Grinstein, F
Ein Mehrgitter/Gleitkomma-Zusatz für den	**Knotenprozessor** eines Multiprozessors	Bode, A.
ng von Ableitungen erster Ordnung, variable	**Koeffizienten**	Stephany, P.
zur iterierten Defektkorrektur und zu ihrer	**Kombination** mit Mehrgitterverfahren	Hackbusch, W
as Problem einer ebenen, reibungsfreien und	**kompressiblen** Unterschallströmung am Beispiel des Krei	Schüller, A.
Fourieranalyse zur	**Konstruktion** schneller MGR-Verfahren	Winter, G.
Kritische Betrachtungen und	**konstruktive** Vorschläge zur Frage der Entwicklung eine	Giloi, W.
Ein	**Konzept** zur Nutzung eng gekoppelter Mehrrechnersysteme	Mierendorff,
Zur SUPRENUM-	**Konzeption**	Trottenberg,
eis und Ringgebiet unter Verwendung lokaler	**Koordinaten**	Linden, J.
itterverfahren für die Poisson-Gleichung in	**Kreis** und Ringgebiet unter Verwendung lokaler Koordina	Linden, J.
ssiblen Unterschallströmung am Beispiel des	**Kreisprofils**	Schüller, A.
	Kritische Betrachtungen und konstruktive Vorschläge zu	Giloi, W.
Flux-vector splitting and Runge-	**Kutta** methods for the Euler equations	Turkel, E.;
a multigrid finite volume method with Runge-	**Kutta** time integration for solving the Euler and Navie	Wagner, B.;
ations by finite volume methods using Runge-	**Kutta** time-stepping schemes	Jameson, A.;
A	**Lagrangian** fractional step method for the incompressib	Börgers, C.
The prediction of compressible	**laminar** viscous flows using a time marching control-vo	Davis, R.L.
, and the use of a multigrid technique, for	**laminar** source-sink flow between corotating discs	Lonsdale, G.
bal parabolized Navier-Stokes solutions for	**laminar** and turbulent flow	Rubin, S.G.;
s with a multiple grid and a preconditioned	**Lanczos** method	Wesseling, P
	Language for processes of numerical solutions to diffe	Brandt, A.;
	Language for processes of numerical solutions to diffe	Ophir, D.

Use of the multigrid method for	**Laplacian** problem in three dimensions	Nowak, Z.P.
Techniques for	**large** sparse systems arising from continuation methods	Chan, T.F.
The fast numerical solution of very	**large** elliptic difference schemes	Hackbusch, W
Multiple grid acceleration of	**Lax** Wendroff algorithm	Johnson, G.M
Multiple-grid acceleration of	**Lax**-Wendroff algorithms	Johnson, G.M
larity of the principal value of the double	**layer** potential	Schippers, H
	Lecture notes of a seminar on multiple grid methods	Hemker, P.W.
Multi-	**level** adaptive computations in fluid dynamics	Brandt, A.
Introduction -	**levels** and scales	Brandt, A.
Algol 68 multigrid	**library**	Hemker, P.W.
Multigrid technique applied to	**lifting** transonic flow using full potential equation	Arlinger, B.
Multigrid methods for calculating the	**lifting** potential incompressible flows around three-di	Hackbusch, W
solver for transonic potential flows around	**lifting** airfoils	Boerstoel, J
tigrid method for determining the transonic	**lifting** flows around airfoils	Nowak, Z.P.
On the comparison of	**Line**-Gauss-Seidel and ILU relaxation in multigrid algo	Hemker, P.W.
Multigrid analysis of	**linear** elastic stress problems	Barrett, K.E
Numerical solution of	**linear** and nonlinear parabolic control problems	Hackbusch, W
Multi-grid solutions to	**linear** and nonlinear eigenvalue problems for integral	Hackbusch, W
Multi-grid methods for non-	**linear** boundary-value problems	Amini, S.
Some implementations of multigrid	**linear** system solvers	Hemker, P.W.
Multigrid algorithms for the solution of	**linear** complementarity problems arising from free boun	Brandt, A.;
chemes - part II: regularity estimates for	**linear** and nonlinear problems	Hackbusch, W
ethod as an iterative technique for solving	**linear** systems	Greenbaum, A
ive method for symmetric, positive definite	**linear** complementarity problems	Mandel, J.
n-multi-grid method for the solution of non-	**linear** partial differential equations	Fuchs, L.
t le V-cycle et le W-cycle; applications au	**lissage** de Gauss-Seidel	Musy, F.
	Local and multi-level parallel processing mill	Brandt, A.
	Local defect correction method and domain decompositio	Hackbusch, W
	Local mesh refinement multilevel techniques	Bai, D.; Bra
A	**local** convergence proof for the iterative aggregation	Mandel, J.;
On	**local** mode analysis in multi-grid methods	Reimers, M.
A priori	**local** grid refinement in the multigrid method	Dendy, J.E.
A-posteriori error estimates, adaptive	**local** mesh refinement, and multigrid iteration	Bank, R.E.
algorithms and data structures for regular	**local** mesh refinement	Bank, R.E.;
Efficient multigrid algorithms for	**locally** constrained parallel systems	Kolp, O.; Mi
pid solution of finite element equations on	**locally** refined grids by multi-level methods	Rosendale, J
MGR - ein blitzschneller elliptischer	**Löser**	Ries, M.; Tr
che Simulation auf der Basis superschneller	**Lösungsverfahren** I	Trottenberg,
che Simulation auf der Basis superschneller	**Lösungsverfahren** II	Trottenberg,
stems für numerische Anwendungen (schnelle	**Lösungsverfahren**)	Giloi, W.
	Lokal verfeinerte Mehrgitter-Methoden für Gebiete mit	Ritzdorf, H.
ng in Kreis und Ringgebiet unter Verwendung	**lokaler** Koordinaten	Linden, J.
Algorithm 621. Software with	**low** storage requirements for two-dimensional, nonlinea	Sommeijer, B
The incomplete	**LU**-decomposition as a relaxation method in multi-grid	Hemker, P.W.
Incomplete block	**LU**-factorisation	Meijerink, J
linear FAS multi-grid method and incomplete	**LU** decomposition	Wees, A.J. v
	L3AMG - an algebraic multigrid routine	Dorr, M.R.
	Magnetohydrodynamic equilibrium calculations using mul	Braams, B.J.
treatment of convective flow in the earth's	**mantle**	Baumgartner,
	Marching multigrid solutions to the parabolized Navier	Israeli, M.;
An improved time	**marching** method for turbomachinery flow calculation	Denton, J.D.
ressible laminar viscous flows using a time	**marching** control-volume and multi grid technique	Davis, R.L.
Three-dimensional	**mass**- and momentum- consistent Helmoltz-equation in te	Schumann, U.

ntraction number of a multigrid method with	**mesh** ratio 2 for solving Poisson's equation	Verfürth, R.
thods and developments of three dimensional	**mesh** generation procedures	Steger, J.L.
ic potential flow calculations on arbitrary	**meshes** by the multiple grid method	Jameson, A.
s in the relaxation method on a sequence of	**meshes**	Langer, U.
Une	**methode** d'elements finis multigrille pour le calcul d'	Bredif, M.
Mehrskalen-	**Methoden** der Statistischen Physik: Ausgangspunkte zu M	Hahn, H.
	MG techniques for staggered differences	Favini, B.;
	MGR (Upsilon) multigrid methods	Kamowitz, D.
	MGR - ein blitzschneller elliptischer Löser	Ries, M.; Tr
A note on	**MGR** methods	Ries, M.; Tr
Fourieranalyse zur Konstruktion schneller	**MGR**-Verfahren	Winter, G.
	MG00-Implementierungen auf EGPA-Multiprozessorsystemen	Geus, L.; He
f elliptic problems on rectangular domains:	**MG00** (Release 1)	Foerster, H.
	MG01: A multi-grid program to solve Delta U - c(x,y)U	Stüben, K.
Eine Erweiterung des Mehrgitterprogramms	**MG01**: Einbeziehung von Ableitungen erster Ordnung, var	Stephany, P.
weiterungsmöglichkeiten des Programmpaketes	**MG01** auf nichtlineare Aufgaben	Schwichtenbe
Numerical software for fixed point	**microprocessor** applications and for fast implementatio	McCormick, S
Local and multi-level parallel processing	**mill**	Brandt, A.
läge zur Frage der Entwicklung eines großen	**MIMD**-Multiprozessorsystems für numerische Anwendungen	Giloi, W.
ür anisotrope 3D-Aufgaben: Anforderungen an	**MIMD**-Rechnerarchitekturen	Thole, C.-A.
	Mixed defect correction iteration for the accurate sol	Hemker, P.W.
	Mixed defect correction iteration for the solution of	Hemker, P.W.
A	**mixed** variable finite element method for the efficient	Axelsson, O.
Two algorithms for	**mixed** problems	Verfürth, R.
Numerical solution of	**mixed** finite element problems	Verfürth, R.
A multilevel algorithm for	**mixed** problems	Verfürth, R.
Analysis and multigrid solutions of	**mixed** finite element and mixed difference equations	Hackbusch, W
igrid solutions of mixed finite element and	**mixed** difference equations	Hackbusch, W
Multi-level adaptive technique (	**MLAT**) for fast numerical solution to boundary value pr	Brandt, A.
Multi-level adaptive techniques (	**MLAT**) for partial differential equations: ideas and so	Brandt, A.
Multi-level adaptive techniques (	**MLAT**) for singular-perturbation problems	Brandt, A.
Multi-level adaptive techniques (	**MLAT**). I. The multi-grid method	Brandt, A.
Implementation of the multigrid method "	**MLAT**" in the finite element method	Tal-Nir, A.;
theory for multi-level adaptive techniques (	**MLAT**)	McCormick, S
Multigrid methods: Fundamental algorithms,	**model** problem analysis and applications	Stüben, K.;
with application to a geothermal reservoir	**model**	Gary, J.
ation for some finite element subspaces and	**model** problems	Maitre, J.F.
ergence of an iterative method applied to a	**model** elliptic difference equation	Nicolaides,
hod to a two dimensional nonlinear elliptic	**model** equation	Asselt, E.J.
methods for certifying the accuracy of PDE	**modeling**	Forester, C.
d numerical aspects of semiconductor device	**modeling**	Bank, R.E.;
n of the diffusion equation in VLSI process	**modeling**	Seidl, A.
	Modellierung von parallelen Mehrgitteralgorithmen und	Mühlenbein,
Numerical	**modelling** and fast-solver calculation of approximately	Boerstoel, J
r ein zweidimensionales, singulär gestörtes	**Modellproblem**: Grobgitter- und Glättungsoperatoren	Debus, B.
An efficient	**modular** algorithm for coupled nonlinear systems	Chan, T.F.
rtable vector-code for autonomous multigrid	**modules**	Hemker, P.W.
Three-dimensional mass- and	**momentum**- consistent Helmoltz-equation in terrain-foll	Schumann, U.
der Strömungs- und Verbrennungsvorgänge im	**Motorzylinder**	Ruttmann, B.
grid method for solving the two-dimensional	**multigroup** diffusion equation	Alcouffe, R.
	Multiprocessor-Architekturkonzept für Mehrgitterverfah	Hofmann, F.;
Multigrid methods for	**multiprocessor** computers	McCormick, S
Multigrid algorithms on the Hypercube	**multiprocessor**	Chan, T.F.;

n method for elliptic problems with turning	**points**	Bolstad, J.H
Vectorized multigrid	**poisson** solver for the CDC Cyber 205	Barkai, D.;
Performance analysis of	**Poisson** solvers on array computers	Grosch, C.E.
Fast iterative solution of	**Poisson** equation with Neumann boundary conditions in n	Rieger, H.;
Mehrgitterverfahren für die	**Poisson**-Gleichung in Kreis und Ringgebiet unter Verwen	Linden, J.
A fast iterative method for solving	**Poisson**'s equation in a general region	Hackbusch, W
Application of the multigrid method to	**Poisson**'s equation in boundary-fitted coordinates	Ohring, S.
Experiments with a vectorized multigrid	**Poisson** solver on the CDC Cyber 205, Cray X-MP and Fuj	Lemke, M.
High order multi-grid methods to solve the	**Poisson** equation	Schaffer, S.
multigrid solver for the three-dimensional	**Poisson** equation	Holter, W.H.
analysis of multigrid techniques applied to	**Poisson**'s equation in three dimensions	Merriam, M.L
method with Gauss-Seidel relaxation for the	**Poisson** equation	Braess, D.
method with Gauss-Seidel relaxation for the	**Poisson** equation (revised)	Braess, D.
tigrid method with mesh ratio 2 for solving	**Poisson**'s equation	Verfürth, R.
umber of a multigrid method for solving the	**Poisson** equation	Braess, D.
ehrgitterverfahren für die dreidimensionale	**Poissongleichung**	Bannasch, F.
ren für eine Mehrstellendiskretisierung der	**Poissongleichung** und für eine zweidimensionale singulä	Börgers, C.
and experimental study of driven flow in a	**polar** cavity	Fuchs, L.; T
and step technique for atmospheric chemical	**pollutant** transport and diffusion	Lee, H.N.
Berechnung von stationären Strömungen durch	**poröse** Medien mit freiem Rand	Bollrath, C.
A	**portable** vector-code for autonomous multigrid modules	Hemker, P.W.
f the multigrid method in the symmetric and	**positive** definite case - a convergence estimation for	Maitre, J.F.
multi-level iterative method for symmetric,	**positive** definite linear complementarity problems	Mandel, J.
A-	**posteriori** error estimates, adaptive local mesh refine	Bank, R.E.
A-	**posteriori** error estimation, adaptive mesh refinement	Craig, A.W.;
Some a	**posteriori** error estimators for elliptic partial diffe	Bank, R.E.;
A multigrid full	**potential** transonic code for arbitrary configurations	Luntz, A.L.;
An embedded-mesh	**potential** flow analysis	Brown, J.L.
Comparison of the full	**potential** and Euler formulations for computing transon	Flores, J.;
Calculation of transonic	**potential** flow past wing-tail-fuselage combinations us	Shmilovich,
Acceleration of transonic	**potential** flow calculations on arbitrary meshes by the	Jameson, A.
Threedimensional transonic	**potential** flow about complex 3-dimensional configurati	Reyhner, T.A
Robust calculation of 3d transonic	**potential** flow based on the nonlinear FAS multi-grid m	Wees, A.J. v
Field panel/finite difference method for	**potential** unsteady transonic flow	Hounjet, M.H
A multigrid method for the transonic full	**potential** equation discretized with finite elements on	Deconinck, H
A multigrid algorithm for steady transonic	**potential** flows around aerofoils using Newton iteratio	Boerstoel, J
calculation of three-dimensional transonic	**potential** flows	Caughey, D.A
fast-solver algorithm for steady transonic	**potential**-flow computations with Newton iteration and	Boerstoel, J
of the principal value of the double layer	**potential**	Schippers, H
ative method with applications to transonic	**potential** flow	Nowak, Z.P.;
d solver for two-dimensional transonic full	**potential** flow calculations	Becker, K.
d strongly implicit procedure for transonic	**potential** flow problems	Sankar, N.L.
ficient solution of nonlinear diffusion and	**potential** flow equations	Axelsson, O.
g technique for three-dimensional transonic	**potential** flow analysis	Brown, J.L.
grid methods with applications to transonic	**potential** flow	Streett, C.L
hing: A robust fast solver for 3D transonic	**potential** flow	Wees, A.J. v
id code for the three-dimensional transonic	**potential** flow about inlets	McCarthy, D.
ion into a robust fast-solver for transonic	**potential** flows around lifting airfoils	Boerstoel, J
ltigrid methods for calculating the lifting	**potential** incompressible flows around three-dimensiona	Hackbusch, W
pplied to lifting transonic flow using full	**potential** equation	Arlinger, B.
rid finite element method for the transonic	**potential** equation	Deconinck, H
ti-grid method to calculations of transonic	**potential** flow about wing-fuselage combinations	Shmilovich,
ultiple grid methods for the calculation of	**potential** flow around 3-d bodies	Wolff, H.

n Mehrgitterverfahren zur Lösung der vollen	**Potentialgleichung** im Falle transsonischer Strömungen	Becker, K.
gramm zur Mehrgitterbehandlung subsonischer	**Potentialströmungen** um Tragflächenprofile	Becker, K.
rgitterprogramm zur Berechnung subsonischer	**Potentialströmungen** um Tragflächenprofile	Becker, K.
is multigrille pour le calcul d'ecoulements	**potentiels** transsoniques	Bredif, M.
Theoretical and	**practical** aspects of a multigrid method	Wesseling, P
Theoretical and	**practical** aspects of multigrid methods in boundary ele	Schippers, H
On some theoretical and	**practical** aspects of multigrid methods	Nicolaides,
	Praktische und theoretische Aspekte der ILU-Glättung b	Oertel, K.-D
A	**preconditioned** conjugate residual algorithm for the St	Verfürth, R.
Successive overrelaxation, multigrid, and	**preconditioned** conjugate gradients algorithms for solv	Gary, J.; Mc
relaxation schemes in robust multigrid and	**preconditioned** conjugate gradient methods	Kettler, R.
ical experiments with a multiple grid and a	**preconditioned** Lanczos method	Wesseling, P
chastic Helmholtz equation with a multigrid	**preconditioner**	Jordan, K.E.
	Preconditioners for the spectral multigrid methods	Phillips, T.
	Preconditioning and coarse grid corrections in the sol	Houwen, P.J.
	Preconditioning and two-level multigrid methods of arb	Axelsson, O.
ation and the role of squaring matrices for	**preconditioning**	Braess, D.;
The	**prediction** of compressible laminar viscous flows using	Davis, R.L.
for solving finite element equations in the	**presence** of singularities	Yserentant,
Some iterative methods: past,	**present**, future	Jacobs, D.A.
The	**pressure** correction method, and the use of a multigrid	Lonsdale, G.
Multi-grid solution of the	**pressure** equation in reservoir simulation	Behie, A.; F
Multigrid computation of the	**pressure** of an incompressible fluid in a rotating sph	Linden, J.;
plication of the multigrid technique to the	**pressure**-correction equation for the SIMPLE algorithm	Pau, V.; Lew
grid solution of Navier-Stokes equations in	**primitive** variables	Vanka, S.P.;
On the regularity of the	**principal** value of the double layer potential	Schippers, H
Extensions of the defect correction	**principle**	Hemker, P.W.
onstruction of visual surfaces: Variational	**principles** and finite element representation	Terzopoulos,
A	**priori** local grid refinement in the multigrid method	Dendy, J.E.
Multigrid iteration solution	**procedure** for solving two-dimensional sets of coupled	Vondy, D.R.
A multigrid strongly implicit	**procedure** for transonic potential flow problems	Sankar, N.L.
Analysis of a multi-level inverse iteration	**procedure** for eigenvalue problems	Bank, R.E.
Basic smoothing	**procedures** for the multigrid treatment of elliptic 3D-	Thole, C.-A.
Navier-Stokes and parabolized Navier-Stokes	**procedures** and computational techniques	Rubin, S.G.
pments of three dimensional mesh generation	**procedures**	Steger, J.L.
An optimal order	**process** for solving finite element equations	Bank, R.E.;
The speed of convergence of an iterative	**process**	Fedorenko, R
solution of the diffusion equation in VLSI	**process** modeling	Seidl, A.
Language for	**processes** of numerical solutions to differential equat	Brandt, A.;
Language for	**processes** of numerical solutions to differential equat	Ophir, D.
A note on defect correction	**processes** with an approximate inverse of deficient ran	Hemker, P.W.
n overview with emphasis on grid generation	**processes**	Stüben, K.;
A signal	**processing** approach to the numerical solution of parab	Rabenstein,
Concurrent	**processing** adaptation of a multiple-grid algorithm	Johnson, G.M
Local and multi-level parallel	**processing** mill	Brandt, A.
The multigrid method on parallel	**processors**	McBryan, O.A
omputation of flows around a Karman-Trefftz	**profile**	Mol, W.J.A.
MG01: A multi-grid	**program** to solve Delta U - c(x,y)U = f(x,y) (on Omega)	Stüben, K.
Algol 68 Fourier analysis	**program**	Hemker, P.W.
PLTMGC: A multi-grid continuation	**program** for parameterized nonlinear elliptic systems	Bank, R.E.;
ridpack: Toward unification of general grid	**programming**	Brandt, A.;
Erweiterungsmöglichkeiten des	**Programmpaketes** MG01 auf nichtlineare Aufgaben	Schwichtenbe
Vectorization of computer	**programs** with applications to computational fluid dyna	Gentzsch, W.

A mesh	**refinement** method for A*x=Lambda*B*x	McCormick, S
Local mesh	**refinement** multilevel techniques	Bai, D.; Bra
A priori local grid	**refinement** in the multigrid method	Dendy, J.E.
Automatic adaptive grid	**refinement** for the Euler equations	Berger, M.J.
The use of adaptive grid	**refinement** for badly behaved elliptic partial differen	Bank, R.E.;
and data structures for regular local mesh	**refinement**	Bank, R.E.;
-posteriori error estimation, adaptive mesh	**refinement** and multigrid methods using hierarchical fi	Craig, A.W.;
eriori error estimates, adaptive local mesh	**refinement**, and multigrid iteration	Bank, R.E.
Algorithms for	**refining** triangular grids suitable for adaptive and mu	Rivara, M.C.
for solving Poisson's equation in a general	**region**	Hackbusch, W
r solving Helmholtz's equation in a general	**region**	Hackbusch, W
lliptic solvers for general two-dimensional	**regions**	Chan, T.F.;
finement algorithms and data structures for	**regular** local mesh refinement	Bank, R.E.;
On the	**regularity** of difference schemes	Hackbusch, W
On the	**regularity** of difference schemes - part II: regularity	Hackbusch, W
On the	**regularity** of the principal value of the double layer	Schippers, H
regularity of difference schemes - part II:	**regularity** estimates for linear and nonlinear problem	Hackbusch, W
ittermethoden auf das Problem einer ebenen,	**reibungsfreien** und kompressiblen Unterschallströmung a	Schüller, A.
On multiple grid and	**related** techniques for solving discrete elliptic syste	Nicolaides,
r integral equations of the second kind and	**related** problems	Mandel, J.
A	**relaxation** method for solving a biharmonic difference	Otroshcenko,
A	**relaxation** method for solving elliptic difference equa	Fedorenko, R
A	**relaxation** method in a sequence of grids for elliptic	Astrakhantse
New	**Relaxation** methods for incompressible flow problems	Fuchs, L.
General	**relaxation** schemes in multigrid algorithms for higher	Oskam, B.; F
Multigrid	**relaxation** for the Euler equations	Mulder, W.A.
Integrating multigrid	**relaxation** into a robust fast-solver for transonic pot	Boerstoel, J
A preliminary study of	**relaxation** methods for the inviscid conservative gasdy	Steger, J.L.
On the convergence of a	**relaxation** method with natural constraints on the elli	Bakhvalov, N
Analysis and comparison of	**relaxation** schemes in robust multigrid and preconditio	Kettler, R.
The incomplete LU-decomposition as a	**relaxation** method in multi-grid algorithms	Hemker, P.W.
andard multigrid techniques using checkered	**relaxation** and intermediate grids	Foerster, H.
ate of a multigrid method with Gauss-Seidel	**relaxation** for the Poisson equation	Braess, D.
ate of a multigrid method with Gauss-Seidel	**relaxation** for the Poisson equation (revised)	Braess, D.
n the choice of iterative parameters in the	**relaxation** method on a sequence of meshes	Langer, U.
tations with Newton iteration and multigrid	**relaxation**	Boerstoel, J
the comparison of Line-Gauss-Seidel and ILU	**relaxation** in multigrid algorithms	Hemker, P.W.
tigrid methods for the Euler equations: the	**relaxation** technique	Jespersen, D
Introductory	**remarks** on multigrid methods	Brandt, A.
: Variational principles and finite element	**representation**	Terzopoulos,
Algorithm 621. Software with low storage	**requirements** for two-dimensional, nonlinear, parabolic	Sommeijer, B
Multigrid methods for oil	**reservoir** simulation in three dimensions	Scott, T.
i-grid solution of the pressure equation in	**reservoir** simulation	Behie, A.; F
id methods with application to a geothermal	**reservoir** model	Gary, J.
rid methods and adaptive techniques for oil	**reservoir** simulation	McCormick, S
A preconditioned conjugate	**residual** algorithm for the Stokes problem	Verfürth, R.
l de l'ecoulement dans une entree d'air par	**resolution** numerique des equations d'Euler	Koeck, C.
nalysis of gridfunctions, prolongations and	**restrictions**	Hemker, P.W.
on-accelerated, Neutral-Particle Transport,	**Rev.** 1	Alcouffe, R.
ution of axisymmetric flows about bodies of	**revolution** using a multigrid method	Woan, C.J.
Analysis of	**Richardson** iteration in multigrid methods for nonlinea	Houwen, P.J.
A multigrid method for Cauchy-	**Riemann** and steady Euler equations based on flux-diffe	Dick, E.
hren für die Poisson-Gleichung in Kreis und	**Ringgebiet** unter Verwendung lokaler Koordinaten	Linden, J.

Mathematical	**software** for hyperbolic equations and two point bounda	Keller, H.B.
Algorithm 621.	**Software** with low storage requirements for two-dimensi	Sommeijer, B
On efficient multigrid	**software** for elliptic problems on rectangular domains	Foerster, H.
f fully adaptive, multigrid, finite element	**software**	Rivara, M.C.
oration of multigrid methods in accelerator	**software**	Steffen, B.
r partial differential equations: ideas and	**software**	Brandt, A.
Implementation of multigrid in	**SOLA** algorithm	Brockmeier,
f a coupled pair of elliptic equations from	**solid** state electronics	Phillips, T.
Multigrid	**solutions** of the steady Euler equations	Hemker, P.W.
On a multigrid method to	**solve** the integral equations of 3-D Stokes flow	Hebeker, F.K
Eigenvalue and near eigenvalue problems	**solved** by Brandt's multigrid method	Guderley, K.
A fast	**solver** for nonlinear eigenvalue problems	Mittelmann,
A fast	**solver** for the Stokes equations using multigrid with a	Maitre, J.F.
A fast-	**solver** algorithm for steady transonic potential-flow c	Boerstoel, J
A multigrid	**solver** for a stabilized finite element discretization	Abdalass, E.
A multigrid	**solver** for the in-cylinder turbulent flows in engines	Ruttmann, B.
A multigrid	**solver** for two-dimensional transonic full potential fl	Becker, K.
A vectorized multigrid	**solver** for the three-dimensional Poisson equation	Holter, W.H.
A multi-grid optimal order	**solver** for elliptic boundary-value problems	Douglas, C.C
Numerical modelling and fast-	**solver** calculation of approximately normal shocks	Boerstoel, J
Vectorized multigrid poisson	**solver** for the CDC Cyber 205	Barkai, D.;
Application of a single equation MG-FAS	**solver** to elliptic grid generation equations (sub-grid	Roache, P.J.
employing ILU/SIP smoothing: A robust fast	**solver** for 3D transonic potential flow	Wees, A.J. v
cation of data flow concepts of a multigrid	**solver** for the Euler equations	Merriam, M.L
e of parallelism in a highly concurrent PDE	**solver**	Gannon, D.B.
ing multigrid relaxation into a robust fast-	**solver** for transonic potential flows around lifting ai	Boerstoel, J
riments with a vectorized multigrid Poisson	**solver** on the CDC Cyber 205, Cray X-MP and Fujitsu VP	Lemke, M.
Multigrid	**solvers** for non-elliptic and singular-perturbation ste	Brandt, A.
Multigrid	**solvers** on parallel computers	Brandt, A.
A comparison of elliptic	**solvers** for general two-dimensional regions	Chan, T.F.;
Highly parallel multigrid	**solvers** for elliptic PDE's: an experimental analysis	Gannon, D.B.
On the construction of fast	**solvers** for elliptic equations	Stüben, K.;
Performance analysis of Poisson	**solvers** on array computers	Grosch, C.E.
Multigrid methods: development of fast	**solvers**	Hemker, P.W.
implementations of multigrid linear system	**solvers**	Hemker, P.W.
structures for adaptive multigrid elliptic	**solvers**	Rosendale, J
Analysis of a	**SOR**-like multi-grid algorithm for eigenvalue problems	Hofmann, G.
e use of a multigrid technique, for laminar	**source**-sink flow between corotating discs	Lonsdale, G.
the multi-level splitting of finite element	**spaces** for indefinite elliptic boundary value problems	Yserentant,
Techniques for large	**sparse** systems arising from continuation methods	Chan, T.F.
Algebraic multigrid (AMG) for	**sparse** matrix equations	Brandt, A.;
	Spectral methods for inviscid, compressible flows	Hussaini, M.
	Spectral multi-grid methods for elliptic equations	Zang, T.A.;
	Spectral multigrid methods for Dirichlet problems	Phillips, T.
	Spectral multigrid methods for elliptic equations II	Zang, T.A.;
	Spectral multigrid methods with applications to transo	Streett, C.L
Improved	**spectral** multigrid methods for periodic elliptic probl	Brandt, A.;
Iterative	**spectral** methods and spectral solutions to compressibl	Hussaini, M.
Preconditioners for the	**spectral** multigrid methods	Phillips, T.
Iterative spectral methods and	**spectral** solutions to compressible flows	Hussaini, M.
The	**speed** of convergence of an iterative process	Fedorenko, R
onjugate gradient-type methods: A multigrid	**speed** of convergence	Yserentant,
Ansatz	**spezieller** Mehrgitterkomponenten für ein zweidimension	Debus, B.

lver to elliptic grid generation equations (	**sub**-grid and super-grid coefficient generation)	Roache, P.J.
Ein Mehrgitterprogramm zur Berechnung	**subsonischer** Potentialströmungen um Tragflächenprofile	Becker, K.
erimentierprogramm zur Mehrgitterbehandlung	**subsonischer** Potentialströmungen um Tragflächenprofile	Becker, K.
Multigrid method in	**subspace** and domain partitioning in the discrete solut	Vassilevski,
an exact evaluation for some finite element	**subspaces** and model problems	Maitre, J.F.
Multigrid calculation of	**subvortices**	·Gustafson, K
	Successive overrelaxation, multigrid, and precondition	Gary, J.; Mc
tic grid generation equations (sub-grid and	**super**-grid coefficient generation)	Roache, P.J.
Multigrid algorithms run on	**supercomputers**	Hemker, P.W.
für die numerische Simulation auf der Basis	**superschneller** Lösungsverfahren I	Trottenberg,
für die numerische Simulation auf der Basis	**superschneller** Lösungsverfahren II	Trottenberg,
Zur	**SUPRENUM**-Konzeption	Trottenberg,
Steady state reactive kinetics on	**surfaces** exhibiting defect structures	Grinstein, F
Multi-level reconstruction of visual	**surfaces**: Variational principles and finite element re	Terzopoulos,
	Survey of convergence proofs for multigrid iterations	Hackbusch, W
A	**survey** of multi-grid methods for nonlinear problems	Asselt, E.J.
Algebraic multigrid (AMG) for geodetic	**survey** problems	Ruge, J.
ultigrid methods and applications - a short	**survey** and one result on a special nearly singular pro	Becker, K.;
Multigrid methods for	**symmetric** variational problems: A general theory and c	Maitre, J.F.
Algebraic multigrid theory: The	**symmetric** case	Brandt, A.
A multi-level iterative method for	**symmetric**, positive definite linear complementarity pr	Mandel, J.
Algebraic study of multigrid methods for	**symmetric**, definite problems	Mandel, J.
Comparison of fast iterative methods for	**symmetric** systems	Behie, A.; F
ormalisation of the multigrid method in the	**symmetric** and positive definite case - a convergence e	Maitre, J.F.
Solution of coupled	**system** of PDE by the transistorized multi-grid method	Shieh, A.S.L
Some implementations of multigrid linear	**system** solvers	Hemker, P.W.
Convergence of an iterative method for the	**system** Ax + y = x using aggregation	Mandel, J.
Transportleistung und Größe paralleler	**Systeme** bei speziellen Mehrgitteralgorithmen	Mierendorff,
für die Navier-Stokes-Gleichungen auf EGPA-	**Systemen**	Geus, L.
	Systems of nonlinear algebraic equations arising in si	Sheikh, Q.M.
Bus coupled	**systems** for multigrid algorithms	Kolp, O.; Mi
Black box multigrid for	**systems**	Dendy, J.E.
Techniques for large sparse	**systems** arising from continuation methods	Chan, T.F.
Efficient solution of elliptic	**systems**	Hackbusch, W
algorithms for locally constrained parallel	**systems**	Kolp, O.; Mi
ation and multi-grid for nonlinear elliptic	**systems**	Bank, R.E.;
ccelerated convergence of structured banded	**systems** using constrained corrections	Kneile, K.
ed techniques for solving discrete elliptic	**systems**	Nicolaides,
ent modular algorithm for coupled nonlinear	**systems**	Chan, T.F.
On the solution of nonlinear finite element	**systems**	Mansfield, L
ons in nonorthogonal curvilinear coordinate	**systems** by a multiple grid method	Rieger, H.;
rogram for parameterized nonlinear elliptic	**systems**	Bank, R.E.;
s an iterative technique for solving linear	**systems**	Greenbaum, A
son of fast iterative methods for symmetric	**systems**	Behie, A.; F
	Systemunabhängige Organisation von Mehrgitterverfahren	Kolp, O.; Mi
ation of transonic potential flow past wing-	**tail**-fuselage combinations using multigrid techniques	Shmilovich,
tionel: theorie generale et estimations sur	**taux** de convergence	Musy, F.
Computation of anomalous modes in the	**Taylor** experiment	Bolstad, J.H
dimensional sets of coupled equations (high	**temperature** gas cooled reactors)	Vondy, D.R.
Numerische Behandlung von	**Template**-Matching in der automatischen Bildverarbeitun	Verhuven, T.
	Termination strategies for Newton iteration in full mu	Asselt, E.J.
d momentum- consistent Helmoltz-equation in	**terrain**-following coordinates	Schumann, U.
Vektorisierung von Mehrgitterverfahren:	**Tests** auf einer Cray X-MP	Weidner, P.;

	Theoretical and practical aspects of a multigrid metho	Wesseling, P
	Theoretical and practical aspects of multigrid methods	Schippers, H
On some	**theoretical** and practical aspects of multigrid methods	Nicolaides,
Praktische und	**theoretische** Aspekte der ILU-Glättung bei Mehrgitterve	Oertel, K.-D
Multigrid methods: convergence	**theorie** in a variational framework	Maitre, J.F.
ultigrilles pour les problemes variationel:	**theorie** generale et estimations sur taux de convergenc	Musy, F.
An algebraic	**theory** for multigrid methods for variational problems	Mandel, J.;
A variational	**theory** for multi-level adaptive techniques (MLAT)	McCormick, S
Algebraic multigrid	**theory**: The symmetric case	Brandt, A.
Multi-grid convergence	**theory**	Hackbusch, W
d methods for variational problems: general	**theory** for the V-cycle	McCormick, S
Fast adaptive composite grid (FAC) methods:	**Theory** for the variational case	McCormick, S
r symmetric variational problems: A general	**theory** and convergence estimates for usual smoothers	Maitre, J.F.
A multigrid scheme for the	**thermal**-hydraulics of a blocked channel	Lacroix, M.;
Multigrid scheme for	**thermohydraulic** flow	Lacroix, M.;
	Three-dimensional mass- and momentum- consistent Helmo	Schumann, U.
	Three-dimensional treatment of convective flow in the	Baumgartner,
Analysis of	**three**-dimensional viscous internal flows	Ghia, K.N.;
Solution of	**three**-dimensional viscous incompressible flows by a mu	Fuchs, L.; Z
Computation of	**three** dimensional vortex flows past wings using the Eu	Koeck, C.; C
Computation of	**three**-dimensional flow using the Euler equations and a	Koeck, C.
Multigrid scheme for	**three**-dimensional body-fitted coordinates in turbomach	Camarero, R.
Multi-grid solution of	**three**-dimensional problems with discontinuous coeffici	Behie, A.; F
Numerical simulation of	**three**-dimensional flowfields using the Cyber 205	Swisshelm, J
A multigrid code for the	**three**-dimensional transonic potential flow about inlet	McCarthy, D.
Multi-grid calculation of	**three**-dimensional transonic potential flows	Caughey, D.A
Multigrid applications to	**three**-dimensional elliptic coordinate generation	Thames, F.C.
Multiple-grid solution of the	**three** dimensional Euler and Navier-Stokes equations	Johnson, G.M
A vectorized multigrid solver for the	**three**-dimensional Poisson equation	Holter, W.H.
Block implicit multigrid calculation of	**three** dimensional recirculating flows	Vanka, S.P.
A multigrid mesh-embedding technique for	**three**-dimensional transonic potential flow analysis	Brown, J.L.
e multigrid method for Laplacian problem in	**three** dimensions	Nowak, Z.P.
fting potential incompressible flows around	**three**-dimensional bodies	Hackbusch, W
ithms for zonal methods and developments of	**three** dimensional mesh generation procedures	Steger, J.L.
rid methods for oil reservoir simulation in	**three** dimensions	Scott, T.
spects of Multigrid Methods for Problems in	**Three** Dimensions	Kettler, R.;
techniques applied to Poisson's equation in	**three** dimensions	Merriam, M.L
	Threedimensional transonic potential flow about comple	Reyhner, T.A
A	**time**-accurate multiple-grid algorithm	Jespersen, D
On	**time** dependent multi-grid numerical techniques	Lee, H.N.; M
On the	**time** integration of parabolic differential equations	Houwen, P.J.
An improved	**time** marching method for turbomachinery flow calculati	Denton, J.D.
Multigrid solution of	**time**-dependent incompressible flow	Karlsson, A.
Fast and accurate solution of	**time**-dependent incompressible flow	Karlsson, A.
The fast numerical solution of	**time** periodic parabolic problems	Hackbusch, W
On special multigrid methods for the	**time**-dependent Navier-Stokes equations	Zang, T.A.;
The multigrid method in the solution of	**time**-dependent nonlinear partial differential equation	Vries, H.B.
by finite volume methods using Runge-Kutta	**time**-stepping schemes	Jameson, A.;
compressible laminar viscous flows using a	**time** marching control-volume and multi grid technique	Davis, R.L.
several MG-methods for the solution of the	**time**-dependent Navier-Stokes equations	Schröder, W.
d for fluid dynamic problems with disparate	**time** scales	Israeli, M.;
igrid finite volume method with Runge-Kutta	**time** integration for solving the Euler and Navier-Stok	Wagner, B.;
Gridpack:	**Toward** unification of general grid programming	Brandt, A.;

andlung subsonischer Potentialströmungen um	**Tragflächenprofile**	Becker, K.
echnung subsonischer Potentialströmungen um	**Tragflächenprofile**	Becker, K.
Solution of coupled system of PDE by the	**transistorized** multi-grid method	Shieh, A.S.L
	Transonic flow calculations for aircraft	Jameson, A.
	Transonic flow calculations using triangular finite el	Pelz, R.B.;
	Transonic flow computation by a multi-grid method	Fuchs, L.
Viscous	**transonic** airfoil flow simulation	Longo, J.M.;
Axisymmetric	**transonic** flow computations using a multigrid method	Arlinger, B.
Calculation of	**transonic** potential flow past wing-tail-fuselage combi	Shmilovich,
Acceleration of	**transonic** potential flow calculations on arbitrary mes	Jameson, A.
Calculations of	**transonic** flows around single and multi-element airfoi	Nowak, Z.P.
Threedimensional	**transonic** potential flow about complex 3-dimensional c	Reyhner, T.A
Multi-grid method for	**transonic** wing analysis and design	Raj, P.
Robust calculation of 3d	**transonic** potential flow based on the nonlinear FAS mu	Wees, A.J. v
A multigrid full potential	**transonic** code for arbitrary configurations	Luntz, A.L.;
A multigrid method for the	**transonic** full potential equation discretized with fin	Deconinck, H
A multigrid algorithm for steady	**transonic** potential flows around aerofoils using Newto	Boerstoel, J
A fast-solver algorithm for steady	**transonic** potential-flow computations with Newton iter	Boerstoel, J
Vectorizable multigrid algorithms for	**transonic** flow calculations	Melson, N.D.
A multigrid solver for two-dimensional	**transonic** full potential flow calculations	Becker, K.
Applications of multi-grid methods for	**transonic** flow calculations	Schmidt, W.;
Multigrid technique applied to lifting	**transonic** flow using full potential equation	Arlinger, B.
A multigrid finite element method for the	**transonic** potential equation	Deconinck, H
A multigrid code for the three-dimensional	**transonic** potential flow about inlets	McCarthy, D.
of the Euler equations for two dimensional	**transonic** flow by a multigrid method	Jameson, A.
/SIP smoothing: A robust fast solver for 3D	**transonic** potential flow	Wees, A.J. v
A multigrid strongly implicit procedure for	**transonic** potential flow problems	Sankar, N.L.
Application of a multi-level grid method to	**transonic** flow calculations	South, J.C.
ential and Euler formulations for computing	**transonic** airfoil flows	Flores, J.;
h-embedding technique for three-dimensional	**transonic** potential flow analysis	Brown, J.L.
hod to solve the Euler equations for steady	**transonic** flow	Rizzi, A.; E
id relaxation into a robust fast-solver for	**transonic** potential flows around lifting airfoils	Boerstoel, J
ifference methods for plane steady inviscid	**transonic** flows	Fuchs, L.
mesh multigrid treatment of two-dimensional	**transonic** flows	McCarthy, D.
Multi-grid calculation of three-dimensional	**transonic** potential flows	Caughey, D.A
Newton multigrid method for determining the	**transonic** lifting flows around airfoils	Nowak, Z.P.
of an iterative method with applications to	**transonic** potential flow	Nowak, Z.P.;
of the multi-grid method to calculations of	**transonic** potential flow about wing-fuselage combinati	Shmilovich,
te difference method for potential unsteady	**transonic** flow	Hounjet, M.H
tral multigrid methods with applications to	**transonic** potential flow	Streett, C.L
Numerical solution of semiconductor	**transport** equations in two dimensions by multi-grid me	Gaur, S.P.;
al, Diffusion-accelerated, Neutral-Particle	**Transport**, Rev. 1	Alcouffe, R.
echnique for atmospheric chemical pollutant	**transport** and diffusion	Lee, H.N.
	Transportleistung und Größe paralleler Systeme bei spe	Mierendorff,
lle pour le calcul d'ecoulements potentiels	**transsoniques**	Bredif, M.
sung der vollen Potentialgleichung im Falle	**transsonischer** Strömungen	Becker, K.
Computation of flows around a Karman-	**Trefftz** profile	Mol, W.J.A.
Algorithms for refining	**triangular** grids suitable for adaptive and multigrid t	Rivara, M.C.
Transonic flow calculations using	**triangular** finite elements	Pelz, R.B.;
hree-dimensional body-fitted coordinates in	**turbomachine** applications	Camarero, R.
An improved time marching method for	**turbomachinery** flow calculation	Denton, J.D.
A multigrid solver for the in-cylinder	**turbulent** flows in engines	Ruttmann, B.
zed Navier-Stokes solutions for laminar and	**turbulent** flow	Rubin, S.G.;

INDEX